Fundamentals of Materials Science for Technologists

Properties, Testing, and Laboratory Exercises

Second Edition

LARRY HORATH

California University of Pennsylvania

WAVELAND

PRESS, INC.

Long Grove, Illinois

Dedication

This work has resulted from the advice, support, and encouragement of many individuals. Therefore, it is dedicated to my friends and family, without whose support it could not have been completed.

For information about this book, contact:
Waveland Press, Inc.
4180 IL Route 83, Suite 101
Long Grove, IL 60047-9580
(847) 634-0081
info@waveland.com
www.waveland.com

Preface

This book and its accompanying materials are meant as an introduction for learning about the importance of materials and materials testing. This introduction is appropriate for two-year and four-year colleges and universities preparing future technicians, technologists, and engineers.

The objective of this volume and its supplemental materials is to provide an awareness of the theory, manufacturing, processing, properties, applications, and common test procedures relating to common materials. These materials include the most common materials that a learner is most likely to encounter, including metals, plastics, glass and ceramics, wood and composites, fuels, and adhesives. Also included are the important advantages, disadvantages, and importance of these materials.

The first section of this book is a background on the chemistry of materials, including their structure, properties, and selection. This includes their common properties that are important to material applications and how these properties may be created, reduced, and altered for various applications. The second section deals with the destructive and nondestructive evaluation of material properties, focusing on the mechanical properties. Further support for the learner is included in the accompanying materials found on the book's web page at waveland.com.

Each chapter contains an introduction, summary, and questions and problems. These are intended to assist learners in identifying the overall objectives of the chapter and section and reinforce their understanding and learning. Examples and photos are included as visual aids for the concepts presented. Laboratory assignments and reference materials have been included for hands-on learning in a supervised environment. Thus, it helps promote understanding of the "why" of studying materials and materials testing.

It is beyond the scope of this book to cover every material and test available. However, the intent of this material is to provide a basic understanding of the materials available and the tests made on these materials to identify their properties. There are endless combinations of materials available today. This book provides a discussion and examples of the most common. This information can then be applied in other situations and applications.

ACKNOWLEDGMENTS

It is a daunting task to try and acknowledge everyone involved in this book and its additional materials. However, I would like to take this opportunity to thank Bob and Lori Olson, multimedia development; Dan Arnold, photographer; Bethany, Elyse, and Rachael; and all of the various contributors to this work. I would also like to thank the following reviewers of the manuscript: Thomas R. Clark, Rock Valley College (IL), Dr. Greg E. Maksi, State Technical Institute at Memphis (TN), and Dr. Christian Unanwa, South Carolina State University.

Thank you all for your patience, support, and understanding.

Contents

SECTION ONE
Fundamentals of Materials Technology

CHAPTER 1
Introduction to Materials Technology

Objectives
- State reasons for studying materials.
- List and describe common terms related to the study of materials.
- Describe and define terms and conditions associated with atomic structure and atomic theory.
- Recognize and describe how the periodic table of elements is used and the structure of the table.
- List and describe how the various bonding forces act to hold atoms together.
- Define the various structures of materials in crystal lattice formation.
- Describe the solidification process.
- List and recognize various standards organizations and their purposes.

1.1 INTRODUCTION

The general purpose of this text is to introduce the reader to the nature of various materials, including their applications, advantages and disadvantages, and properties. Later, testing procedures and specifications are provided so that students can conduct their own tests and analyze the data they collect.

Materials technology deals with the specifications, properties, selection, and testing of engineering materials. *Engineering materials* consist primarily of those materials used in the construction of various structures and machines and the many different products manufactured or produced from these materials throughout the world. To introduce the subject of materials technology, aspects of the origins, content, components, specifications, and standards regarding materials are presented in this chapter.

1.2 ATOMIC THEORY

Why study engineering materials? These materials surround us. To understand the function of machines and the actions and components of structural assemblies, we must first identify and understand the characteristics of materials and their structures. Once we understand the structure of materials, we are able to ask questions about how they perform in service. This latter aspect of materials technology relies on testing to collect data on a material's performance. New materials are constantly researched and developed, putting a greater emphasis on the understanding of the structure and properties of these materials.

All materials are made of *atoms*. Atoms are the smallest parts of an element that still retain the properties of that element. These atoms are bonded together in different patterns using different methods to form different materials.

It is often necessary to classify or categorize materials according to their attributes. We may base this classification on available empirical data or on some theoretical basis yet to be proved or disproved. At present, there are several ways to classify materials, but they all refer to four important categories:

- Chemical composition
- The material's natural state
- The refining or manufacturing processes the material must undergo before it reaches its final useful or applicable state
- Atomic structure

All materials can be classified into one of the following areas based on their chemical composition: elements, compounds, or mixtures.

Elements cannot be broken down into chemically simpler substances. The periodic table of elements is shown in Figure 1–1. Dmitri Mendeleyev produced the first table of elements more than a century ago. Many of the elements in Figure 1–1 were not discovered until long after Mendeleyev's table. All of the elements listed are in their most basic form and cannot be simplified. Oxygen (O) and hydrogen (H) are two common basic elements. By studying the elements, scientists discovered repeti-

IA	IIA	IIIB	IVB	VB	VIB	VIIB	VIII	VIII	VIII	IB	IIB	IIIA	IVA	VA	VIA	VIIA	INERT GASES
1 H 1.00797																	2 He 4.0026
3 Li 6.939	4 Be 9.0122											5 B 10.811	6 C 12.01115	7 N 14.0067	8 O 15.9994	9 F 18.9984	10 Ne 20.183
11 Na 22.9898	12 Mg 24.312											13 Al 26.9815	14 Si 28.086	15 P 30.9738	16 S 32.064	17 Cl 35.453	18 Ar 39.948
19 K 39.102	20 Ca 40.08	21 Sc 44.956	22 Ti 47.90	23 V 50.942	24 Cr 51.996	25 Mn 54.9380	26 Fe 55.847	27 Co 58.9332	28 Ni 58.71	29 Cu 63.54	30 Zn 65.37	31 Ga 69.72	32 Ge 72.59	33 As 74.9216	34 Se 78.96	35 Br 79.909	36 Kr 83.80
37 Rb 85.47	38 Sr 87.62	39 Y 88.905	40 Zr 91.22	41 Nb 92.906	42 Mo 95.94	43 Tc (99)	44 Ru 101.07	45 Rh 102.905	46 Pd 106.4	47 Ag 107.870	48 Cd 112.40	49 In 114.82	50 Sn 118.69	51 Sb 121.75	52 Te 127.60	53 I 126.9044	54 Xe 131.30
55 Cs 132.905	56 Ba 137.34	57 *La 138.91	72 Hf 178.49	73 Ta 180.948	74 W 183.85	75 Re 186.2	76 Os 190.2	77 Ir 192.2	78 Pt 195.09	79 Au 196.967	80 Hg 200.59	81 Tl 204.37	82 Pb 207.19	83 Bi 208.980	84 Po (210)	85 At (210)	86 Rn (222)
87 Fr (223)	88 Ra (226)	89 †Ac (227)															

*Lanthanum Series

58 Ce 140.12	59 Pr 140.907	60 Nd 144.24	61 Pm (147)	62 Sm 150.35	63 Eu 151.96	64 Gd 157.25	65 Tb 158.924	66 Dy 162.50	67 Ho 164.930	68 Er 167.26	69 Tm 168.934	70 Yb 173.04	71 Lu 174.97

†Actinium Series

90 Th 232.038	91 Pa (231)	92 U 238.03	93 Np (237)	94 Pu (242)	95 Am (243)	96 Cm (247)	97 Bk (247)	98 Cf (249)	99 Es (254)	100 Fm (253)	101 Md (256)	102 No (256)	103 Lr (257)

( ) Numbers in parentheses are mass numbers of most stable or most common isotope.

Atomic weights corrected to conform to the 1963 values of the Commission of Atomic Weights.

Figure 1–1
Periodic table of elements.

tive patterns, which allowed them to predict the nature and properties of elements not discovered until much later. This periodic (repetitive pattern) nature of elements is determined in part by the nuclear particles and also by the behavior and configuration of the electrons.

Looking at the table of elements, note that the element hydrogen (H) has an atomic number of 1. The *atomic number* is the number of protons in the nucleus or center of the atom. Protons are positively charged particles. Other elements have various numbers of protons in the nucleus, characterized by their atomic number. Also included in the nucleus are neutrons, or neutral particles. Neutrons remain at the center or nucleus of the atom. Electrons are negatively charged particles that orbit the nucleus at velocities approaching the speed of light. Atoms in a free state (balanced electronic charge) must contain the same number of electrons as protons. In the free state, the negative charges balance the positive. Electrons do not follow neat, well-defined orbits. Instead they randomly orbit the nucleus in what have been described as clouds. Electrons orbit so quickly that, if you could see an atom, the orbit of the electrons would appear to be a random pattern typified by a cloud structure (Figure 1–2).

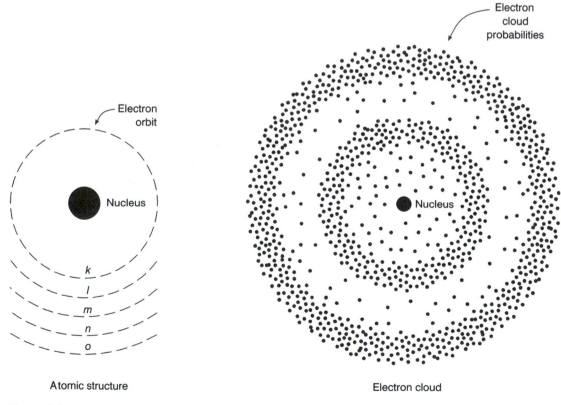

Atomic structure

Electron cloud

Figure 1–2
Atomic structure and electron cloud.

An atom consists of the central cluster called the *nucleus* and the orbiting electrons. The number of protons in the nucleus determines which element the atom represents. For example, carbon (C) has an atomic number of 6, because there are six protons in the nucleus of a carbon atom. Neutrons and protons are much heavier than the other subatomic particles. Based on this fact, scientists often ignore the mass of the other particles and define the *atomic mass* of an atom as the sum of the masses of the protons and neutrons. *Mass* is a property of a body, expressed in pounds or kilograms. *Weight* is the gravitational force exerted on a body by the earth. The terms mass and weight are often confused and erroneously used interchangeably.

An electron, the largest of the remaining particles, weighs about 1/2000 as much as a proton. An arbitrary mass of 1 is assigned to protons and neutrons. For example, carbon has six protons and six neutrons in the nucleus. Therefore, its atomic mass is 12. Refer to the periodic table in Figure 1-1 and compare the atomic masses to the atomic numbers of several elements. The difference between the two numbers is roughly the number of neutrons in the nucleus.

Isotopes are atoms that vary from the normal atomic mass found in naturally occurring forms of the element. For instance, an atom may contain more or fewer neutrons in the nucleus. Neutrons are neutral particles (without electrical charge); therefore, they do not alter the atomic number or the chemical properties of the element. However, some isotopes are unstable and radioactive. These isotopes are used as chemical tracers and nuclear fuels. For example, some elements are known to collect in certain cancerous tumors. Doctors inject radioisotopes into the bloodstream of a patient to detect or to monitor change in tumors using sensitive radioactivity monitors.

Electrons orbit the nucleus of the atom. This tiny solar system behaves in ways that cannot be predicted by the common laws of physics. The speeds involved and the size of the particles make it necessary to use a branch of physics called *quantum mechanics* to study atoms. Quantum mechanics is a field of study that uses energy levels, motion analysis, and probability theories to describe and predict the motion and behavior of electrons. Quantum mechanics makes a basic assumption related to the electrons: Electrons behave in a wavelike fashion rather than like individual particles. Waves can be diverted by reflection or diffraction. This makes it difficult to predict the exact locations or paths of electrons. It is more accurate or descriptive to use energy levels or energy contents of electrons in describing their location in space. Again using the idea of the solar system, the higher the energy level or energy content of an electron, the farther it is from the central core or nucleus.

It is common to have more than one electron in an orbit. Two electrons in the same orbit or energy level have the same energy content. In addition to orbiting the nucleus, electrons also spin or rotate while orbiting. Electrons spin in a clockwise or counterclockwise direction. For two electrons to orbit at the same energy level, they must spin in opposite directions, one positive and one negative. The two electrons are virtually identical except for spin direction.

Energy levels appear at predictable intervals. Electron levels cannot exist in the gaps between these major levels. The first orbit, or electron shell, called 1s, can hold up to two electrons. The second orbit contains two subshells, called 2s (lower energy) and 2p (higher energy), that are close to each other in energy values. The subshell designations are *s, p, f*, and *d*, and they increase in energy from *s* to *d*. Referring to the periodic table in Figure 1–1, the repetitive pattern of the elements becomes clearer. Looking at Group I, the hydrogen contains one proton in the nucleus and one electron at the 1*s* level. Hydrogen, also in the first period, has two electrons, both in the 1*s* level but with opposite spin directions. The second period begins with lithium, which has one electron in the outer shell, like hydrogen. Lithium (Li) has two electrons in the 1*s* level and one electron in the 2*s* level. The second and third periods (rows) continue to fill the third energy level. Starting with the fourth period, lower subshells must be filled before the outermost energy shell is filled. Electron shells are shown in Figure 1–3.

Quantum mechanics has come up with four quantum numbers to describe an electron. The total energy of an electron is designated by the principal quantum number, *n*. The letter *n* is used to designate the shell: 1, 2, 3, 4, and so forth. The angular momentum of an electron, *l*, is the second quantum number. The angular momentum of the electron ranges from 0 to $n-1$, and the subshell is $2(2l+1)$. This expression is used to determine the subshell inhabited by the electron. Subshells are designated as *s, p, d, f, g*, and *h*. The magnetic moment of the electron, designated *M*, is the third quantum number. *M* has values from +1 to 0 to –1 and designates the orbit within the subshell that corresponds to its energy level. M_s is the spin direction of the electron. It can have values from +1/2 to –1/2. No two electrons have the same four quantum numbers.

The vertical groupings in the periodic table are based on similar electron configurations and similarities in both chemical and physical properties. For example, the members of Group IA are called the alkali metals; Group IIA are the alkaline-earth metals; Groups IIIA through VA and VIIA are mostly nonmetals; and the last group contains the inert gases. The groups listed as transition elements are mostly metals with incomplete subshells. The atomic size of the elements decreases from left to right within a period and increases from top to bottom within a group.

Elements are roughly divided into two categories: metals or nonmetals. Metals tend to be solid at room temperature and to be good conductors. Nonmetals may be solids, liquids, or gases and tend to be insulators. These are general considerations and are meant to be rough guidelines, not absolutes. About 92 elements occur naturally in the earth. With the addition of synthesized or manufactured elements, the total number of elements is around 120. Many of these synthetic elements are short-lived, and many even remain unnamed.

Compounds are combinations of two or more elements. These combinations may have properties that are different from any of the component elements. Sodium chloride (NaCl), or salt, is an example of a compound that has properties different from either constituent. Sodium is a metal that burns at room temperature, and chlorine is a poisonous gas. *Alloys* are metals combined with one or more other elements. The smallest part of a compound that retains the properties of that compound is a *molecule*. Molecules are generally made up of different types of atoms in varying combinations.

Mixtures are constructed of two or more pure substances that have been mechanically mixed together. The pure substances can be elements or compounds. Mixtures

Element	Symbol	Atomic Number	Shell and Subshell Configuration
Hydrogen	H	1	$1s$-1
Helium	He	2	$1s$-2
Lithium	Li	3	$1s$-2, $2s$-1
Beryllium	Be	4	$1s$-2, $2s$-2
Boron	B	5	$1s$-2, $2s$-2, $2p$-1
Carbon	C	6	$1s$-2, $2s$-2, $2p$-2
Nitrogen	N	7	$1s$-2, $2s$-2, $2p$-3
Oxygen	O	8	$1s$-2, $2s$-2, $2p$-4
Fluorine	F	9	$1s$-2, $2s$-2, $2p$-5
Neon	Ne	10	$1s$-2, $2s$-2, $2p$-6
Sodium	Na	11	$1s$-2, $2s$-2, $2p$-6, $3s$-1
Magnesium	Mg	12	$1s$-2, $2s$-2, $2p$-6, $3s$-2
Aluminum	Al	13	$1s$-2, $2s$-2, $2p$-6, $3s$-2, $3p$-1
Silicon	Si	14	$1s$-2, $2s$-2, $2p$-6, $3s$-2, $3p$-2
Phosphorus	P	15	$1s$-2, $2s$-2, $2p$-6, $3s$-2, $3p$-3
Sulfur	S	16	$1s$-2, $2s$-2, $2p$-6, $3s$-2, $3p$-4
Chlorine	Cl	17	$1s$-2, $2s$-2, $2p$-6, $3s$-2, $3p$-5
Argon	A	18	$1s$-2, $2s$-2, $2p$-6, $3s$-2, $3p$-6
Potassium	K	19	$1s$-2, $2s$-2, $2p$-6, $3s$-2, $3p$-6, $4s$-1
Calcium	Ca	20	$1s$-2, $2s$-2, $2p$-6, $3s$-2, $3p$-6, $4s$-2
Scandium	Sc	21	$1s$-2, $2s$-2, $2p$-6, $3s$-2, $3p$-6, $3d$-1, $4s$-2
Titanium	Ti	22	$1s$-2, $2s$-2, $2p$-6, $3s$-2, $3p$-6, $3d$-2, $4s$-2
Vanadium	V	23	$1s$-2, $2s$-2, $2p$-6, $3s$-2, $3p$-6, $3d$-3, $4s$-2
Chromium	Cr	24	$1s$-2, $2s$-2, $2p$-6, $3s$-2, $3p$-6, $3d$-5, $4s$-1
Manganese	Mn	25	$1s$-2, $2s$-2, $2p$-6, $3s$-2, $3p$-6, $3d$-5, $4s$-2
Iron	Fe	26	$1s$-2, $2s$-2, $2p$-6, $3s$-2, $3p$-6, $3d$-6, $4s$-2
Cobalt	Co	27	$1s$-2, $2s$-2, $2p$-6, $3s$-2, $3p$-6, $3d$-7, $4s$-2
Nickel	Ni	28	$1s$-2, $2s$-2, $2p$-6, $3s$-2, $3p$-6, $3d$-8, $4s$-2
Copper	Cu	29	$1s$-2, $2s$-2, $2p$-6, $3s$-2, $3p$-6, $3d$-10, $4s$-1
Zinc	Zn	30	$1s$-2, $2s$-2, $2p$-6, $3s$-2, $3p$-6, $3d$-10, $4s$-2
Gallium	Ga	31	$1s$-2, $2s$-2, $2p$-6, $3s$-2, $3p$-6, $3d$-10, $4s$-2, $4p$-1
.			
.			
.			
Silver	Ag	47	$1s$-2, $2s$-2, $2p$-6, $3s$-2, $3p$-6, $3d$-10, $4s$-2, $4p$-6, $4d$-10, $5s$-1
.			
.			
.			
Cesium	Cs	55	$1s$-2, $2s$-2, $2p$-6, $3s$-2, $3p$-6, $3d$-10, $4s$-2, $4p$-6, $4d$-10, $5s$-2, $5p$-6, $6s$-1
.			
.			
.			
Gold	Au	79	. . . $3s$-2, $3p$-6, $3d$-10, $4s$-2, $4p$-6, $4d$-10, $4f$-14, $5s$-2, $5p$-6, $5d$-10, $6s$-1
.			
.			
.			
Lead	Pb	82	. . . $3s$-2, $3p$-6, $3d$-10, $4s$-2, $4p$-6, $4d$-10, $4f$-14, $5s$-2, $5p$-6, $5d$-10, $6s$-2, $6p$-2

Figure 1–3
Electron shells and orbits for selected elements.

differ from compounds in that, theoretically, mixtures can be separated back into their component parts. One example of a mixture is oil and water, which can be mixed into an emulsion and later separated.

A naturally occurring material exists in nature in the form in which it will be used. These materials include wood, stone, and water, which occur naturally in the form in which we use them. The majority of materials used today are manufactured or refined in some way. The form in which we use them is based on some manufacturing or refinement process. Thus, materials may be classified as natural or manufactured and further classified according to the principle or process used to refine them.

Materials may also be classified as *organic* or *inorganic*. Organic materials are carbon-based materials. All biological systems are organic. Some materials, such as limestone, have biological origins but are not organic. Most plastics are organic, as are most petroleum-derived products. Inorganic materials are not derived from living things. Sand, rocks, and metals are all inorganic materials.

1.3 BONDING OF MATERIALS

Many materials consist of atoms bonded together through some means. This bonding can occur directly between atoms or can occur between previously grouped atoms called *molecules*. The molecules then combine to form the material. The bonding forces help determine the chemical and physical properties of the material.

The properties of a material are affected by the willingness of the material to combine or react with other materials. This willingness is related to the chemical structure of the material. Of particular importance is the outermost shell, or *valence shell*, which has an overall positive or negative charge. The electrons in that shell have the highest amount of energy but are the least tightly held electrons in the atom. These high-energy electrons, or *valence electrons*, are responsible for the chemical activity of the atom. The activity stems from the ease with which these valence electrons attract other electrons or are given up to another atom to combine with other materials.

For instance, the inert gases are not chemically active. Their outermost shells are full and, therefore, they do not have free electrons to attract or to give up. Their valence shells are satisfied. The most reactive elements are the alkali metals and the halogens, Groups II and VII. These elements have a shortage of one electron in a filled valence shell or have a valence shell that contains only one electron. Therefore, they have a surplus or shortage of an electron. The activity levels of the other groups decline in comparison with these two situations.

The two general types of elements, metals and nonmetals, act differently in combining with other atoms. Metals tend to give up, or donate, their valence electrons when bonded with other atoms. Nonmetals are those that tend to gain or attract other electrons. Metals are generally those with fewer than four valence electrons, and nonmetals have more than four valence electrons. The transition metals are difficult to understand in valence terms and do not act in a clearly metallic or nonmetallic manner.

The physical state of an element depends on the type of bonding by which the atoms are attracted and the relative strength of those bonds. For example, atoms of gases are loosely bonded together and are free to move independently of one another. A solid material represents the most tightly bound substance; the atoms are rigidly held in place by other atoms.

Liquids and gases exhibit random arrangements. When several unlike atoms combine, they quickly separate and reform. Due to the activity level and the excitedness of the atoms, the action of random grouping, scattering, and regrouping is a characteristic of liquids. This is also true in the case of gases.

As the energy level and activity decrease, the random groupings and scattering become less frequent. Bonds become stronger and begin to form ordered patterns called *lattices*. These lattices form *crystals*, which in turn form larger patterns we know as *solids*. These patterns are recognized as the grain structure of the material. Exceptions to this order of events are the *amorphous* materials. Amorphous materials retain the random order and do not exhibit regular patterns, order, or groupings in their solid form. Glass is an amorphous material. Amorphous materials may exhibit short-run chains but no definite, recognizable pattern.

Bonding also gives materials density. For instance, tighter bonding increases the viscosity of liquids. Liquids are heavier than gases because they contain more atoms. "Denser," in this situation, means more atoms per unit volume. Solids are more dense than liquids and maintain their shape better. Liquids and gases tend to take the shape of their containers because they are free to move around. Solids maintain their shape until acted upon by a force greater than that holding them together.

When materials gain or lose an electron, they become *ions*. Ions are versions of the original element that deviate from the normal neutral state—as, for instance, when sodium and chlorine combine to form salt. The sodium atom relinquishes its valence electron and becomes a positive sodium ion. Chlorine, with its seven outer electrons, absorbs this electron and becomes a negative ion. The charges on these

Ionic bonding of
sodium chloride (NaCl)

Covalent bonding
of methane (CH₄)

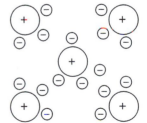

Metallic bonding
where positive ions
donate electrons;
electron cloud bonds
positive ions together.

Figure 1–4
Bonding.

atoms are dependent on the original number of protons in the nucleus. The net charge is dependent on their original state and the gains made in bonding. Sodium and chlorine share an electron. The attraction of the ions for the electron holds the two together. This results in a very tight *ionic bond*. The resulting molecule (NaCl) is electrically neutral and very stable after bonding.

Large numbers of ionically bonded atoms arrange themselves so that the positive ions are surrounded by negative ions of other pairs. Again, the attractions between pairs hold the entire arrangement together (Figure 1–4). Each ion is rigidly held in place by the attraction of the other ions that surround it. This bonding produces a definite, regular structure known as a *lattice structure,* or *crystalline structure.* Ionically bonded materials, such as sand and salt, tend to be hard and have high melting points. This makes them useful in applications such as ceramics and refractories, where these properties are extremely significant.

Covalent bonding is another type of bond mechanism. Many gases are *diatomic;* that is, they consist of two atoms bonded together in the natural state, such as hydrogen (H_2), nitrogen (N_2), and oxygen (O_2). Oxygen, for example, has six electrons in its outer shell. It needs two more to complete its valence shell. When two atoms combine, as in O_2, the two atoms combine and share enough electrons to complete their outer shells. This, effectively, completes each of the atoms' outer shells. The two atoms are bonded quite strongly. Many gases and all organic materials are covalently bonded.

Van der Waals forces, named after the Dutch physicist who explained their function, are another type of binding force. These forces result from the influence that the action of the valence electrons in one atom exerts on the valence electrons in another atom. For example, when two atoms have valence electrons that move in compatibility, they produce a small attractive force. These forces are very weak and are able to act over considerable distances, but they are often masked by the larger bonding forces. To liquefy a pure gas where only these forces are at work, high pressures are applied to the gas to squeeze the atoms together so that the stronger bonding needed for liquefaction can take place.

Metallic bonding is a fourth type of bonding, in which a localized exchange of electrons takes place between a small number of atoms. This is more likely to take place in metals, where all of the atoms have similar valences and cannot directly exchange electrons to complete their outer shells. Metallic bonds are based on the donation of electrons by the constituent atoms to a common pool of electrons known as an *electron cloud.* This results in a structure where positively charged ions with completed outer shells are surrounded by a negatively charged electron cloud, which moves randomly and fluidly throughout the localized structure. The cloud bonds the atoms together and can be shaped or resized as needed.

We emphasize here that the chemical makeup and activity of the element(s) of which a material is composed help determine the physical, chemical, mechanical, and other properties in application. The bonding mechanism also helps determine the properties of a material. In addition, the lattice and crystalline structure of the material, which are examined soon, also affect the properties of a material. It is important to note that the properties of these materials depend on these characteristics.

1.4 CRYSTALLINE STRUCTURES

If we consider an atom as a solid sphere, two atoms will approach each other until the forces of attraction equal the forces of repulsion between like charges. This distance is different for different materials. It is related to the number of shells and number of valence electrons present. The greater the number of shells, the greater the distance between atoms. The greater the number of valence electrons, the less the distance between atoms. Imaginary lines that connect the centers of atoms in a pattern are called lattice structures. Through the use of X-ray diffraction analysis, it has been determined that common metals crystallize into one of seven types of lattice structures: cubic, tetragonal, orthorhombic, rhombohedral, hexagonal, monoclinic, or triclinic.

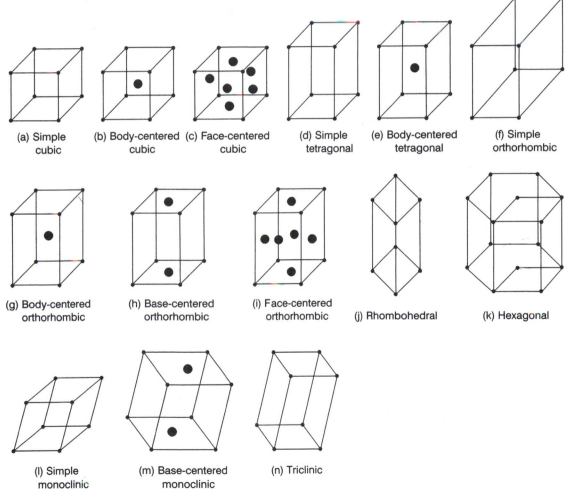

(a) Simple cubic

(b) Body-centered cubic

(c) Face-centered cubic

(d) Simple tetragonal

(e) Body-centered tetragonal

(f) Simple orthorhombic

(g) Body-centered orthorhombic

(h) Base-centered orthorhombic

(i) Face-centered orthorhombic

(j) Rhombohedral

(k) Hexagonal

(l) Simple monoclinic

(m) Base-centered monoclinic

(n) Triclinic

Figure 1–5
Crystal lattice structures.

Every crystal lattice structure has its own *unit cell*. A unit cell is the smallest unit into which a lattice structure can be broken down that still retains the properties of the whole structure. Figure 1–5 illustrates the seven basic crystalline structures.

Figure 1–5(a) shows a *simple cubic* structure. In it, eight atoms form a cube whose sides are equal in size. When the cube has an atom centered in the cube, such as the common material iron, it is called a *body-centered cubic* structure, as in Figure 1–5(b). If the simple cubic has an atom centered on each of its faces, the resulting structure is called a *face-centered cubic* structure [Figure 1–5(c)]. Body-centered cubic structures are found in metals such as barium, cesium, potassium, lithium, molybdenum, sodium, tantalum, and tungsten. Face-centered cubic structures are found in aluminum, copper, gold, lead, nickel, platinum, and silver.

In the *simple tetragonal* and *body-centered tetragonal* structures [Figs. 1–5(d) and (e), respectively], all of the atomic planes are still at right angles to each other, as in the cubic structure. However, one of the dimensions is longer than the other two dimensions. A good example of a natural tetragonal structure is pure tin.

Figure 1–5(f) is an example of the *simple orthorhombic* structure, in which the lengths of all three axes are all different. However, the planes of atoms are perpendicular. Figure 1–5(g) is an example of a *body-centered orthorhombic* structure. Notice the atom centered within the tetrahedron. The *base-centered* and *face-centered* structures [Figs. 1–5(h) and (i), respectively] are the two remaining orthorhombic structures.

The *rhombohedral* structure differs from the three previous structures because none of the planes are perpendicular. However, as illustrated in Figure 1–5(j), the distances between the crystal planes are equal.

Figure 1–5(k) illustrates the *close-packed hexagonal* structure. The distances between atoms in the bases are equal in the hexagonal structure. Although the bases are perpendicular to the sides, the angle between the sides is 120°. When three unit cells are put together, the resulting structure forms a hexagon formation, which is the source of the name. The cells of a crystal structure may overlap when assembled. This results in the material packing more closely together than normal. When this happens, the structure is termed *close-packed*. Graphite is a common material that is close-packed hexagonal in structure. Diamond, however, is carbon that has formed in the *face-centered close-packed cubic* structure, or *complex cubic* form. This form is often referred to as the *diamond structure*. Materials that have a close-packed hexagonal structure include beryllium, cadmium, cobalt, magnesium, titanium, zinc, and zirconium.

In a *monoclinic* structure, two of the atomic planes are perpendicular, but the third angle is not 90°. Also, the atoms in the crystal plane are not the same distance apart. There are two variations: *simple monoclinic* and *base-centered monoclinic* structures. Figure 1–5(l) illustrates the monoclinic structure, and Figure 1–5(m) shows the base-centered monoclinic.

The last crystalline structure, the *triclinic*, is shown in Figure 1–5(n). No two planes are perpendicular to each other, the distances between the atoms are all different, and the angles between atom planes are not equal. The result is an elongated, thin crystal formation. Figure 1–6 gives the crystal lattice structure for several common elements.

Figure 1–6
Lattice structures for common metals.

Material	Lattice Structure at 68°F (20°C)
Aluminum	FCC
Antimony	Rhombohedral
Arsenic	Rhombohedral
Barium	BCC
Beryllium	HCP
Boron	Orthorhombic
Bromine	Orthorhombic
Cadmium	HCP
Calcium	FCC
Carbon	Hexagonal
Cesium	BCC
Chlorine	Tetragonal
Chromium	BCC
Cobalt	HCP
Copper	FCC
Germanium	Cubic
Gold	FCC
Helium	HCP
Hydrogen	Hexagonal
Iodine	Orthorhombic
Iron	BCC
Lead	FCC
Lithium	BCC
Magnesium	HCP
Manganese	Cubic
Mercury	Rhombohedral
Molybdenum	BCC
Neon	FCC
Nickel	FCC
Oxygen	Cubic
Phosphorus	Cubic
Platinum	FCC
Potassium	BCC
Silicon	Cubic
Silver	FCC
Sodium	BCC
Sulfur	Orthorhombic
Tantalum	BCC
Tin	Tetragonal
Titanium	HCP
Tungsten	BCC
Uranium	Orthorhombic
Vanadium	BCC
Zinc	HCP
Zirconium	HCP

Note: Examine the structures of these elements and try to determine correlations between their structures and their properties.

Figure 1–7
Grain boundary.

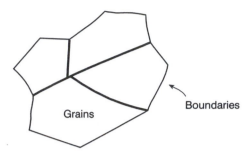

    Many of a material's physical and mechanical properties depend on the crystal structure of that material. Generally, soft metals have a face-centered cubic structure, whereas mild steels and less ductile metals have body-centered cubic structures and brittle metals are often simple cubic structures. Most common metals are face-centered cubic (FCC), body-centered cubic, or hexagonal in structure. Refer to Figure 1–6 for a partial list of the structure of many metals.

    Some elements can exist in two or more crystal structures. These alternative structures are known as *allotropes*. Allotropes depend on temperature and pressure changes. For example, pure iron will assume a body-centered cubic structure at normal temperature and pressure conditions. It undergoes an allotropic change to FCC if the temperature is raised above 1670°F (910°C). Allotropes are also known as *polymorphs*.

    As the energy in a liquid system decreases, the forces that are grouping, scattering, and regrouping atoms tend to develop distinct patterns, which become the characteristic lattice structure for that material. The temperature or energy level at which this begins is called the *freezing point* for that material.

Figure 1–8
Supercooling process.

Figure 1–9
Dendrites.

The formation of these lattice crystals produces heat. As the heat produced is drawn off, the lattice grows until stopped by some energy block, such as another lattice or a container. Two lattice structures that collide exhibit what is known as a *grain boundary.* The space between these lattice structures is the boundary that separates them. Figure 1–7 illustrates a grain boundary.

For lattice growth to start, a nucleus, or seed, must be present. For very pure metals, rapid cooling may not allow time for the nucleus to grow. If the temperature falls below the freezing temperature, it is called *supercooling* the liquid. The temperature increases as the nucleus forms and levels off as the lattice structure evolves. A graph of the supercooling process is given in Figure 1–8.

Crystal size depends on the particular metal, the temperature, and the cooling rate. Rapid cooling produces smaller crystals. Slow cooling produces larger crystals. As the material cools and crystals form, heat energy is released ahead of the lattice growth. This heat release blocks the formation of crystals in that direction. Growth then continues in a perpendicular direction, and the process repeats. Thus, lattice growth occurs more rapidly in directions perpendicular to each other. This is called the *dendritic* nature of crystal growth. Dendrites are shown in Figure 1–9.

When a liquid solidifies, the temperature at which solidification occurs is constant. As soon as all of the liquid freezes, the temperature then continues to decrease. Plotting the temperature of a liquid against time yields the plot in Figure 1–10. The linear portion of the curve illustrates the latent heat of fusion period. Latent heat opposes solidification.

1.5 SPECIFICATION OF MATERIALS

Specifications are an attempt by the customer to specify or define what he or she wants. The quality of the specifications depends on the ability of the customer to define the product needed and the accuracy and ability with which the object can be specified. Thus, the specification becomes a statement of the minimum quality that the customer will accept from the producer. Several characteristics are used to specify materials, such as physical, mechanical, electrical, and other properties, and crystalline structure.

Figure 1–10
Solidification curve.

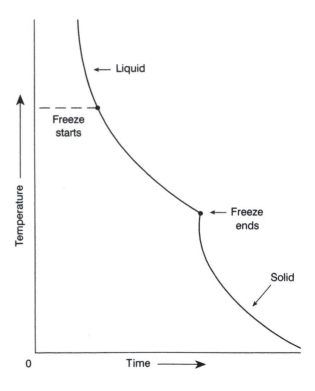

Specifications are often criticized for being too loose or too tight. In the case of being too loose, inferior materials may be used. If the specification is too tight, the producer may not be able to meet the specification economically, if at all. Specifications can also exclude certain materials or specify a certain brand or type of product based on tradition, personal prejudice, or inadequate or improper criteria. With these facts in mind, a specification should be written that is both practical and economical without bias or prejudice. Items that are typically specified include the following:

- Method of manufacture or processing
- Forms, dimensions, and finish (physical attributes)
- Desired chemical, physical, or mechanical properties

Universal testing machine.

Courtesy of Tinius Olsen Testing Machine Co., Inc.

These items should be periodically reviewed to determine whether these specifications continue to be the most practical, economical, and satisfying for the producer and the consumer.

Table 1–1 lists general properties of engineering materials that are the main concern of materials technologists.

1.6 STANDARDS ORGANIZATIONS

Standards represent the efforts of many people and groups of people to come to some reasonable understanding based on common definitions and commonly defined procedures. However, standards work properly only when everyone adopts and follows them. The following organizations are involved in setting standards for materials and material applications. Further information on their purpose, their procedure, and the standards that they set can be obtained directly from the organization, from your local library, or from an Internet search. Further information is available on the accompanying CD-ROM.

Organizations involved in developing standards include the following:

SAE	Society of Automotive Engineers
API	American Petroleum Institute
ACI	American Concrete Institute
AISI	American Iron and Steel Institute
ASCE	American Society of Civil Engineers
GSA	General Services Administration
DOT	Department of Transportation
DOD	Department of Defense
ANSI	American National Standards Institute
ASME	American Society of Mechanical Engineers
IEEE	Institute of Electrical and Electronic Engineers
NEMA	National Electrical Manufacturers Association
ISO	International Standardization Organization
EIA	Electronic Industries Association

One organization, the American Society for Testing and Materials (ASTM), is of particular interest to those concerned with materials testing. ASTM is concerned with developing methods of testing materials, constructing new standard definitions, drafting new materials specifications, and designing new practices for the processes involved in the utilization of materials.

Each year, ASTM publishes a revised set of standards. This set of standards includes specifications, test methods, definitions, classifications, and suggested practices. Each of these standards must be reviewed and revised or reapproved at least every five years, or they are removed. Standards are constantly changing, and some of those discussed in this text may be changed by the time you read about them.

Table 1–1
General properties of materials.

Class	Property
General	Density, porosity, moisture content, macrostructure, microstructure
Chemical	Composition, pH, resistance to corrosion, weatherability
Physiochemical	Liquid absorption/repulsion, shrinkage and swell due to hygroscopic action
Mechanical	Strength in: tension, compression, shear, torsion and flexure; static, impact, and endurance, stiffness, toughness, elasticity, plasticity, ductility, hardness
Thermal	Specific heat, thermal expansion, thermal conductivity
Electrical/magnetic	Conductivity, permeability, galvanic action
Acoustic	Transmission, reflection, absorption
Optical	Color, transmission, reflection, collimation

Each ASTM standard has a letter and number designation. The letter designates the group to which it belongs:

A. Ferrous metals
B. Nonferrous metals
C. Cement, ceramic, and masonry materials
D. Miscellaneous materials
E. Miscellaneous subjects
F. Specific applications of materials
G. Corrosion, deterioration, and degradation of materials
ES. Emergency standards

The number designates the order in which it was adopted within the group.

1.7 SUMMARY

The study of materials and materials testing is not a new concept. People have been using materials since the beginning of time. Modern methods of study started with the development of the scientific method of study. Numerous improvements have been made in testing methods, and the fields of materials science and materials testing have gained in importance.

Early building materials were wood, stone, bone, leather, and similar readily available materials. These materials proved deficient for tooling and weapons. The discovery of metals (gold, silver, and copper) brought on a new age in materials science.

Engineering materials can be classified as elements, compounds, or mixtures. Elements can be further defined as metals or nonmetals. All materials can be categorized as being organic or inorganic. Organic materials are carbon-based. The properties of materials can be generally classified into eight categories: general, chemical, physiochemical, mechanical, thermal, electrical/magnetic, acoustic, and optical.

Materials testing involves the identification of the properties of materials through well-established procedures. These procedures are conducted in a manner that is both accurate and precise. How well a test is able to predict the performance of a material in service determines the significance of the test.

Specifications are based on the demands of the customer. These are the desired characteristics of a material that define what the customer will accept. Through materials testing, the producer and customer can determine if the specifications are being met.

Standards organizations are individuals or groups of individuals who work jointly to develop agreements concerning how testing procedures should be conducted and to define the methods, procedures, and definitions that should be utilized in materials testing. One group in particular, ASTM, issues periodic updated material on standards for testing and materials.

QUESTIONS AND PROBLEMS

1. Using the periodic table in Figure 1–1, categorize each element as metal or nonmetal. On what basis did you make your decision for each?
2. List five naturally occurring engineering materials.
3. List three common compounds found in the home.
4. List three common mixtures.
5. What is meant by the terms *organic* and *inorganic*?
6. Give the electron shell structure for (a) hydrogen, (b) helium, (c) sodium, (d) copper, (e) iron, (f) gold, and (g) silver.
7. List and define the four types of bonding forces.
8. List and describe the four quantum numbers.
9. Define the terms (a) isotope, (b) ion, and (c) allotrope.
10. Describe the seven types of crystal structures discussed.
11. Describe the process of solidification from a liquid to a solid state.
12. Describe the process of slip.
13. Write a specification for a pencil.
14. Why are there standards for materials and testing procedures?

15. List the standards organizations and their abbreviations.
16. Using current ASTM standards as a reference, determine to what the following ASTM standards relate: C 120, ES 10, F 22, B 200, A 57, D 33, and G 17.
17. List common properties that are important in selecting materials and why you think these properties are important.
18. How does atomic bonding affect properties?
19. Describe five applications that owe their development to new materials.
20. List five applications where a traditional material has been replaced by a newer material or process. Provide the reason(s) that you believe the material was replaced.
21. Provide a complete specification for a bookcase, particularly materials, hardware, manufacturing processes, coatings and finishes, and other such details.
22. Describe the differences between quality and value.

CHAPTER 2
Ferrous Metals

Objectives
- List the various ingredients of cast iron, steels, and stainless steels.
- Recognize and use the nomenclature associated with steels, stainless steels, and cast iron.
- Describe the processes used to refine iron ore in the production of cast iron, steels, and stainless steels.
- Describe the processes used to produce steels, cast iron, and stainless steels.
- List the major shapes in which ferrous metal products are available.
- List and describe the various alloying elements in ferrous metals and the purposes of each.
- Describe the process of galvanic corrosion in metals.

2.1 INTRODUCTION

Ferrous metals, mainly steels, are important for any industrialized, high-technology civilization. In many applications, there are no substitutes or alternatives for steels. *Steel* is not an element; rather, it is an iron-carbon alloy that contains less than 2% carbon. If the alloy contains more than 2% and less than 4% carbon, it is called *cast iron*.

Steel has been around for many centuries, dating back to about 2000 B.C., when charcoal was packed with iron bars and heated to approximately 1000°C. This is known as the *cementation process*. Quality varied widely in the early processes. Some iron mongers were able to produce harder steels than others, and no one knew why. Although crude and difficult to control accurately, the cementation process allowed the carbon in the charcoal to diffuse into the iron bar to produce steel.

An improvement on the cementation process was the *crucible process*. In the crucible process, the steel bars produced by the cementation process were melted together in a large pot, or crucible. Once melted, the steel was cast into bars. The crucible process produced a more uniform quality steel.

The modern steelworking process really began in 1856, when Sir Henry Bessemer introduced the *converter process* for steelmaking. The *open hearth* process followed approximately a decade later. These two processes were predominant for roughly a century.

Steel is one of the basic building materials in any industrialized nation. Although its use dates back to about 2000 B.C., steel was not widely used until the latter part of the nineteenth century when it became available in larger quantities at lower costs. Prior to this, two ferrous metals were used, cast iron and *wrought iron*. Wrought iron is almost pure iron that includes silicate slag. It was used for decorative pieces such as railings and ornamentation. Wrought iron is no longer produced in the United States in any quantity, although soft steel products in these applications are still termed "wrought iron." Cast iron is still used for castings, but it is generally too hard and brittle for general use.

Iron ore mine.
Courtesy of Inland Steel Company. Photo by Bill Staniec.

For modern engineering purposes, steel is the cheapest and most versatile of metals. Steels are available in thousands of types, ranging from hard to soft, can be magnetic or nonmagnetic, are heat-treatable and weldable, and have various resistances to heat, corrosion, impact, and abrasion. Although other materials, such as plastics, composites, and ceramics, have replaced metals in certain applications, steel production continues to be a predominant industry.

2.2 PRODUCTION OF IRON

Only limited amounts of pure iron, usually ingot iron or iron powders, are used in modern industry. Steels of iron and alloying elements, such as carbon, silicon, nickel, chromium, and manganese, are more widely used. A *plain-carbon steel*, which contains carbon, silicon, and manganese, rarely contains more than 1% of an alloying element. A *low-alloy steel* contains carbon, silicon, and manganese, together with small quantities of nickel, chromium, molybdenum, and other alloying elements that alter the properties of steel. *High-alloy steels* have higher quantities (more than 5%) of alloying elements.

Iron ore hauler.
Courtesy of Inland Steel Company.
Photo by Bill Staniec.

Three primary ores are (1) *magnetite*, (2) *hematite*, and (3) *taconite*. These are among the several kinds of iron that are generally refined. The first of these, magnetite, is a combination of ferric oxide (Fe_2O_3) and ferrous oxide (FeO) that is black in color. It contains approximately 65% iron and is highly magnetic. *Lodestone* is a form of magnetite found in nature. Hematite is typically ferric oxide (Fe_2O_3), or what we typically refer to as *rust*. It is blood red in color and contains approximately 50% iron. Taconite is a green-colored, low-grade ore that contains less than 30% iron. It often contains much silica. Other ores that are available include *limonite* [hydrated ferrous oxide ($FeO \cdot H_2O$)], *siderite* [ferrous carbonate ($FeCO_3$)], and *iron pyrites*, which are iron sulfides (FeS). These are such low-grade, low-yield ores that they are seldom used today.

The first iron mill in the United States began production in Hammersmith in the early 1600s. This site is now located in Saugus, MA. This mill produced iron used in tools, implements, pots and pans, nails, and other products for the community.

The earliest smelting of iron ore used charcoal. Carbon was obtained from wood or coal. The charcoal was mixed with the iron ore and placed in the furnace. Bellows were used to blow hot air from the fire under or outside the furnace through the charcoal-iron mixture. This process is often depicted in the blacksmith shops in classic movies and old westerns. This process produces a spongy mass called a *bloom*, which must be hammered to force out slag and impurities.

Modern practice involves heating coal in a furnace from which most of the air (oxygen) has been removed. This furnace is often called a *coking oven*. Hydrogen and other elements are driven from the coal, leaving carbon in the form of *coke*. The iron ore is cleaned and layered in a large *blast furnace* (Figure 2–1), along with coke and limestone. In metal-melting operations, a *slag* is often necessary to flux or combine with impurities. The slag and impurities then float on the molten metal and are separated and removed after the metal is tapped from the furnace. Limestone is used as a blast furnace slag to remove such impurities as sulfur and silica. Limestone, coke, and iron ore are continuously charged into the top of the furnace, so that the furnace is always full.

Air is blown in from the bottom of the furnace at 1100°F through water-cooled nozzles. At this temperature, the carbon in the coke reacts with the oxygen in the ore and starts to burn off the oxygen from the iron oxides. This reaction causes the temperature to rise above the melting point of iron, about 3000°F. After about 5 to 6 hours, the iron is tapped from the furnace and poured into ingots. These ingots are called *pigs*, and each one weighs about a ton. Pig iron contains many impurities, such as silica and sulfur. The carbon content of pig iron is 4%, roughly that of cast iron. It can be remelted and used in that state, or alloying elements can be added to enhance the properties, if cast iron is required. To make 3000 tons of pig iron a day would require approximately:

6,000 tons of iron ore
3,000 tons of coke
1,500 tons of limestone
90,000 ft^3/min of hot air

(a)

(b)

(c)

Figure 2–1

A. Blast furnace. B. Blast furnace control room. C. Smelting process.

Parts A. and B. courtesy of Inland Steel Company.

Ore pellets: Note blast furnace in background.

Courtesy of Inland Steel Company. Photo by Bill Staniec.

Slabbing mill.

Courtesy of Inland Steel Company. Photo by Bill Staniec.

2.3 PRODUCTION OF STEEL

Pig iron contains too much carbon and too many impurities to be used directly in most applications. It must be converted to steel in one of several types of *converters*. These converters may differ in appearance, but they do the same thing: burn off the carbon in the iron.

Basic oxygen furnace-combination caster.

Courtesy of Inland Steel Company. Photo by Bill Staniec.

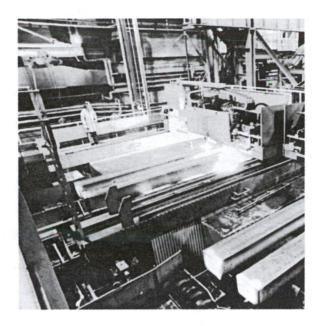

The oldest of the converters is the *open hearth converter* (Figure 2–2). The open hearth converter is basically a large bowl made of firebrick, which typically ranges from 15 to 50 ft across and stands up to 40 ft high. The pigs from the blast furnace process are mixed with scrap steel and melted in the converter. Hot air is blown across the surface, and the excess carbon and impurities are burned off the melt. Typical melt sizes, which range from 150 to 250 tons, take approximately 12 hours to refine. At the end of the refining period, the exact amounts of carbon and the desired alloying elements are added, and the steel is drawn off into crucibles and cast into ingots.

Around the same time that the open hearth furnace was developed, Henry Bessemer was searching for a method of producing better metal for gun barrels. Bessemer developed a process using a pear-shaped converter lined with firebrick and mounted on trunnions, called a *Bessemer converter* (see Figure 2–3). The pigs and scrap metal are placed into this converter, and hot air is blown through the melt. The carbon is burned off, the desired carbon content and alloying elements are added, and the steel is drawn off or poured into ingots. It takes approximately 12 to 15 min to refine

Figure 2–2
Open hearth converter.

Hot air →   ← Raw materials

Molten iron

→ Iron

Figure 2–3
Bessemer converter.

an average melt of 25 tons of steel. The converter can be tipped on its trunnions and the metal poured into a large ladle, which is used to transfer the steel into ingots.

One problem with both the open hearth and Bessemer converters is that the resulting steel contains many impurities, including oxygen, nitrogen, sulfur, phosphorus, and other unwanted elements, which weaken the steel. About 1900, the *electric arc furnace* (Figure 2–4) was invented. It is basically an open hearth furnace mounted on trunnions. Pigs and scrap metal are loaded into the furnace. The electrodes are lowered from the top of the furnace close to the charge, and an arc is struck. This produces enough heat to melt the metal and burn off the carbon. Because no air is required, no air bubbles are introduced into the melt, which lowers the oxygen and nitrogen content of the melt. Electric arc furnaces generally range in capacity from 1 to 100 tons.

Figure 2–4
Electric arc furnace.

Figure 2–5
Oxygen lance converter.

The main drawback of the electric arc furnace is the amount of power that it requires. A typical furnace requires 200 V, three-phase power, which means a total supply requirement of 30,000 kW. This equates to about 400 kW/ton, or about 200 W/lb, of steel for an average production day. The electrodes for the furnace range from 12 to 24 in. in diameter and may draw up to 25,000 A each. In consumable electrode processes, the composition of the electrodes is carefully controlled to reduce contamination. Each ton of steel consumes approximately 15 lb from the electrodes, so the electrodes must constantly be fed into the melt. The electric arc furnace is often used in conjunction with an open hearth or a Bessemer converter to further refine the steel and thus produce high-quality steel. The less expensive methods are used to convert the iron into steel, and the electric arc furnace is used to finish the product.

The *oxygen lance method* is the newest conversion method. It was invented in Europe and first used in the United States about 40 years ago. The oxygen lance method uses an apparatus similar to the Bessemer converter (Figure 2–5). The premelted iron and scrap metal mixture is lowered into the furnace, along with a water-cooled pipe or lance. Pure oxygen is blown through the lance and, therefore, through the liquid metal at pressures around 150 lb/in^2. The oxygen burns off the carbon very rapidly, taking approximately 45 min per average melt. Once the carbon is removed, the lance is withdrawn, and the desired carbon content and alloying elements are added. The steel is then cast into ingots.

Figure 2–6
Continuous-pour process.

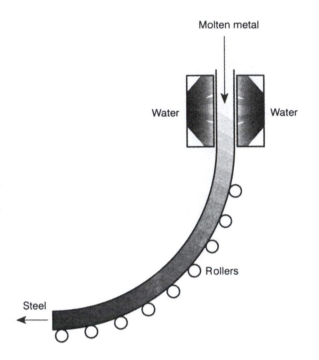

When steel is cast into ingots, the size of the products that can be made from the ingot is limited. Realizing this, a *continuous-pour process* has been developed, which removes the size restriction on the ingots. Figure 2–6 illustrates the continuous-pour process. Molten steel is poured through a form, which is water-cooled on the outside. As the steel passes through the form, it cools and solidifies as gravity pulls it downward, and it is cut to length. The entire melt from a converter can be formed using this process and cut into desired lengths while hot. Later, the metal can be formed or rolled into any desired shape. This process is not limited to steel. Many nonferrous metal products are being manufactured using the continuous-pour process.

Once the conversion process is complete, the steel is then formed into one of the many shapes available. The steel can be rolled in a rolling mill while red hot (hot-rolled, at approximately 2200°F) or formed later after cooling (cold-rolled or cold-drawn at room temperature). Hot-rolling is generally used to form shapes, whereas cold-rolling is used for finishing thin, flat products. Other shapes can be formed using cold-drawing. Ingots can be formed into preliminary shapes, depending on the final desired product. These preliminary shapes are then formed into some of the following commonly available commercial shapes:

Angles	Angles can have legs of equal or unequal length. Equal-leg lengths range from 1×1 up to 8×8 in. Unequal-leg lengths go up to 9×4 in.
Bars	Solid shapes, hot- or cold-drawn, in sizes ranging from 3/4 to 12 in. thick.

Beams	Standard I and H beams. I beams range from 3×2.25 in. up to 24×8 in. H, or wide-flange, beams range from 8×5.25 in. up to 36×16.5 in.
Billets	Section of ingot suitable for rolling.
Blooms	Slab of steel in which width and depth are generally equal.
Channels	U-shaped in cross section. Sizes range from 1.25×3 in. up to 4×18 in.
Plates	Large, flat slabs thicker than 1/4 in.
Sheets	Generally less than 1/4 in. in thickness, hot- or cold-rolled; may be available in coils as well as flat sheets.
Tubing	Square, rectangular, and round tubing and pipe made from sheet stock. Rectangular tubing ranges in size from 2×3 in. up to 12×20 in. Square tubing ranges from 2×2 in. to 16×16 in. Round tubing and pipe are available from 1/4 in. up to several feet in diameter in several thicknesses.
Wires	Drawn from bars that have been rolled down to small diameters. Steel wire is further fabricated into nails, cable, fence, screens, and similar products.

If a special shape is required, it may be requested as a special-order product. If the order is economically feasible, the steel mill will accommodate the job. However, nonstandard shapes are generally expensive and are not usually ordered.

2.4 CARBON CONTENT IN STEELS

The most important alloying element in steel is carbon. Its presence is very important, and its effect is very potent, considering that most steels contain less than 1%

Steel span bridge.

carbon. Steels in which carbon is the only significant alloying element are termed *plain-carbon steels. Low-carbon*, or *mild, steels* are produced in greater tonnages than all other steels combined. Mild steel is cheap, soft, ductile, and readily welded. However, it cannot be heat-treated. Mild steels are used to produce car bodies, appliances, bridges, tanks, and pipe. The following steel classifications are based on carbon content:

Name	Content (% C)	Examples
Low carbon (mild)	0.05–0.32	Sheet, structural
Medium carbon	0.35–0.55	Machinery
High carbon	0.60–1.50	Machine tools
Cast iron	>2.00	Castings

Medium-carbon steels are used for reinforcing bars in concrete, farm implements, machine tool gears and shafts, and the automobile and aircraft industries. High-carbon steels are used in knives, files, machine tooling, hammers, chisels, axes, mauls, and similar applications. Notice that as the amount of carbon increases, the products become less ductile, harder, and stronger.

A small increase in the carbon content (0.1%) has a significant impact on the properties of steel. As the carbon content of steel increases, steel

- Becomes more expensive to produce
- Becomes less ductile, i.e., more brittle
- Becomes harder
- Loses machinability
- Has a higher tensile strength
- Has a lower melting point
- Becomes easier to harden
- Becomes harder to weld

In addition to carbon content, cold-working of steels is used to enhance their properties. For example, a reduction in the thickness of 4% by cold-working raises the tensile strength of the product by 50%. Cold-working of this type is plastic deformation at room temperature. This working produces a large number of dislocations in the metal's structure. These dislocations block each other as they try to slide along slip planes. In order to drive these dislocations through the blocks, higher stresses must be applied. Therefore, cold-working effectively increases the strength of the metal. Cold-rolled sheet steel and cold-drawn tubing are examples of products that use this principle. Cold-working also has its drawbacks. As the yield strength increases, higher loads are required to size the material. In addition, as the material is being formed, a process called *work-hardening* occurs. Materials become harder due to the strengthening that occurs during cold-working. This may not always be desirable. To relieve the blocks to dislocation motion and to relieve internal stresses, metals can be heat-treated. Heat treatment is covered in a later chapter.

2.5 OTHER ELEMENTS IN STEELS

Besides carbon, steels are often classified by their alloying elements. Among these elements are aluminum, manganese, molybdenum, nickel, and silicon. Each element has a specific purpose as an alloying element of steel.

Certain undesirable elements are left in the steel when it is converted. These elements can be removed by expensive processing or rendered harmless by adding other elements that rely on the chemical properties of undesirable elements. For instance, oxygen makes steel brittle. Oxygen can be rendered harmless by adding aluminum or silicon to the melt. As the hot air blows over the melt, oxygen reacts with the iron and produces sparks, which shoot out of the converter. Aluminum is added to the melt because aluminum is more chemically active than iron, so the oxygen reacts with the aluminum rather than the iron and no sparks are produced. Steel that has aluminum added is called *killed steel*. Aluminum is also added to steels to promote smaller grain size, which adds to the toughness of the steel. Another example of an alloying element is manganese. Manganese increases the strength, hardenability, and hardness of the steel. Sulfur in the limestone and iron ore is introduced into the melt. Sulfur accumulates at the grain boundaries as iron sulfide. This causes a condition known as *hot short*; that is, the steel loses its strength at high temperatures. Small amounts of manganese tie up the sulfur and increase the malleability of the steel. Boron is added to increase the hardenability of the steel. Boron can be used only with aluminum-killed steels. Small amounts of copper are added to increase the corrosion resistance of steels. Other elements used as alloying elements include chromium (increases corrosion resistance and hardenability), niobium (increases tensile strength), titanium (gives high strength at high temperatures), tungsten carbide (gives high hardness), and vanadium (increases toughness and impact resistance). Notice that most of these alloying elements are metals. There are many more alloying elements that alter, enhance, or reduce certain characteristics of the steels containing them. The ones given here are just some of the more common.

2.6 NOMENCLATURE OF STEELS

The SAE and the AISI have developed a method of cataloging steels that uses their carbon content and alloying elements. An AISI 8620 steel is the same as an SAE 8620 steel. Steels are referred to by a four-digit designation that is descriptive of the alloying elements, contents, and carbon percentage of that sample. The numbering system has an implied decimal for carbon percentage, e.g., (0.18%, 0.22%). The second digit designates the actual percentage of the alloying element. For example, a steel is labeled as 1018 steel. The first two numbers, 10, mean that it is a plain-carbon steel, whereas the last two numbers, 18, mean that it has 0.18% carbon. Another example is a 4030 steel, which is a molybdenum steel of 0.15% to 0.30% molybde-

Number	Type of Steel	Alloying Elements (%)*
10--	Plain carbon	None, 0.4 Mn
11--	Sulfurized free-machining	0.7 Mn, 0.12 S
12L--	Leaded free-machining	Pb added
13--	Manganese	1.60–1.90 Mn
2---	Nickel steels	3.5–5.0 Ni
3---	Nickel-chromium	1.0–3.5 Ni, 0.5–1.75 Cr
40--	Molybdenum	0.15–0.30 Mo
41--	Chrome-molybdenum	0.80–1.1 Cr, 0.15–0.25 Mo
43--	Nickel-chromium-molybdenum	1.65–2.0 Ni, 0.4–0.9 Cr, 0.2–0.3 Mo
46--	Nickel-molybdenum	1.65 Ni, 1.65 Mo
5---	Chromium	0.4 Cr
6---	Chromium-vanadium	
7---	Unused	
8---	Low nickel-chromium-molybdenum	
9---	Nickel-chromium with small percentage of molybdenum	

Figure 2–7
Steel nomenclature.

*All steels contain a certain percentage of manganese, phosphorous, silicon, and sulfur.

num and 0.30% carbon. A partial table of the numbering system for steels is given in Figure 2–7.

When only the first digit is given, as in 2 --- for nickel steels, the second digit gives the actual full percentage of the alloy. For example, 22 -- is a nickel steel with 2% nickel. Carbon percentages have an implied decimal point, i.e., 0.18%, 0.30%, and so on. In addition, numbers such as 10100, with five digits, indicate 1.00% carbon or more. The letter *B* in the middle of a number, as in 81B40, indicates that a minimum of 0.0005% boron has been added for hardenability.

A sample of the uses of the various common steels includes the following:

1010	Steel tubing for general purposes
1040	Connecting rods for automobiles
4140	Sockets and socket wrenches
52100	Ball and roller bearings
8620	Shafts, gears, and machinery parts

The carbon content of a steel has a specified range of 0.05%. This variation is due to various environmental and control factors such as electrode erosion, air quality, oxidation reactions, and similar factors. Thus, a steel that has a stated carbon content of 0.5% may range from 0.48% to 0.53%, inclusive. Similar ranges can be given for other alloying elements. Greater allowances are made for high-carbon steels and for certain elements in high-alloy steels.

2.7 TOOL STEELS

Tool steels are special types of steel produced to make tooling used to cut or shape other materials. These steels are generally produced by electric furnaces, where conditions are more closely controlled. All tool steels may be hardened and vary from high carbon to high alloy, depending on the application. These steels have high wear and heat resistance, have high strength, and exhibit good hardenability. In addition to carbon, tool steels may contain alloying elements such as chromium, cobalt, copper, manganese, molybdenum, nickel, silicon, tungsten, and vanadium.

Tool steels have an additional nomenclature. The several classes of tool steels are classified by letter designations:

Type	Purpose
A	Air-hardening, medium-alloy steels
H	Hot-working steels
M	High-speed steels containing molybdenum
O	Oil-hardening, low-alloy steels
S	Shock-resisting, medium-carbon, low-alloy steels
T	High-speed steels containing tungsten
W	Water-hardening, high-carbon steels

Specific alloys are designated by number along with the type. For example, W-1 is a plain-carbon steel that is 0.9% carbon. *High-speed steels*, such as those used in machine tools, typically contain 0.75% carbon, 18% tungsten, 4% chromium, and 1% vanadium. This is just one example, of which there are many combinations. These ingredients are adjusted, depending on the intended use of the machine tool.

Tool steels have properties that make them especially useful in the cutting and shaping of other materials. Applications focus on four basic categories: shock-resistant, such as hammers, chisels, and other tools used for striking; hot-working, such as forging equipment; cold-working, such as dies and hand tools; and high-speed, for applications such as lathe tools, drills, and mill cutters. Each of these categories has a separate prefix designation. Shock-resistant tool steels have an *S* prefix. Hot-working has the *H* prefix. Cold-working includes the water-hardening *(W)*, air-hardening *(A)*, oil-hardening *(O)*, and high-carbon, high-chromium *(D)*. High-speed tool steels have either an *M* or *T* prefix. These are the most common of the tool steel types. Specific applications and compositions for tool steels are provided as a guide in their selection:

Tool Steel Compositions

The following includes nominal percentages of various alloying elements for the tool steel examples. Iron makes up the difference in the following percentages:

Cold-Working: Air-Hardening

| Designation | %C | %Co | %Cr | %Mn | %Mo |
%Ni	%Si	%V	%W		
A2	1.00	—	5.00	—	1.00
—	—	—	—		
A4	1.00	—	1.00	2.00	1.00
—	—	—	—		
A6	0.70	—	1.00	2.00	1.25
—	—	—	—		
A7	2.25	—	5.25	—	1.00
—	—	4.75	—		
A8	0.55	—	5.00	—	1.25
—	—	—	1.25		
A9	0.50	—	5.00	—	1.50
1.50	—	1.00	—		
A10	1.35	—	—	1.80	1.50
1.80	1.25	—	—		

Cold-Working: High-Carbon, High-Chromium

| Designation | %C | %Co | %Cr | %Mn | %Mo |
%Ni	%Si	%V	%W		
D2	1.50	—	12.00	—	1.00
—	—	1.00	—		
D3	2.25	—	12.00	—	—
—	—	—	—		
D4	2.25	—	12.00	—	1.00
—	—	—	—		
D5	1.50	3.00	12.00	—	1.00
—	—	—	—		
D7	2.35	—	12.00	—	1.00
—	—	4.00	—		

Note: A2 is the most commonly used tool steel.
Note: D-series tool steels are split between air- and oil-hardening. Type D2 is widely used for cold-work tooling.

Cold-Working: Oil-Hardening

Designation %Ni	%C %Si	%Co %V	%Cr %W	%Mn	%Mo
01	0.90	—	0.50	1.00	—
—	—	—	0.50		
02	0.90	—	—	1.60	—
—	—	—	—		
06	1.45	—	—	0.90	0.30
—	1.00	—	—		
07	1.20	—	0.75	—	—
—	—	—	1.75		

Cold-Working: Water-Hardening

Designation %Ni	%C %Si	%Co %V	%Cr %W	%Mn	%Mo
W1	0.6–1.4	—	—	—	—
—	—	—			
W2	0.6–1.4	—	—	—	—
—	0.25	—			
W5	1.0	—	0.50	—	—
—	—	—	—		

Hot-Working

Designation %Ni	%C %Si	%Co %V	%Cr %W	%Mn	%Mo
Chromium Type					
H10	0.40	—	3.25	—2.50	—
1.00	0.40	—			
H11	0.35	—	5.00	—	1.50
—	1.00	0.40	—		
H12	0.35	—	5.00	—	1.50
—	1.00	0.40	1.50		
H14	0.40	—	5.00	—	—
—	1.00	—	5.00		
H19	0.40	4.25	4.25	—	—
—	—	2.00	4.25		

Note: These are the lowest-cost tool steels because of their low alloy content. They are primarily carbon steels with low percentages of single alloying elements.

Tungsten Type

Designation %Ni	%C %Si	%Co %V	%Cr %W	%Mn	%Mo
H21	0.35	—	3.50	—	—
—	—	0.50	9.00		
H23	0.30	—	12.00	—	—
—	—	1.00	12.00		
H24	0.45	—	3.00	—	—
—	—	0.50	15.00		
H26	0.50	—	4.00	—	—
—	—	1.00	18.00		

Molybdenum Type

Designation %Ni	%C %Si	%Co %V	%Cr %W	%Mn	%Mo
H42	0.60	—	4.00	—	5.00
—	—	2.00	6.00		

Shock-Resistant

Designation %Ni	%C %Si	%Co %V	%Cr %W	%Mn	%Mo
S1	0.50	—	1.50	—	—
—	—	—	2.50		
S2	0.50	—	—	—	0.50
—	1.00	—	—		
S4	0.55	—	—	0.80	—
—	2.00	—	—		
S6	0.45	—	1.50	1.40	0.40
—	2.25	—	—		
S7	0.50	000	3.25	0.60	1.40
—	—	—	—		

High-Speed

Designation %Ni	%C %Si	%Co %V	%Cr %W	%Mn	%Mo
M1	0.85	—	4.00	—	8.50
—	—	1.00	1.50		
M2	0.85	—	4.00	—	5.00
—	—	2.00	6.00		
M4	1.30	—	4.00	—	4.50
—	—	4.00	5.50		
M7	1.00	—	4.00	—	8.75
—	—	2.00	1.80		
M42	1.10	8.00	3.75	—	9.50
—	—	1.20	1.50		
T15	1.50	5.00	4.00	—	—
—	—	5.00	12.00		

Note: M designates molybdenum as the major alloying element; T designates tungsten.

Tool steels have unique properties that make them especially adaptable for shaping and cutting other materials. Among these properties are high temperature stability, wear resistance, machinability, and toughness. Machine tools are an example of why these tool steel properties are important. These tools tend to heat up at the point of contact. If the surface temperature exceeds the softening temperature of the tool steel, the machine tool will melt, rendering it unusable. Therefore, a tool steel must be able to maintain its properties at elevated temperatures.

Wear resistance is another important quality in a tool steel. It is an indication of how long the machine tool will last in service. Wear can occur through abrasions, where the rough surface contacts remove pieces of the surface of the machine tool. Machinability, a subjective quality, is an indication of how well the tool steel can be shaped into a usable form. The easier a material is to form, the lower the cost of producing tools. Finally, toughness indicates how well a tool steel can withstand sudden impacts without fracture. Most tool steels are brittle, unless specially treated to increase their toughness. In fact, simply dropping a hardened tool steel on the floor may cause it to fracture, unless it has been properly tempered. "Hard" does not always mean brittle. Alloying elements are used to increase hardenability with toughness.

2.8 CAST IRON

Other ferrous metals include *cast iron* (grey and white), *ductile cast iron, malleable cast iron*, and *wrought iron*. A steel containing more than 2% carbon is called an *iron*. Steels with more than 2% carbon have almost no ductility and must be formed by casting. The most common type of cast iron is grey cast iron, which contains approximately 3% to 3.5% carbon and more than 1% silicon. Carbon exists interstitially in iron up to concentrations of 0.8%. Beyond that, carbon appears in the form of iron carbide or graphite flakes. Excess carbon forms minute cracks of differing sizes in the metal. This network of cracks lowers the strength of the metal. These graphite flakes produce a dark appearance, which can be seen in fresh breaks in cast iron. The high percentage of silicon aids in the formation of the graphite. *Grey cast iron* has almost no ductility and will break if heated or cooled too rapidly. ASTM has devised several classes for grey cast iron, depending on its minimum tensile strength. Tensile strength is the resistance of the metal to being pulled apart. A class 20 grey cast iron has a minimum tensile strength of 20,000 lb/in.^{2}, whereas a class 40 has a minimum tensile strength of 40,000 lb/in^2. Grey cast iron is used in furnace doors, machine bases, and crankshafts due to its compression strength, machinability, and vibration-damping qualities.

White cast iron contains between 2.5% and 3.5% carbon, with 0.5% to 1.5% silicon. The difference in grey and white cast iron can be seen from a freshly fractured surface. Grey cast iron, where all the carbon content is in the graphite form, appears grey and dull. White cast iron, with the carbon as iron carbide or cementite, appears bright and silvery at the fracture. White cast iron parts are hard and brittle, which allows them to be used in rolling and crunching equipment where wear resistance is necessary.

When small amounts (around 0.05%) of calcium, cerium, lithium, magnesium, or sodium are added to cast iron and the metal is allowed to cool slowly, the excess carbon forms small balls, or spheres, called *nodules*, or *spherulites*, instead of the flat plates in ordinary cast iron. The additive and casting process removes the stress risers, which produce the cracks in the cast product. The product is termed *nodular*, or *ductile,* cast iron, because the ductility is increased in products made from this metal. It still contains approximately 4% carbon, with 2.5% silicon. ASTM has classifications of these ductile cast irons according to the tensile strength in thousands of pounds per square inch, the yield strength, and the percent elongation of the metal. For example, a 60:40:20 ductile cast iron has a minimum tensile strength of 60,000 lb/in.^{2}, a yield strength (the point at which the material "yields," or elongates, without a corresponding increase in load) of 40,000 lb/in.^{2}, and a percent elongation of 20%. Percent elongation is the ratio of the final length prior to fracture divided by the original length, given as a percentage. Ductile cast iron is often used for engine blocks, pistons, crankshafts, machine parts, and similar pieces.

Malleable cast irons are heat-treated versions of white cast iron. A cast iron containing 2% to 3% carbon is heated to approximately 1750°F (962°C), where the iron carbide or cementite is allowed to form spherulites. This produces a metal similar to ductile cast iron. There are two types of malleable cast iron: pearlitic malleable iron and ferritic malleable iron. The ferritic malleable iron casting is heated to 1700°F (925°C) for several hours. It is then cooled to just above the eutectoid temperature and cooled through the eutectoid temperature range at a rate of 10° to 25°F (6° to 14°C) per hour. This takes about 24 hours, allowing the excess carbon to form into graphite spheroids. The casting is then allowed to cool to ambient temperature. Pearlitic malleable iron castings are also heated to 1700°F (925°C) but are then quenched in oil or air to room temperature. The casting is then reheated to 850° to 900°F (458° to 486°C) for several hours to allow the pearlite to form into spheroids. This special heat-treatment process gives the malleable cast irons a minimum percent elongation of between 10% and 20%.

2.9 STAINLESS STEELS

Normal steels corrode rapidly when left uncoated. At higher temperatures, corrosion in the form of oxidation occurs even more rapidly. This process can be slowed with the addition of chromium and, sometimes, nickel. These steels are termed *stainless steels*. When chromium is present in excess of 12%, the steel becomes highly corrosion-resistant due to the formation of chromium oxide on the surface.

Stainless steels are divided into three major categories: *ferritic stainless steels,* *austenitic stainless steels*, and *martensitic stainless steels*. Ferritic stainless steels typically contain between 12% and 25% chromium and 0.1% and 0.35% carbon. They are ferritic in structure up to their melting temperature. Therefore, austenite cannot form, and these steels cannot form the hard martensitic structure. They may be strengthened by work-hardening. Because the ferritic variety of stainless steel is very formable, it is often used in applications not requiring high strength, such as jewelry, decorations, utensils, and automotive trim pieces.

When chromium and nickel are added to steels, the result is a face-centered cubic structure, or austenite. These austenitic stainless steels are nonmagnetic; therefore, by placing a magnet on the surface of an austenitic stainless steel, you can readily identify it. The magnet will not be attracted to the steel. They are also low-carbon steels, generally below 0.15% carbon. They contain between 16% and 26% chromium and from 6% to 23% nickel. One particular austenitic stainless steel is 18/8 stainless, which contains 18% chromium and 8% nickel. These stainless steels are used in the chemical processing industry, for food utensils, for architectural products, and for other generally low-strength applications. They are machinable and weldable, to a certain extent, but are not heat-treatable. Austenitic stainless steels to be welded typically contain small stabilizing amounts of niobium or titanium.

Martensitic stainless steels contain between 6% and 18% chromium, with carbon concentrations of 0.1% to 1.5%. These can have up to 2% nickel, but most do not contain nickel. These steels can be hardened by rapid cooling (quenching) the steel from the austenitic range. This property, along with the corrosion resistance, makes martensitic stainless steels an excellent material for knives and cutlery. However, this property also lowers their machinability and weldability. Martensitic stainless steels are magnetic.

Other high-strength stainless steels can be made through the addition of certain alloying elements and hardening. These high-strength steels are referred to as *super-alloys,* or *maraging steels*. They contain 18% to 25% nickel, 7% cobalt, and small amounts of other elements, such as molybdenum, titanium, and zirconium. These steels are heated to 1500°F (822°C) for 1 hour per inch of thickness in order to produce a uniform structure. They are then allowed to air-cool to room temperature. Once cooled, they are reheated to about 900°F (486°C) for approximately 3 hours. Once the heat treatment is over, the steels are cold-worked to achieve greater strengths. One advantage of the maraging steels is that they can be comparatively easily worked and machined. They are used in large structures such as bridges, buildings, and aircraft components.

The AISI has developed a three-digit numbering system for stainless steels. The 200- and 300-series stainless steels are austenitic; the 400-series are ferritic and martensitic. Figure 2–8 illustrates the AISI nomenclature for stainless steels.

2.10 CORROSION

Ferrous metals have a tendency to rust, because the iron in them reacts with oxygen in the environment to form iron oxide, or rust. This process is called *corrosion*. The process of corrosion can involve the metal coming into contact with water, which contains oxygen or another element that is strongly negative. Two different metals in

Figure 2–8
AISI nomenclature for stainless steels.

No.	Composition	Uses
Austenitic		
201	0.15 C, 17 Cr, 7.5 Mn, 3.5–5.5 Ni	General-purpose
202	0.15 C, 18 Cr, 10 Mn, 4–6 Ni	General-purpose
301	0.15 C, 17 Cr, 2 Mn, 6–8 Ni	Trim-general
302	0.15 C, 18 Cr, 2 Mn, 8–10 Ni	General-purpose
303	0.15 C, 18 Cr, 2 Mn, 8–10 Ni	Free machining
304	0.08 C, 19 Cr, 2 Mn, 8–12 Ni	Weldable
308	0.08 C, 20 Cr, 2 Mn, 10–12 Ni	Corrosion resistant
316	0.08 C, 17 Cr, 2 Mn, 2 Mo, 10–14 Ni	Chemical resistant
Ferritic		
405	0.08 C, 11–15 Cr, 0.1–0.3 Al	Weldable
430	0.12 C, 14–18 Cr	General-purpose
445	0.2 C, 23–28 Cr	High temperature
Martensitic		
403	0.15 C, 11–13 Cr, 0.5 Si	Turbine blades
410	0.15 C, 12 Cr, 1.0 Si	General-purpose
420	0.15 C, 12–15 Cr	Heat treatable

contact with an electrolyte, typically water, compose an electrolytic cell, or galvanic cell. One metal acts as the positive electrode, or anode, and the other acts as the negative electrode, or cathode. This process is called *galvanic corrosion*. Galvanic corrosion is an electrochemical process, which erodes the anode. The oxide or iron forms a larger crystal than the steel itself, causing the rust to buckle away from the surface of the metal and to flake off, eroding the surface. If allowed to continue, this flaking will eventually eat through the metal. Just look at old car fenders to witness what eventually happens in the process of corrosion. Metals are arranged in a galvanic series, shown in Figure 2–9. The further apart the two metals, the more severe the corrosion, due to the greater electrode potential difference.

For example, take two metals, aluminum and steel, joined together. Look at the relative positions of the two metals in the galvanic series. Aluminum is higher than steel. Therefore, aluminum is the more positive, or anodic. Aluminum will give up electrons to the conducting liquid and will corrode. In some situations, this fact is used as an advantage. For instance, brass and bronze contain copper. The copper will react with chlorine, oxygen, or sulfur to form a green or black coating of copper chloride, copper oxide, or copper sulfate, respectively. These coatings stick to the surface of the metal and protect it from further attack. This can be observed in statues where the artist artificially corrodes the metal for visual appeal and to protect the base metal.

Figure 2–9
Galvanic series.

Magnesium
Aluminum
Zinc
Iron
Steel
Cast iron
Lead
Brass
Copper
Bronze
Nickel
Stainless
steel
Silver
Graphite

There are several factors that may speed the corrosion process. Among these factors are an increase in temperature, the presence of certain gases, environmental factors such as acid rain, metal fatigue, cold working/forming, and other similar factors. Alloying elements tend to retard the corrosion process by protecting the metal against oxide formation. In addition, various coatings are used to protect the surface of the metal.

Corrosion occurs in other materials, such as plastics and elastomers, which are described in more detail later. Corrosion in these materials tends to deteriorate them, reducing their mechanical properties and making them more brittle. Corrosion in old tires is apparent as weathering and rot, generally caused by environmental factors. Other materials such as glass and ceramics are oxides, sulfides, and other natural compounds that resist deterioration.

Corrosion is often an electrochemical process. Therefore, poor conductors such as ceramics are corrosion-resistant. The process requires an anode, a cathode, and some form of electrolyte. If you were to immerse two metals, such as aluminum and steel, in water, galvanic corrosion will occur. One metal acts as an anode and the other as the cathode. Oxidation or the loss of electrons takes place at the anode, while reduction or the gaining of electrons takes place at the cathode. The water serves as the conductor or electrolyte. As mentioned earlier, the corrosion process is electrochemical. A potential and a path for current flow must exist. The anode and the cathode can exist in the same material. For example, two alloying elements in the same material within a humid environment may produce the necessary conditions for corrosion.

There are several ways to protect against corrosion. Many of these will be discussed later in the Chapter 10, "Adhesives and Coatings." Corrosion can be prevented or lessened by coatings, design considerations, environmental control, and alloying, among others.

2.11 SUMMARY

Steel is not an element; rather, it is an iron-carbon alloy, which contains less than 2% carbon. If the alloy contains more than 2% and less than 4% carbon, it is called cast iron. There are several ores of iron that are generally refined. The major ores are (1) magnetite, (2) hematite, and (3) taconite. The first iron mill in the United States was built at Hammersmith (today at the site of Saugus, MA) in the 1600s. This mill was used to produce tools, implements, pots and pans, and other products for the community.

Ferrous metals contain iron. Iron must be extracted from iron ore. This requires high heat, limestone, coke, and iron ore. The molten iron is extracted from the oven and poured into pigs. Pig iron contains many impurities. The metal is reheated and alloying elements are added to produce various steels. Cast iron, steels, and stainless steels have their own nomenclatures or numbering systems. These systems describe the properties and constituents of the metal.

Through hot-working or cold-working, products such as angles, bars, beams, billets, blooms, channels, plates, sheets, tubing, and wires are made from the metal.

Stainless steels are highly corrosion-resistant metals that contain certain alloying elements to increase these properties. Depending on the amount of carbon, chromium, nickel, and other elements available in the steel, these stainless steels may be ferritic, austenitic, or martensitic.

All ferrous metals are susceptible to corrosion. Galvanic corrosion is the erosion of the anodic or positive electrode. The relative placement in the galvanic series determines the degree of corrosion and which metal will undergo anodic displacement.

QUESTIONS AND PROBLEMS

1. What are the major differences between iron and steel?
2. What are the major differences between steel and cast iron?
3. Give the carbon content and alloying elements for the following steels:
   a.   1020
   b.   1060
   c.   1340
   d.   3350
   e.   4120
   f.   4340
   g.   6140
   h.   8640
4. Describe the differences between plain-carbon, low-alloy, and high-alloy steels.
5. List the three primary iron ores.
6. Define the terms *coke* and *slag*.
7. Describe the operation of a blast furnace.
8. Describe the operation of the open hearth converter.
9. Describe the operation of the Bessemer converter.
10. Describe the operation of the electric arc furnace.
11. Describe the operation of the oxygen lance method.
12. List 10 common shapes in which steel products are produced.
13. List five effects that increasing carbon content in steels produces.
14. List and describe the purpose of five common alloying elements in steels.
15. Define killed steel.
16. Define the term *hot short*.

17. What is ferrite?
18. What is austenite?
19. What is pearlite?
20. What is cementite?
21. What is martensite?
22. Describe the difference between grey and white cast iron.
23. List the three major categories of stainless steels.
24. Describe maraging steels.
25. Describe the process of galvanic corrosion.
26. List the best ferrous metal(s) for a machine gear.
27. List the best ferrous metal(s) for aircraft construction.
28. List the best ferrous metal(s) for knives and cutlery.
29. List the best ferrous metal(s) for jewelry.
30. List the best ferrous metal(s) for porch railings.
31. List the best ferrous metal(s) for shock-dampening castings.
32. List, by appropriate number, an austenitic stainless steel.
33. List, by appropriate number, a martensitic stainless steel.
34. List, by appropriate number, a ferritic stainless steel.
35. Give the appropiate number for a steel with the following components:
    a. No alloying elements, 0.2% carbon
    b. 1.75% manganese, 0.4% carbon
    c. 1% chromium, 0.2% molybdenum, 0.2% carbon
    d. 0.3% nickel, 0.5% chromium, 0.1% molybdenum, 0.4% carbon
    e. 4.0% nickel, 0.4% carbon
36. Give the appropriate designation for a ductile cast iron with 60,000 lb/in. tensile strength, 40,000 lb/in. yield strength, and a percent elongation of 10%.
37. List five applications that you encounter every day that use ferrous metals and the reason(s) why these metals were chosen for each application. Can you think of a better material for the application?
38. Select a common product made from cast iron, steel, or stainless steel and produce a flowchart that illustrates how that product was produced from raw material to finished goods.
39. Why is time important in the transformation of steels both in hardening and softening procedures?

CHAPTER 3
Nonferrous Metals

Objectives • Describe the refinement process, major alloys, uses, and properties of copper, brasses, bronzes, magnesium, chromium, titanium, lead, tin, zinc, gold, silver, and platinum.
• Explain the refinement process, major alloys, uses, designation system, and properties of aluminum and nickel.
• Describe the uses and properties of the major refractory metals.

3.1 INTRODUCTION

There are a number of nonferrous metals (those that do not contain iron) and alloys that are widely applied in modern products. The radioactive metals uranium, thorium, and plutonium are used as nuclear fuels. Zirconium is an alloying element and is also used in the nuclear field. The light metals aluminum, beryllium, calcium, lithium, magnesium, potassium, titanium, and sodium also have their particular uses. Aluminum, beryllium, and titanium are used as structural metals, whereas the remaining light metals are too soft and chemically reactive; these metals are used to extract metals from their ores. Sodium and potassium are used in the nuclear field as coolants. Nickel and lead are versatile metals used in many applications, whereas copper is used primarily for its thermal and electrical conductivity. Cadmium, tin, and zinc are often used in electrical applications and for bearings. Cobalt and manganese are used as alloying elements in ferrous and nonferrous metals. Silver is used as a decorative metal and in brazing alloys, whereas gold, silver, and platinum are used for electrical contacts and jewelry. Finally, the refractory metals, those with melting points above 3600°F (2000°C), such as columbium, titanium, tungsten, vanadium, and zirconium, are used in applications requiring high strength, hardness, and high temperatures. This chapter deals with metals other than those derived from iron. Please refer to the periodic table of elements (Figure 1–1) for more information on these metals.

3.2 ALUMINUM

Aluminum is not only one of the most abundant elements in the earth's crust (after oxygen and silicon), but it is also the most abundant metal in the earth's crust. Most clay contains approximately 8% *alumina*, pure aluminum oxide (Al_2O_3). For nonferrous metal ores, this percentage is considered to be very high. Although the content of clay is high in aluminum oxide, extraction costs are lower for *bauxite* ore ($Al_2O_3 \cdot 3\ H_2O$), which is a hydrated aluminum ore.

The process of extracting aluminum from ore has changed over the years since its discovery. Aluminum was discovered in 1825 by Hans Oersted. It was later isolated by Friedrich Wöhler, professor of chemistry at the University of Heidelberg. In 1854, Wöhler, together with Henri Deville, produced aluminum through a reaction with sodium metal. Aluminum was first produced by reaction with potassium. Both of these processes proved to be very expensive. Because of its price (around $500 per pound, or $1.10 per gram), European royalty once used aluminum instead of other precious metals for cutlery to impress their guests. In 1886, Charles Hall developed a new method for producing aluminum using electrolysis. The *Hall method* involved the electrolysis of a molten solution of alumina in cryolite or sodium aluminum fluoride (Na_3AlF_6) at temperatures around 1745°F (950°C), which separated the aluminum. Hall went on to found the Aluminum Company of America (ALCOA). The Hall method brought the price of aluminum down to as low as 15¢ per pound, or 0.0333¢ per gram.

Bauxite ore is the most economical of the aluminum ores. Bauxite is also known as gibbsite, beohmite, and diaspore. Bauxite was first found near the French town of Le Baux; hence its name. The majority of domestic bauxite comes from the state of Arkansas; however, the majority of bauxite used in the United States comes from foreign sources in Surinam, Jamaica, and Guyana. Bauxite normally contains iron oxides and other minerals. This presents some difficulty in extracting the aluminum metal. In general, the easier a metal is to oxidize, the more difficult it is to decompose. Because aluminum is a very active metal, it will not react with carbon to reduce the oxide, as do iron and copper. The bauxite must be purified prior to electrolysis. Pure aluminum oxide is extracted using the *Bayer process*. Bauxite is mined near the surface in an open-pit or strip mine. The ore is crushed and washed to remove undesirable materials and then dried. The dried powder is then mixed with soda ash (Na_2Co_3), lime (CaO), and water(H_20, where it forms sodium aluminate ($Na_2Al_2O_4$). After filtering, aluminum hydrate [AlO(OH)] is precipitated from the solution and heated to temperatures of about 2000°F (1100°C) to form aluminum oxide (Al_2O_3). This product is approximately 99.6% pure. The aluminum oxide is then electrolyzed using the Hall method and placed in a container containing cryolite at 1800°F (990°C). Large carbon electrodes are lowered into the molten solution, and a large direct current is applied (around 100,000 A). The electrodes are positively charged, whereas the lining of the container is the negative electrode. The product of the Hall method is metallic aluminum, which is drained off the container and cast into ingots. The aluminum produced by the Hall method is approximately 99.5% to 99.8% pure, with the major impurities being iron, manganese, and silicon. The aluminum production process is given in Figure 3–1.

Figure 3–1
Aluminum production process.

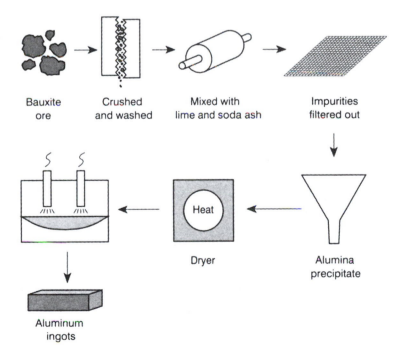

Bauxite
ore

Crushed
and washed

Mixed with
lime and soda ash

Impurities
filtered out

Heat

Dryer

Alumina
precipitate

Aluminum
ingots

Aluminum has a number of properties that make it a very useful engineering material. It is corrosion resistant and lightweight, with a conductivity approximately 60% that of copper. However, the per-pound conductivity of aluminum is twice that of copper and makes it attractive for use in electrical lines that span long distances. Pure aluminum has low strength, but alloying elements are used to increase its strength and strength-to-weight ratio. Due to its face-centered cubic structure, aluminum is ductile and easily shaped through a wide variety of common methods.

Because aluminum is very chemically active, it attracts oxygen, and any fresh surface oxidizes rapidly. Aluminum trim, windows, and structural frames often appear slightly dull-grey in appearance. This color is due to the oxidized surface layer of the aluminum. Unlike ferrous metals, the oxidized surface of aluminum sticks to the metal and provides a protective coating, which prevents further attack. This fact is the basis for *anodizing* of aluminum. The anodizing process involves placing an anode of aluminum in an *electroplating* cell with oxalic, sulfuric, or chromic acid as the plating solution, or *electrolyte*. Current is applied to the solution, and the anode is plated with an oxide coating that is hard and wear resistant. Anodized coatings give the metal a better appearance and may be colorized by using different metals in the electrolyte or by dying the metal.

The Aluminum Association has devised a numbering method for cast and wrought aluminum alloys. The system for wrought aluminum alloys is a four-digit system. The first digit indicates the alloying elements in the alloy. The second digit indicates the alloy modifications or degree of control of the impurities. The third and fourth digits are arbitrary numbers that identify a specific alloy or indicate the purity of the alloy over 99%. Figure 3–2 illustrates the four-digit numbering system for wrought aluminum alloys.

Some of the more common aluminum alloys are silicon alloys, used for castings; copper alloys, for machining; magnesium alloys, for welding; pure aluminum, for forming; magnesium and silicon alloys, for extrusion; and copper alloys, for strength. Typical examples of specific alloys are as follows:

- 2011 with 5% to 6% copper is a free-machining alloy.
- 2024 contains between 3.8% and 4.9% copper with 1.5% magnesium. This alloy is a heat-treatable aluminum alloy that is commonly used for aircraft parts.

Figure 3–2
Numbering system for wrought aluminum alloys.

Number	Major Alloying Element
1---	None
2---	Copper
3---	Manganese
4---	Silicon
5---	Magnesium
6---	Magnesium and silicon
7---	Zinc
8---	Other
9---	Unused

- 3003 has 1% to 1.5% manganese, which provides additional strength.
- 4043 contains 4.5% to 6.0% silicon and is commonly used in welding wire.
- 5154 contains 3.1% to 3.9% magnesium; it is weldable and is available in sheets, plates, and many structural shapes.
- 6063 contains approximately 0.5% magnesium and silicon and is used in windows, doors, and trim.

The casting aluminum alloy numbering system is a three-digit numbering system that is not generally standardized. The more common system is the Aluminum Association's designation. This system places silicon casting alloys up to 99, silicon-copper from 100 to 199, magnesium from 200 to 299, and silicon-manganese from 300 to 399. These designations refer to specific alloys whose individual recipes are available from the Aluminum Association. These casting alloys are used for sand castings, permanent mold castings, and die castings.

The internal structure of aluminum alloys can be modified by heat treatment. When heat-treated, alloys are followed by a heat-treatment designation. For example, 2024-T6 is an example of a copper-aluminum alloy that has been solution heat-treated and then aged artificially. Only the copper, zinc, and magnesium-silicon alloys can be age-hardened. Wrought alloys, which are not heat-treatable, are given either an O (annealed) suffix or an F (as-fabricated) designation. Other heat-treatment designations are shown in Figure 3–3.

Designation		Definition
F		As fabricated
O		Annealed
H		Strain-hardened
	H1*	Strain-hardened only
	H2*	Strain-hardened and partially annealed
	H3*	Strain-hardened and then stabilized
W		Solution heat-treated, unstable temper
T		Treated to produce stable temper
	T2	Annealed
	T3	Solution heat-treated and then cold-worked
	T4	Solution heat-treated and then naturally aged
	T5	Artificially aged
	T6	Solution heat-treated and then artificially aged
	T7	Solution heat-treated and then stabilized
	T8	Solution heat-treated, cold-worked, and then artificially aged
	T9	Solution heat-treated, artificially aged, and then cold-worked
	T10	Artificially aged and then cold-worked

Figure 3–3
Aluminum heat-treatment designations.

*Denotes a number following the suffix designation that describes the degree of cold-working or strain-hardening for the material. For example, 1 indicates one-eighth hard, 2 is one-quarter hard, 4 is one-half hard, 6 is three-quarters hard, 8 is full hard, and 9 is extra hard.

Because of aluminum's many properties, it can be used in a wide variety of applications. For example, its electrical conductivity makes it a good conductor in electrical and electronics applications. Its light weight makes it desirable in structural applications that require medium strength and low weight. Its high reflectivity for infrared and visible radiation makes it desirable in headlights, light fixtures, and insulations. In flake form, aluminum is used as a paint pigment. Cast aluminum engine blocks and pistons are a common use for the aluminum casting alloys.

3.3 CHROMIUM

Chromium was discovered in 1797 by Dr. Louis Vauguelin, a professor of chemistry at the College of France. Chromium was named for its "colorful" nature. Chromic oxide (Cr_2O_3) has a dark green color, potassium chromate ($KCrO_4$) is bright yellow, potassium dichromate ($K_2Cr_2O_7$) is orange, chromium trioxide (CrO_3) is red, and lead chromate ($PbCrO_4$) is yellow. Chromium is the third hardest element; only boron and diamond are harder. It is extremely resistant to corrosion and is often used as a corrosion-resistant alloy or as a plating material.

The primary chromium ore is chromite ($FeOCr_2O_3$), typically found in Albania, Rhodesia, Russia, Turkey, and Iran. The reduction process involves grinding and crushing the ore to a powder. It is then reacted with powdered aluminum to release iron and chromium. The chromium can then be refined by electrolysis, but pure chromium is not always required or desired. Most chromium is used in alloy form.

The uses of chromium as an alloy for ferrous metals, such as high-strength steels and stainless steels, have already been discussed. Chromium is also used as a plating material to provide a hard, corrosion-resistant surface over other materials. Chromium will not stick to steel very well, but it will adhere to nickel. The triple-plating process is used to plate steels. First, the steel is degreased and cleaned well. It is then etched with nitric acid; this process roughens the surface of the steel and gives the plating something to hold onto. The steel is first given a thin layer of copper. After washing, the copper is then covered with a thin layer of nickel. The part is then washed again, and a final layer of chromium is deposited on the surface of the part. Coats of about 0.0002 in. provide a shiny, decorative surface, whereas platings of 0.05 in. are used to provide wear resistance. Chromic acid is used as the electrolyte plating solution. Chromium is important as an alloy in stainless steels, nickel alloys, refractories (discussed later), and bronzes, as well as being a plating material for many metals.

3.4 COPPER, BRASS, AND BRONZE

Copper is one of the oldest metals. It appears in the face-centered cubic structure. Early civilizations worked copper by hammering it into desired shapes or by melting it and casting it. There are many copper ores in the oxide (cuprite), sulfide (chalcopyrites, bornite, chalconite, and covellite), carbonate (malachite and azurite), or silicate form (chrysocolla). Many of these ores are found close to the earth's surface.

Copper is known for its high thermal and electrical conductivity. The thermal conductivity of copper is almost 10 times that of steel. This makes it preferable for chill, casting molds, and applications that require the fast removal of heat. The melting point of pure copper is 1981°F (1092°C). However, oxides form when copper is exposed to heat or environmental conditions for prolonged periods. Surface treatments help preserve the appearance of copper in these cases. Copper for electrical applications must be relatively pure. The addition of even small amounts of alloys greatly reduces the conductivity of the copper. Silver, cadmium, and gold are sometimes added to increase the strength of copper without significantly reducing its conductivity. Copper and copper alloys are used for tubing and pipe and in heat-transfer applications. However, many compounds of copper are toxic, which prevents its use in food-related applications. Two particular alloys are used: brass, which is an alloy of copper and zinc, and bronze, which is an alloy of copper and elements other than zinc.

The copper smelting process is a multistep procedure. Ores that contain silica (sand), aluminum oxide (clays), and other unwanted materials must first be cleaned through the *flotation process*. This involves grinding the ores into powder and placing the powder in water. A foaming material such as soap is added to create a froth,

Once installed, this conveyor belt, itself weighing 15 tons, will move millions of tons of crushed copper ore.

Courtesy of Goodyear.

which brings the copper ore to the surface. The ore is then skimmed off, leaving the undesirable materials in the water. These concentrated ores may contain iron sulfide. Iron sulfides must be converted to iron oxides by roasting. This is accomplished by heating the ore in an oven. After roasting, the ore contains copper oxides, iron sulfides, copper sulfide, iron oxides, iron sulfates, silicates, and other impurities. The ores are then placed in a smelting furnace, where they are melted at temperatures of about 2600°F (1438°C). Once melted, the mixture is called *matte copper*, which contains about 30% copper. This mixture is then placed in a converter with a flux (usually silica), where air is blown through the melt. The oxygen in the air reacts with the sulfur in the melt, much like the steel-conversion process, and produces reasonably pure copper. The copper is then drained off and poured into ingots. The sulfur dioxide produced in the conversion process bubbles out as the copper cools. This gives the copper the name *blister copper*. This copper is 98% to 99% pure. The slag drawn off the mixture in the conversion process is further refined to extract other metals, such as gold and silver, which may be present in the ore.

As previously discussed, even small amounts of impurities reduce the conductivity of the copper. A large percentage of the copper produced is used in the electrical and electronics industries. The impurities in the copper must be removed by *electroplating*. Blister copper is remelted and cast into plates called anodes (+). Refined copper cathodes (–) are placed on the other side in staggered pairs, as shown in Figure 3–4. These plates are immersed in a plating solution of copper sulfate. The anode is connected to the positive terminal of a direct current source, and the cathode is connected to the negative terminal. When an electric current is applied, the metal in the anode goes into the solution, and the copper is plated on the cathode. The impurities in the anode metal are left in the copper sulfate solution. The plating on the cathode will be in excess of 99.9% pure copper.

Copper to be used in electrical wire is remelted using an oxidizing flame to prevent sulfur from being reabsorbed into the copper and to keep the oxygen content at less than 0.04%. Copper produced in this manner is called *electrolytic tough pitch*

Figure 3–4
Electroplating process.

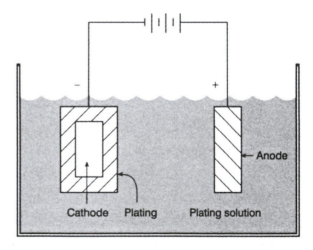

Cathode   Plating   Plating solution

Anode

Table 3–1
Coppers and copper alloys.

Name	Composition (%)	Application
Coppers		
Electrolytic tough pitch copper (ETP)	94.40 Cu, 0.04 O	Roofing, nails, rivets, cotter pins, kettles
Phosphorous deoxidized copper (DHP)	94.40 Cu, 0.02 P	Piping, tubing, heat-transfer equipment, tanks
Free-cutting copper	99.5 Cu, 0.5 Te or 0.6 Se, 1.0 Pb	Electrical connectors, motor and switch parts, oxyacetylene torch tips
Oxygen-free (OF) copper	99.92 Cu	Electrical conductors, bus bars, tubing
Brasses		
Red brass	80 Cu, 20 Zn	Trim, conduit and sockets, fire extinguishers
Cartridge brass	70 Cu, 30 Zn	Radiator cores and tanks, reflectors, munitions
Yellow brass	65 Cu, 35 Zn	Radiators, lamp fixtures, plumbing supplies
Medium-leaded brass	64 Cu, 34.75 Zn, 1 Pb	Engravings, gears, nuts, couplings
High-leaded brass	62.5 Cu, 35.75 Zn, 1.75 Pb	Nuts, gears, wheels, clock parts
Free-cutting brass	61.5 Cu, 35.5 Zn, 3 Pb	Gears and pinions
Muntz metal	60 Cu, 40 Zn	Hardware, panels and sheets, forgings
Naval brass	60 Cu, 39.25 Zn, 0.75 Sn	Nuts and bolts, propeller shafts
Forged brass	60 Cu, 38 Zn, 2 Pb	Forgings, valve stems, plumbing supplies
Bronzes		
Aluminum bronze	95 Cu, 5 Al	Sheets, wire, tubing
Silicon bronze—grade A	95 Cu, 3 Si, 2 Mn	Hydraulic lines, marine hardware
Silicon bronze—grade B	95 Cu, 2 Si, 1 Mn	Hydraulic lines, marine hardware
Phosphor bronze—grade A	95 Cu, 5 Sn	Bearing plates, clutch discs, bushings
Phosphor bronze—grade C	92 Cu, 8 Sn	Springs, brushes, textile machine parts
Aluminum-silicon bronze	91 Cu, 7 Al, 2 Si	Forgings and extrusions
Phosphor bronze—grade D	90 Cu, 10 Sn	Bars and plates, fittings
Commercial bronze	90 Cu, 10 Zn	Munitions, jewelry, plaques and awards
Leaded commercial bronze	89 Cu, 9 Zn, 2 Pb	Screws, hardware
Cupro-nickel	88.5 Cu, 10 Ni, 1.5 Fe	Condensers, ferrules, tanks, valves, auto parts
Nickel-silver	65 Cu, 18 Ni, 17 Zn	Camera parts, plates, fixtures, tableware, zippers
Manganese bronze	59 Cu, 39 Zn, 1.5 Fe, 1 Sn, 0.1 Mn	Shafts, clutch parts, valve stems, forgings, welding rods
Architectural bronze	57 Cu, 40 Zn, 3 Pb	Trim, extrusions, hinges, auto parts

(ETP) copper. Phosphorus (approximately 0.02%) is sometimes added to the melt to control the amount of oxygen in the copper. This product is called *oxygen-free high conductivity* (OFHC) or *phosphorus deoxidized* (DHP) copper.

Copper is very useful in electrical applications. At very low temperatures, approaching absolute zero, copper becomes a *superconductor*. Superconductors have very little resistance to current flow. A current started in a superconducting circuit will continue to flow almost indefinitely. Research continues to find the perfect superconductor. Various alloy combinations have been found to lower the resistance to current flow to astounding levels. The purpose of this research is to find an alloy with superconducting capabilities that can be maintained at reasonable temperatures. This product will provide electrical transmission over long distances with little loss of energy. The applications in communications, transportation, manufacturing, medicine, and other fields are amazing. Magnetic resonance imaging (MRI) devices used in hospitals for diagnosing patients are a current example of a medical application of superconductivity. Magnetic levitation (mag-lev) applications will also increase significantly in the future.

Copper alloys are among the oldest metallic alloys. Alloying elements are added to the copper to increase the strength, hardness, machinability, appearance, and cost. The melting points of copper alloys are lower than that of pure copper. Aluminum, beryllium, lead, manganese, nickel, phosphorus, silicon, tin, and zinc are all common alloying elements of copper. The copper-zinc alloys are referred to as *brasses*. Zinc is added to increase the strength, improve the ductility, and (along with lead) improve the machinability. The copper-tin alloys are called *bronzes*. Tin is added to improve the strength, hardness, and ductility of the alloy while reducing the overall cost. Table 3–1 lists the names, compositions, and typical uses of some of the more common copper alloys generally available.

There are many formulas for brass, as shown in Table 3–1. There are more alloying elements in brasses than copper and zinc alone. Because brass and bronze are multicomponent systems, each has a phase diagram. Figure 3–5 illustrates the copper-zinc phase diagram for brasses. As shown in the copper-zinc phase diagram, up to 36% zinc may dissolve into copper in a single-phase state. These brasses are known as *alpha brasses*. The alpha (α) brasses contain up to 36% zinc and are face-centered cubic (FCC) in structure. The beta phase is body-centered cubic (BCC). Alpha plus beta brasses contain 38% to 46% zinc. Yellow brass and red brass are two common varieties of brasses. Red brasses have less alloy content and offer better corrosion resistance than the yellow variety. Red brasses are among the most ductile and malleable of all brasses.

The term *bronze* refers to those metal alloys that contain copper alloyed with any other metal. There are many different formulas for bronzes, as shown in Table 3–1. Figure 3–6 illustrates the copper-tin phase diagram for bronzes. Traditionally, copper and tin were combined to form tin bronzes. Phosphorus is sometimes added in small amounts to improve the ductility of tin bronzes. These are commonly referred to as *phosphor bronzes*, although they contain only a small percentage of phosphorus (1% to 11%). Phosphorus is also used in small amounts as a deoxidizer. Bronzes that con-

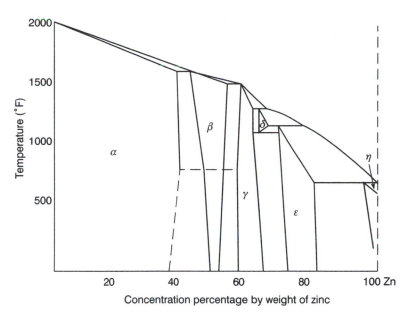

Figure 3–5
Copper-zinc phase diagram.

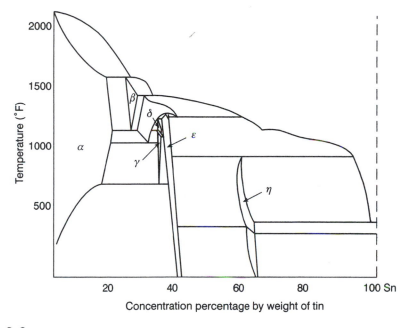

Figure 3–6
Copper-tin phase diagram.

tain more than 90% copper appear reddish in color. Aluminum bronzes are heat-treatable and are among the highest-strength alloys of copper. Their high strength and corrosion resistance make them useful as structural materials. Silicon bronzes are high-strength alloys that are often used in marine applications. Cupronickels are single-phase alloys of copper and nickel, such as monels. They are used primarily for tubing, due to their resistance to attack by fresh and salt water. The beryllium bronzes contain 2% or less beryllium. This combination produces a heat-treatable bronze, which has the highest strength of the copper alloys. Beryllium bronzes are nonsparking, meaning they will not spark when struck with another metal. This feature makes them ideal for working with explosives, fuels, and flammable materials. Nickel silver and German silver are names often given to nickel bronzes. Nickel silver has been used in stamping coins. Zirconium-copper and chromium-copper alloys are heat-treatable combinations used primarily in instrument and electrical hardware.

3.5 MAGNESIUM

Magnesium, discovered in 1808 by Sir Humphrey Davy, is the lightest of the structural metals. Pure magnesium weighs two-thirds as much as aluminum. It is derived primarily from seawater. One pound (454 g) of magnesium can be obtained from every 100 gal (378 L) of seawater. Like most of the light structural metals, magnesium is hexagonal-close-packed in structure. In order to produce magnesium, seawater is filtered through lime [Ca(OH)$_2$] and oyster shells. This process converts the magnesium

Figure 3–7
Magnesium production process.

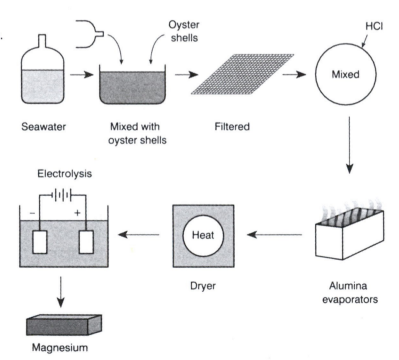

to magnesium hydroxide, which precipitates out of the water. Hydrochloric acid is added to convert the magnesium hydroxide to magnesium chloride. Once the magnesium solution is dried, electrolysis decomposes the magnesium chloride into magnesium metal and chlorine gas. The chlorine gas is recycled into hydrochloric acid, and the magnesium is drawn off. This process is shown in Figure 3–7.

Magnesium is an active metal. It was first used in incendiary bombs because it burns with an extremely hot flame, giving off an intense light. Magnesium chips are readily ignited, which makes it dangerous to try to gas-weld magnesium parts. Its chemical activity is an advantage used in anodes for protecting water tanks, piping, and other applications susceptible to galvanic attack. It is also used as an alloying element in ferrous metals, such as ductile cast iron. Most magnesium is used in lightweight applications where strength is required, such as extension ladders, space vehicles, aircraft, power tools, loading ramps, and similar applications.

The American Society for Testing and Materials (ASTM) has devised a system for designating magnesium alloys. This system uses two letters to designate the primary ingredients, followed by two numbers giving the percentages of the primary ingredients to the nearest whole number. A serial letter follows the digits. Examples are AZ81A, AZ92A, EK42A, and AZ91C. Thus, AZ91C would be used to designate a magnesium alloy containing aluminum and zinc as its primary alloy components. The 9 designates that it contains roughly 9% aluminum, the 1 indicates approximately 1% zinc, and the C indicates that this is the third (A—1, B—2, C—3) alloy qualifying for the AZ91 designation. Figure 3–8 illustrates the designation system for alloying elements of magnesium.

Figure 3–8
Designation system for alloying elements of magnesium.

Letter	Element
A	Aluminum
B	Bismuth
C	Copper
D	Cadmium
E	Rare earths
F	Iron
G	Magnesium
H	Thorium
K	Zirconium
L	Beryllium
M	Manganese
N	Nickel
P	Lead
Q	Silver
R	Chromium
S	Silicon
T	Tin
Y	Antimony
Z	Zinc

Most of the magnesium alloys are of the AZ type. Aluminum-magnesium (AM), zinc-zirconium (ZK), zinc-thorium (ZH), and some of the rare-earth combinations (EK and EZ) are used for casting. These are some of the more common magnesium alloys.

3.6 NICKEL

Nickel is a remarkable metal that closely resembles steel in many properties. Historically, nickel was used in the production of the U.S. 5¢ coin, which is an alloy of 75% copper and 25% nickel. Most of the world's commercial nickel ores are supplied by Canada, Russia, and Australia in the form of pentlandite [$(FeNi)_9S_8$] and pyrrhorite (iron sulfide with nickel). The percentage of nickel in pentlandite is only about 1%. Nickel is relatively expensive to refine.

There are two methods for refining from the sulfide ores. One process, the *Mond process*, does not work for all nickel ores. In this process, carbon monoxide gas is washed over the heated ore, which converts the nickel to nickel carbonyl [$Ni(CO)_4$]. Nickel carbonyl is very volatile; it turns from a solid directly into a gas at a temperature of 1783°F (973°C). The nickel carbonyl is decomposed into metallic nickel dust and carbon monoxide. The other process resembles the method used for extracting copper. In this process, the ore is mined, crushed and ground, washed, and concentrated by the flotation process. The ore is then roasted and smelted in an electric furnace to produce the matte. Next, the matte is placed in the converter, where air is blown through the molten metal to produce blister nickel. The blister is remelted and cast into anodes, which are refined in an electrolytic cell. The nickel is plated on the cathodes, after which it is removed for fabrication or to be used as an alloying element.

Nickel ores typically contain one or more of the following materials, among others: cobalt, gold, selenium, silver, and tellurium. These materials are present in the slag from copper and nickel refining and are extracted when feasible. About 80% of the nickel refined is used in steel alloys; about 10% goes toward nickel-plating, and the remaining 10% is used as nickel and nickel alloys.

Nickel is hard, is moderately corrosion resistant, and polishes well. Nickel is face-centered cubic in structure and is not prone to brittleness at low temperatures. It is therefore used in cryogenic applications at very low temperatures and in gas turbine and rocket engines. No other metal has such a wide temperature range in application. The most common application is as an alloying element. Most companies produce nickel-alloyed materials under proprietary trade names. Alnico, Inconel, and Hastelloy are all examples of these materials (Figure 3–9). Alnico is used in very powerful permanent magnets, Alumel and Chromel are used in thermocouples and temperature-control devices, Hastelloy is used where high strength and corrosion-resistance are required, and Nichrome is used in wire for heating elements. Many rechargeable batteries contain a nickel-cadmium alloy, which is the origin of the name Nicad batteries. Nickel silver, or German silver, is an alloy of cobalt, copper, nickel, and zinc that is used in coinage and jewelry. There are also many *superalloys* made from nickel. A beryllium-nickel alloy containing 98% nickel, 1.5% beryllium, and 0.5% manganese is one example of a superalloy. Most of these alloys, like most

Designation	Composition (%)
Alnico I	67 Fe, 20 Ni, 12 Al, 5 Co
Alnico VIII	35 Co, 34 Fe, 15 Ni, 7 Al, 5 Ti, 4 Cu
Alumel	95.3 Ni, 1.75 Mn, 1.6 Al, 1.2 Si, 0.1 Fe
Chromel A	80 Ni, 20 Cr
Chromel C	60 Ni, 24 Fe, 16 Cr
Chromel D	47 Fe, 35 Ni, 18 Cr
Chromel P	90 Ni, 9.5 Cr, 0.4 Si, 0.2 Fe
Constantan	55 Cu, 45 Ni
Duranickel	94 Ni, 4.5 Al, 0.5 Ti, 0.25 Mn, 0.15 Fe, 0.15 C
Hastelloy A	53 Ni, 22 Fe, 22 Mo, 2 Mn
Hastelloy C	55 Ni, 17 Mo, 6 Fe, 4 W, 1 Mn
Hastelloy X	45 Ni, 24 Fe, 22 Cr, 9 Mo
Inconel 600	76 Ni, 16 Cr, 7 Fe, 0.2 Mn, balance C, S, Cu
Inconel X-750	73 Ni, 15 Cr, 7 Fe, 2.5 Ti, 0.85 Nb, 0.8 Al
H monel	63 Ni, 31 Cu
K monel	66 Ni, 29 Cu, 2.75 Al, balance Fe, Si, Mn, Ti
R monel	67 Ni, 30 Cu
Z monel	98 Ni, 2 Cu
Nichrome V	80 Ni, 20 Cr
Nickel silver	57 Cu, 25 Zn, 15 Ni, 3 Co
Permanickel	98.6 Ni, 0.5 Ti, 0.35 Mo, 0.25 C, 0.1 Fe, 0.1 Mn

Figure 3–9
Common alloys of nickel and their designations.

of the nickel alloys, are available under trade names. For example, Chromel is a product of Hoskins Manufacturing Company, Inconel is produced by the International Nickel Company, and Hastelloy belongs to Union Carbide Corporation. These are a few examples of the more common nickel alloys.

3.7 PRECIOUS METALS

The precious metals traditionally include gold, silver, and platinum. They are called "precious" because of their value and use in jewelry and decoration. In the past, the face value of a coin was the value of the metal from which it was made. For example, a $20 gold piece contained $20 worth of gold. Today, the coin represents only the value we place on the coin, and the coin itself is made of cheaper metals. Precious metals have found limited applications in industry.

Gold is found as nuggets, as dust, and in quartz rock. Although there have been periods where many people have focused their efforts the world over on prospecting for gold, most of today's gold comes from South Africa. Gold may be extracted from quartz ores by reacting them with mercury or cyanide.

The properties of gold, including its electrical conductivity, corrosion resistance, and malleability, make it a valuable material. Gold is also used as a plating material, being electroplated from a gold chloride solution. Gold is too soft to be used in a pure form, so it is often alloyed with other metals. Copper, nickel, and platinum are common alloying elements of gold. Gold exists in a face-centered cubic structure. In addition to jewelry and coinage, gold is used in dental work as caps, crowns, and fillings. Typical dental gold alloys contain up to 70% gold, 5% platinum, 5% palladium, 25% silver, 18% copper, 3% nickel, and 1% zinc. These percentages are maximum percentages of the dental alloys. The purities of gold and gold alloys are given on a carat scale. Twenty-four carat gold is pure gold, 18 carat gold is three-fourths gold, 12 carat gold is 50% gold, and so forth.

The chemical name for silver comes from the Latin *argentum*. Silver also occurs in nature and in argentite (Ag_2S) and horn silver ($AgCl$). It is also found in other metallic ores. It is face-centered cubic in structure, exhibiting excellent malleability and ductility. Until 1964, silver was used in U.S. coins. When the price of silver exceeded the face value of the coins, it was replaced. The British "sterling" silver is actually an alloy of silver and copper.

Silver is used to plate other metals, as electrical conductors, in producing jewelry, and in the making of light-sensitive compounds for photographic materials. Approximately 30% of all silver goes toward photographic films and papers. Silver was also used in making mirrors. Silver iodide is used for seeding clouds to induce rain. Silver is also used in the manufacture of photochromic (light-sensitive) lenses for glasses. Photochromic glasses darken when exposed to light. Other uses for silver include brazing alloys and silver-cadmium batteries. Silver fulminate ($Ag_2C_2N_2O_2$) is used as an explosive. Silver is also used in ointments, salves, and creams for medicinal purposes. Silver and gold are often sold by the *troy ounce*, where 12 oz equal a pound.

The platinum group contains six metals, which are extracted from nickel ores. Of these six metals—*iridium, osmium, palladium, rhodium*, and *ruthenium*—platinum is probably the most useful. Platinum, palladium, iridium, and rhodium are face-centered cubic in structure, whereas osmium and ruthenium are body-centered-cubic. All six have high melting points (greater than 3000°F [1662°C]).

Platinum is found in nature and in the mineral sperrylite ($PtAs_2$). It is used in corrosion-resistant coatings and as a catalyst for many reactions. It is also used in high-resistance wire for furnaces. Platinum is also used in catalytic converters on modern automobiles, where it converts unburned hydrocarbons and carbon monoxide to carbon dioxide and water. A large percentage of platinum is used in laboratory equipment, medical instruments, and fine jewelry. The main disadvantage of using platinum is its cost; it is more expensive than gold.

3.8 REFRACTORY METALS

Refractory metals, by definition, have melting points above 3600°F (2000°C), some of them reaching as high as 6150°F (3426°C). This includes such metals as iridium, osmium, and ruthenium, which are in the platinum group. Common refractory met-

Metal	Melting Point, °F (°C)	Oxide	Melting Point, °F (°C)
Chromium	2,822 (1,562)	Chromia (Cr_2O_3)	4,230 (2,351)
Hafnium	3,866 (2,147)	Hafnia (HfO_2)	5,140 (2,860)
Molybdenum	4,730 (2,631)	Molybdenum oxide (MoO_3)	1,465 (802)
Niobium	4,380 (2,435)	Niobia (Nb_2O_5)	2,500 (1,382)
Rhenium	5,730 (3,191)	Rhenia (Re_2O_7)	565 (298)
Tantalum	5,425 (3,020)	Tantalum oxide (Ta_2O_5)	2,500 (1,382)
Tungsten	6,152 (3,427)	Tungsten oxide (WO_3)	1,830 (1,007)
Vanadium	3,450 (1,914)	Vanadia (V_2O_5)	1,270 (693)

Figure 3–10
Refractory metals, their oxides, and their melting points.

als include *chromium, columbium, hafnium, molybdenum, niobium, rhenium, tantalum, tungsten*, and *vanadium*.

Such metals are used in high-temperature applications, where they exhibit high strength, but they oxidize rapidly at low temperatures. A list of the common refractory metals and the lowest-melting oxide formed from each metal is given in Figure 3–10.

At present, niobium and tantalum are the preferred refractory metals, based on their ductility, high melting point of the oxide, and lower cost. Hafnium is also used to a lesser extent, based on its higher cost.

Hafnium is present in all zirconium ores, and this is its most economical source. It is commonly used in rectifiers and other electronics applications. Rhenium is a high-strength material used in structural applications under vacuum and for filaments in high-power vacuum tubes. Vanadium has many ores and up to 1% is added to steels to enhance the grain structure. Tantalum is used in capacitors and in tantalum-foil rectifiers. Columbium is used in the nose sections and leading-edge components of supersonic aerospace vehicles.

3.9 TITANIUM

Titanium can be readily identified by its light weight, its bluish or silvery color, and its blue grinding sparks. It was first discovered in 1791 by an English clergyman, William Gregor, although it was named by a German chemist Martin Klaproth in 1795. It is found in the minerals rutile (TiO_2), ilmenite ($FeTiO_3$), and sphene (CuTi-SiO_4), among other minerals. Titanium is found in Australia, Canada, Africa, and Russia.

Titanium first became important in the metals industry in the 1950s when research in the aircraft industry demanded higher strengths at higher temperatures. Commercial production of titanium exceeds 15,000,000 lb (68,100,000 kg) annually. Production starts with the mined ore, which is crushed and powdered. The ore is then screened to remove the titanium crystals. The titanium oxides are then reacted with chlorine gas to convert them to titanium tetrachloride. Titanium tetrachloride is used in skywriting and tracer bullets because it forms a dense white cloud when mixed with water. Once the titanium oxide is converted, a stainless steel tank is filled with magnesium or sodium metal and welded shut. Titanium tetrachloride is then run through the tank containing the metal at about 800°F (430°C) for 48 hours. The titanium metal collects on the steel. The tank is cut open, and the titanium is chipped off the tank. The metal is collected, remelted, and cast into ingots.

Titanium has a hexagonal-close-packed structure, in its alpha phase, below 1620°F (882°C). Above this temperature, the titanium goes into a body-centered cubic, beta phase. Titanium is used in high-speed aircraft for its high strength at high temperatures. It is also used in the medical profession. It is lighter than stainless steel when used in medical instruments and is used in artificial replacement joints for two reasons: (1) It is the most biocompatible metal known, and (2) it has a stiffness similar to that of the human bone. Titanium is often used in conjunction with polyethylene and high-purity ceramics, both of which are discussed later. Titanium forms a thin layer of oxide, which protects against further corrosion, much like aluminum and nickel. Titanium may also be colored by passing an electric current through the part and applying an electrolyte to the surface, such as ammonium sulfate. Titanium and tinted titanium are used in watches, pens, cameras, and other similar applications. Titanium oxide is used as a pigment in paints.

3.10 WHITE METALS: LEAD, TIN, AND ZINC

White metals include antimony, bismuth, cadmium, lead, tin, and zinc. These elements have a wide variety of uses. This section deals primarily with lead, tin, and zinc.

Lead has been used since the days of the Romans. The Romans used lead pipes for carrying water. The chemical term for lead, *plumbum,* comes from the Latin name for lead. The terms *plumbers* and *plumbing* are derived from this term. The common minerals from which lead is extracted are anglesite ($PbSO_4O$), cerrusite ($PbCO_3$), minum (Pb_3O_4), and galena (PbS). Of these, galena is the most common ore used. Lead is easily obtained from these ores by roasting to convert the sulfides to oxides and then smelting with carbon to produce lead and carbon dioxide.

Pure lead melts at 621°F (330°C). Lead is versatile and is used for its high density, ductility, low melting point, corrosion resistance, and ability to lubricate. It is a poor conductor of electricity and has been found to be toxic. The greatest use of lead is in storage batteries. Battery plates contain about 10% antimony, with smaller amounts of arsenic, copper, and tin. Lead was used for many years as white, yellow, and red paint pigments. It was also used as an octane booster in automotive fuels as tetraethyl lead [$Pb(C_2H_5)_4$]. It has been replaced because of its toxicity to humans. Due to its density, it has been used as radiation shields for X-rays and other radioactive applications. A summary of the uses of lead is given in Figure 3–11.

Lead is often used to form useful alloys with other metals. It is employed in free-machining steels and brasses, although lead is only one of the alloying elements

Figure 3–11
Common uses for lead and lead alloys.

1. Corrosion-resistant tank linings
2. Lead-acid storage batteries
3. Type metal for the printing trade
4. Bearing metals
5. Biological shield against radiation
6. Pipe and pipe fittings
7. Cable shielding
8. Low-melting solders
9. Alloying element for various metals
10. Thermal fuses

used. To increase the strength of lead, tellurium and antimony are often used as alloying elements. Lead is used in babbitt, an alloy containing 80% lead, 15% antimony, and 5% tin, from which crankshaft bearings are made. Solders are lead-tin alloys where the amount of tin ranges from 5% to 50%. The tin in solders allows the solder to "wet," or bond with, the metals to be joined. Terne steel is a solder-coated steel used in automobile gasoline tanks and other equipment. Terne steel coatings are generally about 20% tin.

The chemical symbol for tin comes from the Roman word *stannum*. Tin is one of the major components of solders and pewter. Pewter is a decorative metal that once contained lead. Modern pewter contains 91% tin, 7% antimony, and 2% copper. One common use of tin is as an alloying element for other metals.

Tin predominantly comes from cassiterite (SnO_2), chiefly found in Bolivia, Indonesia, Malaya, Nigeria, and Thailand. Tin is extracted from the oxide ore by smelting it in the presence of carbon. Pure tin takes on several structures. Below 56°F (13°C), tin is in the alpha state, which is cubic. Above 56°F (13°C), tin converts to a beta phase, which takes on a tetragonal structure.

Tin is often used as a plating material for steels, in solders to help the solder "wet" the base metals, and in babbitts when mixed with antimony and copper. It is also used in the production of tinfoils, which contain approximately 92% tin and 8% zinc.

Zinc is commonly used for galvanizing steels. It is the white coating on the outside of such galvanized metals. It is used as an anodic coating in the hot-dip process or in electrogalvanizing (discussed later). As long as there is zinc available, the zinc will give up electrons to the iron, which protects the iron from corrosion. This process is termed *cathodic protection*. Zinc is used in sheet form as dry-cell casings. It is also used in die castings and as an alloying element for nonferrous metals. Zinc oxide is used in paints, glass, matches, dental cements, and medicines. Perhaps the most recognizable of the die castings of zinc are automobile carburetor, fuel pump bodies, and toys.

Zinc is found in many minerals, including gahnite ($ZnAl_2O_4$), smithsonite ($ZnCO_3$), and wurtzite (ZnS), but it is extracted largely from franklinite ($ZnFe_2O_4$) and zincite (ZnO). These ores are found throughout the world. To produce zinc, the ore is first roasted to convert the sulfides to oxides and then smelted in the presence of carbon. Zinc melts at 787°F (423°C) and boils at 1665°F (914°C). Temperatures used in refinement are about 2200°F (1214°C), which causes the zinc to emerge as a vapor. The vapor is collected and then zinc is reclaimed. Zinc is also extracted by leaching the oxide with sulfuric acid and then using the electrolysis process to refine the metal. Zinc has a hexagonal-close-packed structure.

Antimony, bismuth, and cadmium are also "white" metals. Antimony has a rhombohedral crystalline structure and is used in solders and as an alloying element in lead and tin alloys. Bismuth is also a rhombohedral structure, having the lowest thermal conductivity of any metal except mercury. Bismuth is unique in that it expands upon solidifying, one of only a few that exhibit this characteristic. Cadmium is a hexagonal-close-packed structure. Cadmium is used as a neutron absorber in the control rods found in nuclear power plants.

3.11 SUMMARY

Nonferrous metals are those that do not contain iron as a major alloy. Copper, brass, bronze, aluminum, magnesium, nickel, chromium, titanium, lead, tin, zinc, gold, silver, platinum, and the refractory metals were all covered in this chapter. These metals all come from an ore or are found in nature. These metals are used in a wide variety of applications, depending on their individual properties and characteristics. Many are lightweight, high-strength metals that offer affordable alternatives to the ferrous metals. Others, such as the refractory metals, are used because of their high strength at high temperatures. The precious metals were traditionally used because of their value in coinage and jewelry, but they find other applications. This chapter presented the various refinement processes used for these nonferrous metals, the major alloys of each metal, the various uses and properties of the metals, and the major designation systems used to describe the alloys, where applicable.

QUESTIONS AND PROBLEMS

1. What is a nonferrous metal?
2. What is the difference between brass and bronze?
3. Explain the process of electroplating.
4. Explain the triple-plating process.
5. Of the nonferrous metals discussed, copper, aluminum, magnesium, nickel, chromium, titanium, gold, silver, platinum, lead, tin, and zinc:
   a. What has the highest strength?
   b. Which is the lightest in weight?
   c. Which is the best electrical conductor?
   d. Which is the best thermal conductor?
   e. Which offers the best corrosion resistance?

6. Which of the nonferrous metals are commonly used in plating operations?
7. What is a refractory?
8. For what purpose is most nickel consumed?
9. What element is used as a deoxidizer in copper?
10. Give the alloy composition for the following materials:
   a. Yellow brass
   b. Muntz metal
   c. Cartridge brass
   d. Aluminum bronze
   e. Cupro-nickel

11. Explain the flotation process.
12. What is matte copper?
13. What is blister copper?
14. Define the term *superconductor*.
15. What is the purpose of tin in tin-lead solders?
16. Why are refractory metals not used in low-temperature applications?
17. Explain the Hall method.
18. Describe the Bayer process.
19. Describe the anodizing process for aluminum.
20. Explain the Mond process for nickel.
21. List five common uses for leads and lead alloys.
22. What is meant by the term *cathodic protection*?

23. What factors influence the selection of nonferrous metals?
24. In general, what advantages do nonferrous metals offer over other metals?
25. Select a product made from a nonferrous metal and create a flowchart illustrating how that product was produced from raw material to finished goods.
26. Explain the solution hardening of nonferrous metals. How does solution hardening influence the properties of the metal?
27. What other methods of heat treatment are available for nonferrous metals? Can you think of different methods of influencing nonferrous metal properties?

CHAPTER 4

Heat Treatment

Objectives
- Recognize the major regions and ranges of the iron-carbon phase diagram.
- Define heat treatment and the factors that influence its effectiveness.
- Describe the phases of the quenching process and the various media used.
- Explain the normalizing and annealing processes for metals.
- Describe the purposes and uses of the time-temperature transformation curve.
- Describe the tempering process and its uses.
- Define common terms and processes associated with the hardening of steels.
- List the advantages and disadvantages of the different hardening processes.

4.1 INTRODUCTION

Heat treatment refers to the heating and cooling operations required to alter the properties of metals, plastics, and ceramic materials. Changes in a material's properties result from changes made in the microstructure of the material. In this chapter, we are interested primarily in steels. Because the crystalline structure of steel changes when it is heated, the hardness of steel can be altered by heat treatment. This quality is unique; the crystalline structure of many metals cannot be changed. For example, recall from Chapter 2 that when steel is at room temperature, it is a mixture of ferrite and pearlite. Both of these materials are body-centered cubic (BCC) structures. However, if steel is heated above 1333°F (723°C), it is transformed into austenite, a face-centered cubic (FCC) structure.

Because the rearrangement of steel takes time, if steel is heated into the austenite range and is allowed to cool slowly, it will return to the BCC structure. However, if the steel is heated into the austenite range and is rapidly cooled, the structure does not have enough time to change. This produces a body-centered tetragonal structure called *martensite*, which is the hardest form of steel. The higher the percentage of martensite in steel, the harder the steel. Martensite is formed from pearlite, which has an eutectoid composition of 0.77% carbon. Refer to Figure 4–1, the iron-carbon phase diagram. A steel with 100% pearlite, such as 1080 steel, will transform into 100% martensite if heated into the austenite range and cooled rapidly.

Figure 4–1
Iron-carbon phase diagram.

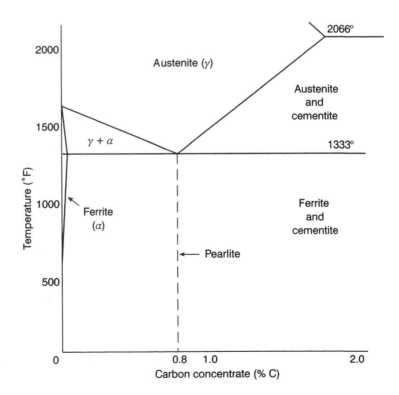

Hypoeutectoid steels (those with less carbon than the eutectoid composition) form less martensite because they contain less pearlite. For example, a 1040 steel has only 50% pearlite and forms only 50% martensite. Steel with higher carbon content can be heat-treated to a much higher hardness.

4.2 IRON-CARBON PHASE DIAGRAMS

Iron and carbon are the two main elements in steels. Their relationship can be plotted in an iron-iron carbide diagram (Figure 4–2). In this diagram, pure iron is plotted along the left side, whereas the carbon, up to 7%, is plotted along the bottom, or x axis. The temperature is plotted along the y axis. The diagram illustrates what occurs when the material is slowly cooled and the resulting transformations.

There are two solid phases of iron and iron in solution of particular interest: ferrite and austenite. Carbon dissolves into iron without separating up to a concentration of 0.025% at a temperature up to 1333°F (723°C). This iron is termed *ferrite*, or *alpha iron* (α iron), and is body-centered cubic and magnetic. At room temperature, ferrite can dissolve only 0.008% carbon. The single-phase region at elevated temperatures between 2552°F (1400°C) and 2802°F (1540°C) is termed *delta iron* (δ iron). These two forms, ferrite and delta iron, are considered to be pure iron. The small amount of carbon available in these metals is dissolved into the crystal lattice interstitially. An *interstitial* is a particle not located at a lattice point in a crystal. It is, therefore, an imperfection in the lattice. An interstitial is shown in Figure 4–3.

A carbon steel of 0.77% carbon (eutectoid) becomes a solid solution at temperatures above 1333°F (723°C) and below 2500°F (1371°C). This single-phase solid solu-

Figure 4–2
Iron-iron carbide diagram.

Figure 4–3
Interstitial.

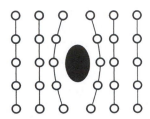

tion of iron is called *austenite*, or *gamma iron* (*γ* iron). All of the carbon is dissolved into the austenite. Austenite is the face-centered cubic, nonmagnetic structure of steel. Between temperatures of 1333°F (723°C) and 1666°F (908°C) and carbon concentrations less than 0.8% is a mixture of austenite and ferrite. Steel must be taken into the austenite region for hardening or softening. When the solid solution is slowly cooled below 1333°F (723°C), the 0.77% carbon separates into two distinct phases: ferrite and cementite. Thus, this point is called the *transformation temperature*. Increasing carbon concentrations up to 0.8% and alloying elements drop the transformation temperature into the austenite region. On the right side of the iron-carbon diagram is a compound called *iron carbide* (Fe_3C), or *cementite*. Because it is a compound and not an alloy, cementite is an extremely hard and brittle substance.

Iron is *allotropic*; i.e., it can exist in more than one crystalline form. The various forms of iron are termed *allotropes*. Ferrite is the name given to the body-centered cubic allotropes of iron and its solid solutions. This includes *α* and *δ* iron. Austenite (*γ* iron) is the name given to the face-centered cubic allotrope of iron and its solid solutions. Cementite is the name given to the carbide of iron. Iron at normal temperatures up to 1666°F (908°C) is body-centered cubic, *α* iron. Upon heating through 1666°F (908°C), the crystalline structure of the iron changes to the face-centered cubic *γ* iron. When heated beyond 2552°F (1388°C), the iron changes back into a body-centered cubic structure, *δ* iron, which is stable up to the melting point of pure iron at 2802°F (1535°C).

Iron is ferromagnetic up to a temperature of 1415°F (768°C). This temperature is called the *Curie temperature*. Ferromagnetism returns when the iron is cooled below 1415°F (768°C). As a point of history, iron between 1415°F (768°C) and 1666°F (908°C) was termed beta iron (ß iron). This is an archaic term, no longer used, because it was found that no change in crystalline form took place at the Curie temperature.

Referring to Figure 4–2, it can be determined that ferrite holds a small amount of carbon in solution. The limits are 0.006% carbon at 392°F (200°C) up to 0.04% at 1333°F (723°C). Austenite, in comparison, can hold considerably more carbon, 0.77% at 1333°F (723°C) up to 1.7% at 2066°F (1130°C). Carbon is held interstitially in both ferrite and austenite.

The lowest temperature at which a single-phase solid exists before it turns into a two-phase solid is called the *eutectoid point*. The eutectoid point occurs at a temperature of 1333°F (723°C) and a carbon concentration of 0.77%. The eutectoid composition of steel is called *pearlite*. Pearlite is the eutectoid mixture of ferrite and cementite formed when austenite decomposes during cooling. It shows up as alternating layers, called *lamellae*, of ferrite and cementite, which is called a lamellar structure.

The terms *hypoeutectoid* and *hypereutectoid* are used to denote steels that contain less carbon and more carbon, respectively, than the eutectoid composition. Steels with less than 0.77% carbon are called hypoeutectoid steels, and steels having more than 0.77% carbon are termed hypereutectoid steels. Hypoeutectoid steels consist of pearlite with an excess of ferrite. A low-carbon steel has much more ferrite than pearlite. Ferrite appears as the lighter grain. Pearlite often appears as a darker grain. Medium-carbon steels have approximately equal amounts of ferrite and pearlite. High-carbon steels are mostly pearlite. Hypereutectoid compositions include cast irons, discussed later.

One other type of steel structure is *martensite*. Martensite is the name given to the very hard and brittle product that is formed when a steel is very rapidly cooled from the austenite state. It comprises ferrite highly saturated with dissolved carbon. Due to the rapid cooling rate, there is not enough time for the austenite solution to separate into the usual ferrite and cementite. What occurs with the rapid cooling is a supersaturated solution of carbon atoms interstitially trapped in a body-centered tetragonal crystal lattice structure. Alloying elements reduce the temperature range at which martensite is formed. A low-alloy steel can begin to transform into martensite at 800°F (430°C), whereas a high-alloy steel may not begin to transform until reaching 600°F (318°C). The hardness of the martensite depends on the amount of carbon in the steel. Martensite is not shown on the iron-carbon diagram because the transformation to martensite is rapid.

4.3 TIME-TEMPERATURE TRANSFORMATION CURVES

Recall that it takes time for the crystalline structure of steel to change from one structure to another. If a piece of steel is heated into the austenite range, removed, and then immediately plunged into a bath of cold water, the crystal structure does not have enough time to change, and the structure is trapped. However, if a piece of steel is heated into the austenite range, removed, and then allowed to cool below the austenite temperature for a few seconds before being plunged into the cold water, the hardness of the heat-treated piece is reduced. The longer the piece is allowed to cool below the austenite temperature, the softer the finished piece, until some minimum hardness is reached. If the piece is allowed to cool long enough, the grain all goes to pearlite, and no martensite is formed.

A graph can be drawn to show the time the steel is allowed to cool versus the temperature. This graph is called a *time-temperature transformation*, or *TTT, curve*. It is also known as a C curve, a Bain-S curve, or an isothermal transformation, or I-T, curve. The curve shows the times and temperatures at which the grain structure of steel changes from austenite to martensite or pearlite. This curve is the basis for all heat treatment of steels. A typical TTT curve is depicted in Figure 4–4.

Martensite.

Courtesy of United States Steel Corp.

Martensite.

Courtesy of United States Steel Corp.

Martensite.

Courtesy of United States Steel Corp.

Martensite.

Courtesy of United States Steel Corp.

Cast austenitic stainless steel.
Courtesy of United States Steel Corp.

Eutectoid steel.
Courtesy of United States Steel Corp.

Ferrite.

Courtesy of United States Steel Corp.

Pearlite.

Courtesy of United States Steel Corp.

Microstructure of the tie rod 30 in. from the failure location showing ferrite and pearlite microconstituents indicating absence of unusual thermal expositor.

Courtesy of United States Steel Corp.

High carbon bainite.
Courtesy of United States Steel Corp.

Large-grain martensite in HAZ
of flame-cut surface.

Courtesy of United States Steel Corp.

Photomicrograph of the roll shaft
material showing ferrite and
pearlite microconstituents.
The material is in hot-finished or
normalized condition, and the
microstructure is consistent with
that of AISI 1045 carbon steel.

Courtesy of United States Steel Corp.

Microstructure of the base metal of the gears exhibiting ferrite and pearlite microconstituents, typical for hot-finished or normalized AISI 1045 carbon steel.

Courtesy of United States Steel Corp.

Photomicrographs of the shaft material showing tempered martensite microstructure.

Courtesy of United States Steel Corp.

Photomicrographs showing graphite size, distribution, and morphology and pearlitic matrix of the housing material. The microstructure is typical for high-strength grey cast iron.

Courtesy of United States Steel Corp.

Photomicrographs *Continued*
Courtesy of United States Steel Corp.

Figure 4–4
TTT curve.

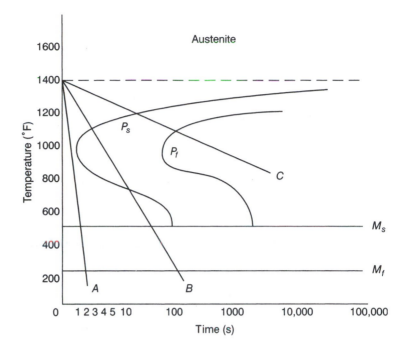

In Figure 4–4, temperature is plotted along the y axis, whereas the time is plotted on a log scale along the x axis. As shown on the graph, lines are drawn from the austenitic temperature down and to the right to designate the rate at which a steel can be cooled. The two curved lines, P_s and P_f, represent the starting and finishing points for pearlite. The two lines marked M_s and M_f represent the starting and finishing points for martensite.

Interpreting the TTT curve may appear difficult, but it really is not. For example, if a steel is cooled along the first cooling-rate line (A) but is cooled rapidly enough to miss the P_s line, it will be 100% martensite. This case results in the hardest possible condition of the steel. However, if a steel follows line C as it cools, pearlite will begin to form when the temperature reaches the P_s line. It will continue to form until the temperature and time reach the P_f line, at which point no further transformation can occur. The steel is then in its softest condition. Different percentages of martensite and pearlite form when the cooling rate lies between A and C. For example, consider line B. Notice that it crosses P_s but not P_f. In other words, pearlite starts to form, but the transformation is not complete. As B crosses M_s, martensite starts to form from the grains not already turned to pearlite. This process continues until B falls below the M_f line. The finished steel contains both martensite and pearlite; the hardness of such a steel lies somewhere between that of steels that are completely pearlitic or completely martensitic.

In cooling the steel, the slowest cooling rate that yields the maximum martensite transformation is termed the *critical cooling rate*. The critical cooling rate differs for each type of steel.

A TTT curve such as the one in Figure 4–4 for plain carbon steel indicates a great deal about the time-temperature relationship that exists for that steel. In order for a completely martensitic steel to form, for example, line *A* indicates that the temperature would have to drop from above 1400°F (766°C) to below 900°F (486°C) in approximately 1 second. Although cooling rates such as this can be obtained for the surface of a steel, they are nearly impossible for an entire piece. Such a cooling pattern would produce a martensitic structure on the surface of the steel; the percent of martensite would drop inside the surface reaching its lowest point at the center. This process is termed *case hardening*, in which a hard outer shell covers a softer central core.

The TTT curve also tells us about the grain structure of the steel. Slower cooling rates, those that are more horizontal, produce coarse-grained pearlite. Slightly faster cooling curves produce finer-grained pearlite. In addition to cooling rate, the mass of the material also affects the hardness. For example, a 1-in.–diameter steel bar can achieve a greater hardness than a 2-in. bar of the same material during the same period. The reasons for this difference are that the heat must escape off a larger surface area, and it takes longer to transfer the heat from the thicker cross section. For this reason, alloy steels are often selected when thicker cross sections need to be hardened. They tend to stretch the curve to the right.

Alloying elements are used to shift the TTT curve to the right, thus allowing longer cooling times for complete martensitic transformation. For instance, the addition of chromium and nickel may produce a curve similar to the one in Figure 4–5. In this steel, 5 seconds or more are allowed for complete transformation to martensite.

Figure 4–5
TTT curve for chromium-nickel steel.

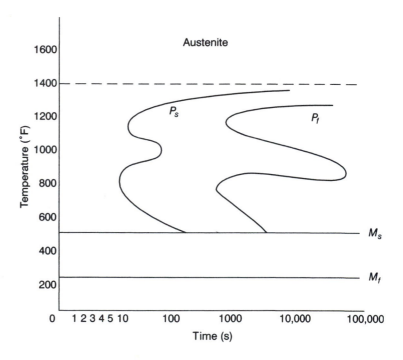

Longer soaking times allow better transformation and also allow transformation deeper into the product rather than yielding a hard surface surrounding a soft core.

4.4 METHODS OF SOFTENING STEELS

Both the iron-phase diagram and the TTT curve provide information on changing the hardness of a steel. In all heat treatments, the steel is heated up and then cooled at a controlled rate. Heat treatments vary in terms of the temperature at which the metal is heated and the time it takes to cool it down.

It is often important for a metal to be more ductile than it is hard and strong. Annealing and normalizing are used to obtain a softer, more ductile, and less distorted metal that is easier to machine and form and is less likely to crack or distort. Both processes involve slowly cooling the heated metal.

Normalizing involves heating the metal well into the austenite range, letting it remain there for 1 hour per inch of thickness, and then letting it cool in still air at room temperature. Normalized metals have an even grain size that makes them easier to machine. Normalizing is often used prior to other heat treatments to provide good machinability.

The slow cooling of a steel from the austenitic temperature range is called *annealing*. Annealing leaves the steel in its softest possible condition. This condition

Heat-treatment oven.

is typical of the steel supplied by mills and warehouses. When annealing, the steel is heated approximately 50° to 100°F (10° to 38°C) into the austenitic range for that steel. The temperature varies according to the carbon content for that steel (refer to Figure 4–1). If the metal is heated too far into the austenitic range, the grain size becomes excessively large. For example, a steel with 0.4% carbon requires about 1500°F (822°C), whereas doubling the carbon content to 0.8% reduces the temperature required to roughly 1400°F (766°C). The steel is held at this temperature for approximately 1 hour per inch of thickness. It is held to allow the steel to come up to an even temperature throughout so the entire piece will be transformed. The temperature is then reduced slowly, between 20° to 40°F (2° to 5°C) per hour until it passes the P_f line on the TTT curve. This process may take 2 to 3 days. The resulting steel will be in its softest possible condition, with a large, uniform grain structure. This process is termed *full annealing*.

Another annealing process involves placing the steel into an oven or box furnace and heating the steel 50° to 100°F (10° to 38°C) into the austenitic range. The temperature remains there for 1 hour per inch thickness. The oven is then turned off. A well-insulated oven, with the door left closed, will hold heat for approximately 24 hours. This process is termed *box annealing*. Steel that has been box-annealed is not quite as soft as full-annealed steel, but the process is cheaper and faster, which makes the steel more economical.

Process annealing is also used when time is more important than full softening. In process annealing, the metal is heated to 1050° to 1300°F and then cooled. This procedure relieves the internal stresses of the metal. Process annealing can also be used as an intermediate step in heat treatment to relieve the internal stresses during manufacture. It gives the metal good malleability.

Spheroidizing is almost identical to process annealing. Parts are heated below the lower transformation temperature of 1333°F (723°C) and are cooled slowly. The cementite in the steel is spheroidized, producing a tougher steel. This process is used with high-carbon steels to make them machinable. Tool steels are heated slightly above the lower transformation temperature for 1 to 4 hours and slowly cooled. Process annealing and spheroidizing depend on the initial condition of the metal prior to these processes.

4.5 METHODS OF HARDENING STEELS

Quenching is simply a controlled cooling process used during the heat treatment of metals. Before quenching, the metal is heated to high temperatures. Several quenching media are commonly used. Among these media are water, brine (salt water), oil, and air. When steel is quenched, four separate actions occur: (1) vapor formation, (2) vapor covering, (3) vapor discharge, and (4) slow cooling. After the initial quenching plunge, the metal starts to cool, but the cooling slows when a vapor film starts to form next to the metal as the liquid boils. As the process enters the second phase, the bubbles formed around the metal stick to the surface of the metal and insulate it, holding the heat in the metal. The metal is gently agitated in the quenching solution to reduce this effect. If these bubbles are not removed, the film of bubbles will form soft spots, warpage, and cracks in the metal. The third phase is apparent. The vapor collapses and explodes off the metal's surface. This is the crackling and sizzling that takes place. The greatest amount of cooling takes place here. Finally, the slow cooling process is continued until the metal reaches room temperature. The full process of conventional hardening followed by annealing is illustrated in Figure 4–6.

Different quenching media are used for different conditions. Water is the most common because it is inexpensive, is convenient, and provides rapid cooling. It is preferred for low-carbon steels, where a very rapid change in temperature is required to obtain high hardness and strength. Because it provides a rapid, drastic quench, water can cause internal stresses, cracks, and distortion. Water as a quenching medium is generally kept at room temperature.

Oil is slower than water as a quenching medium. It is used for thinner parts and those with sharper edges, such as razor blades, knives, and springs. Oil does not provide as sharp a drop in temperature as water. This is due, in part, to the thermal conductivity of oil versus water; i.e., the water dissipates heat faster than oil. Oil provides a compromise in hardness and strength against cracking and distortion. Oil can

Figure 4–6
Quenching process.

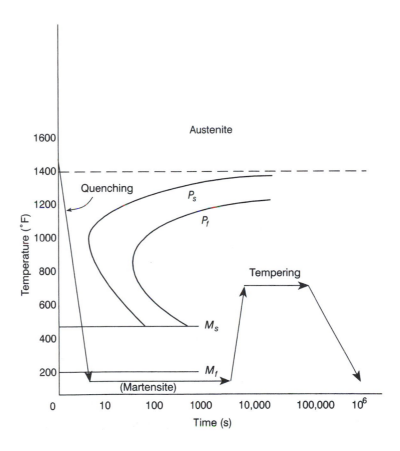

be heated to approximately 100° to 150°F (30 to 66°C) to be more effective. This thins the oil somewhat, allowing it to circulate and flow over the metal more easily.

Air is less abrupt than either water or oil. Air is blown over the heated part. Again, because the metal cools more slowly, the metal's strength and hardness are not as great. Air quenching is generally used in high-alloy metals, where the alloys are used to harden the material.

Brine is the fastest of the quenching media. Brine is a saltwater solution containing between 5% to 10% salt (sodium chloride). The quenching action is slightly faster than water alone but is more drastic. The salt reduces the vapor film phase, causing the outer surface of the metal to explode off.

Of all the quenching media, water and brine are the fastest, followed by oil and then air. Water and brine produce the greatest strength and hardness. Air is the least drastic and, therefore, has the least danger of producing internal stresses, cracks, and distortion.

Rapid drops in temperature cause internal stresses in the metal. To reduce these stresses, the metal must be tempered, or drawn, after quenching. *Tempering* (sometimes called *drawing*, because it supposedly "draws" the hardness from the metal) is the process of reheating the metal immediately after hardening to a temperature below the transformation temperature and cooling it to increase the ductility and toughness of the steel. After hardening, the steel is heated to between 700° and 800°F (375° and 430°C), where it remains for 1 hour per inch of thickness. The steel is then allowed to cool. Tempering softens the steel slightly.

There are three special types of tempering: *martempering, austempering*, and *isothermal quenching and tempering*. Martempering is similar to general tempering, except the part is quenched to a temperature just above the M_s line [between 500° and 600°F (260° and 316°C)] for a few seconds to allow the temperature throughout the part to stabilize. This temperature range is below the knee of the TTT curve but above the M_s line. The part is then quenched through the martensitic range to room temperature. This technique provides a more uniform grain structure in the steel as it enters the martensitic range. As a result, the material is more stress-free than with regular quenching and tempering methods. After quenching through the martensitic range, the part is immediately tempered.

Austempering resembles martempering, except after leveling the temperature at about 700°F (375°C), it is held there for a longer period of time while it passes through the P_s and P_f lines. At this temperature, bainite is formed. *Bainite*, named for Dr. E. C. Bain, who did the initial research on this structure, is the region of transformation between the rapid cooling curve for martensite and the slower cooling rate for pearlite. Bainite is superior to martensite in ductility and toughness but is inferior in hardness and strength. Times for transformation range between 30 minutes to 3 hours, depending on the type of steel. Once the bainite has formed, the steel is quenched to room temperature. Austempering is generally confined to smaller parts, such as springs, lock washers, needles, and other similar parts, where distortion may be a problem.

Isothermal quenching and tempering fit somewhere between martempering and austempering. The result of this process is a steel that is harder and stronger than that produced by austempering, yet more ductile and stress-free than that produced by martempering. The structure in this case is a combination of bainite and tempered martensite. In isothermal quenching and tempering, the metal is heated into the austenitic range and then quenched down to about 50% transformation from austenite to martensite. The temperature [about 300° to 400°F (150° to 205°C)] is then held constant for a few seconds while the remaining austenite transforms to bainite. Finally, the steel is quenched to room temperature.

4.6 SURFACE HARDENING

Case-hardening, also referred to as surface-hardening, involves one of four different methods: *carburizing, nitriding, cyaniding,* or *carbonitriding.* Case-hardening is used on parts such as gear teeth, cutting wheels, and tools. These case-hardened pieces represent a compromise between the hard, wear-resistant brittleness of a high-carbon steel and the softer, more ductile, less wear-resistant low-carbon steels.

Carburizing is accomplished by crowding considerable amounts of carbon into the outer surface of the steel. This can be done by placing the low-carbon steel in a high-carbon atmosphere and heating it into the red-heat range [1400° to 1600°F (766° to 878°C)] (see Figure 4–7). Low-carbon steels are not capable of being hardened appreciably by quenching. The increased carbon increases the steel's ability to be hardened. The atmosphere is a carburizing gas, such as natural gas, which diffuses into the surface of the low-carbon steel and penetrates the crystal structure. The depth of penetration depends largely on the amount of time the low-carbon steel is left in the carbon-rich atmosphere. Typical penetrations for case-hardening range from 0.001 in. (0.025 mm) to 0.125 in. (3.175 mm). Camshafts and piston wrist pins are typical carburized parts.

Figure 4–7
Carburizing.

Part encased in activated charcoal

Sealed, heated container

Carburizing

Figure 4–8
Gas nitriding.

Gas nitriding

Nitriding involves essentially the same process as carburizing. Nitrogen has about the same effect as carbon in steels. Nitrogen also reacts with chromium, molybdenum, and other alloying elements commonly found in steels to form hard, brittle compounds. Nitrogen is added to the outer shell of the part by placing it in a nitrogen-rich atmosphere, much like carburizing. The nitrogen will chemically combine and diffuse into the steel. Ammonia gas (NH_3) and nitrogen gas are commonly used in nitriding (see Figure 4–8). Nitriding offers the hardest and most corrosion-resistant result of all surface-hardening methods.

Cyaniding and carbonitriding both supply carbon and nitrogen to the steel. Sodium cyanide (NaCN) relinquishes the cyanide radical (CN^-) when melted. The hot steel is immersed into the solution for several hours while the two substances, carbon and nitrogen, diffuse into the steel. One precaution when using the cyaniding process: Sodium cyanide reacts with any acid to form hydrogen cyanide gas—an extremely deadly gas when inhaled. Carbonitriding is the same process as cyaniding. The part is immersed in an atmosphere of hot gases containing carbon and nitrogen. Nitrogen also slows down the critical cooling rate of the steel, making it easier to miss the knee of the TTT curve. It also allows the steel to form martensite to a greater depth than normal plain carbon steel. Both processes are used for thin cases.

In addition to case-hardening, *flame-hardening* and *induction-hardening* are used to harden the surface of the metal without changing the structure of the interior (see Figure 4–9). Flame-hardening involves heating the surface of the metal with an oxyacetylene torch or flame source and then immediately quenching it. This produces martensite on the surface of the part. Machine parts such as lathe and mill beds are flame-hardened to prevent wear. Induction-hardening involves using electricity to heat the steel. An electric current passes through a coil wrapped around the part. The passing of an alternating current through the coil builds up and then reduces the magnetic field around the coil. This induces a current in the part many times a second (from 3000 to 1,000,000 cycles per second, or hertz). The resistance in the

Figure 4–9
A. Flame-hardening. B. Induction-hardening.

(a) Flame hardening

metal produces heat when the current flows through the metal. If a high-frequency current is used, only the surface of the metal is affected. This result can be achieved in a few seconds. After the surface is heated, the metal is quenched. Flame-hardening and induction heating differ from the four previous methods in that no carbon or nitrogen is added to the surface, only heat.

4.7   HEAT TREATMENT OF OTHER METALS

Although no nonferrous metal offers the wide range of heat-treatability of steel, some are heat-treatable. The process for nonferrous metals is different, but it has some similarities to that for ferrous metals. A phase change is required, as is an alloying element that will deposit in differing forms or phases, such as carbon in steel. No pure metal can be heat-treated, just as pure iron cannot be heat-treated. Most nonferrous metals do not exhibit a suitable phase transformation when heated. Others exhibit a phase transformation too close to their melting temperature to be practical. The only methods available to change the hardness of pure metals are cold-working and alloying.

Aluminum bronze is one nonferrous alloy that can be heat-treated similarly to steel. It contains approximately 10% aluminum and 90% copper. The more common heat-treatable nonferrous alloys are copper with beryllium, aluminum with copper, aluminum-zinc alloys, aluminum magnesium alloys, and some nickel alloys.

The method primarily used for hardening nonferrous alloys is called *age-hardening*, or *precipitation-hardening*. In addition to being hardened, these alloys can also be annealed. Hardening is a two-stage process: *solution heat treatment* and *aging*. To illustrate the hardening process, consider an aluminum-copper alloy. Up to 5.65% copper will dissolve in the aluminum at the eutectic temperature of 1018°F (548°C), whereas at room temperature, the solubility of copper in aluminum disappears. This forms an intermetallic alloy, $CuAl_2$. Aluminum alloys with less than 2% copper form a large-grain structure when allowed to cool slowly. This grain structure results in a soft material. To harden the material, the alloy is heated into the alpha (α) range, about 950° to 1000°F (510° to 542°C); the $CuAl_2$ then dissolves into the copper and disperses in the solution. Quenching at this point freezes this structure. Quenching does not produce martensite, as in steels. Rather, it freezes the copper as a supersaturated solution, because the copper does not have time to come out of the solution. The quenching process is called a *solution heat treatment*. The process does not harden the alloy; it just prepares it for later hardening. If the copper is left in this condition, an unstable solution of copper in aluminum, the aging process occurs hours or days after the copper aluminide ($CuAl_2$) is forced out of solution. Aging is hastened if the alloy is reheated to about 200° to 400°F (94° to 206°C), which causes the molecules of $CuAl_2$ to precipitate together, effectively hardening the alloy.

4.8 SUMMARY

Heat treatment refers to the heating and cooling operations required to alter the properties of ferrous and nonferrous metals, plastics, and ceramic materials. Changes in a material's properties result from changes made in the microstructure of the material. If a piece of steel is heated into the austenite range, removed, and then immediately plunged into a bath of cold water, the crystal structure does not have enough time to change, and the structure is trapped. Phase changes are an important part of heat treatment.

Time-temperature transformation curves provide feedback on the time and temperature requirements for heat treatment. Using these curves, the heat-treatment parameters can be adjusted to achieve the best results. These parameters also rely on information received from the iron-carbon phase diagram (Figure 4–1).

It is often important for a metal to be more ductile than hard and strong. Annealing and normalizing are used to obtain a softer, more ductile, and less distorted metal that is easier to machine and form and is less likely to crack or distort. Both processes involve slowly cooling the heated metal.

Hardening of metals depends on the quenching process. Quenching is the controlled cooling process used to freeze particular grain structures in metals. Before quenching, the metal is heated to high temperatures. There are several quenching media commonly used. Among these media are water, brine (salt water), oil, and air.

After hardening, metals are often tempered to remove stresses and improve their machinability. In addition, some softer materials are case-hardened to improve their toughness and wearability. There are several methods of case-hardening metals.

Summary of Heat Treatment/Conditioning of Metals

Softening	Hardening	Surface Treatment
Annealing	Flame	*Case-hardening*
Normalizing	Induction	Carburizing
Full	Aging	Nitriding
Process		Cyaniding
Spheroidizing		Carbonitriding
Tempering		
Austemper		
Martemper		
Isothermal		

QUESTIONS AND PROBLEMS

1. Explain the regions and use of the TTT curve.
2. What is normalizing?
3. What is annealing?
4. What is case-hardening?
5. What is spheroidizing?
6. What is tempering?
7. Describe martempering.
8. Describe austempering.
9. Describe the four methods of case-hardening discussed in this chapter.
10. Describe the four phases of quenching.
11. What heat treatment should be used to harden a knife blade?
12. What heat treatment should be used to produce a uniform grain size?
13. What heat treatment should be used to produce the hardest steel?
14. What heat treatment should be used to produce the fastest and hardest surface on a metal?
15. What method is used to heat-treat nonferrous metals?
17. Define the term *allotropic* as it relates to steel.
18. What is ferrite?
19. What is austenite?
20. What is pearlite?
21. What is cementite?
22. What is martensite?
23. Define interstitial as it relates to iron and steel.

CHAPTER 5
Polymers and Elastomers

Objectives • Define the common terms associated with polymers and elastomers.
• List the common properties of a variety of polymer and elastomer families.
• List the various manufacturing processes associated with polymers and elastomers.
• Describe the various properties used to designate or specify various polymers and elastomers.

5.1 INTRODUCTION

The term *polymer* is the correct name for a group of materials commonly referred to as *plastics*. The reason these materials are called plastics is that many exhibit plastic deformation or plastic behavior; in other words, they stretch before failure. Not all polymers exhibit significant plasticity. So, the terms polymer (plastic materials) and elastomer (elastic materials, such as rubber) are used in this chapter to help distinguish between polymeric or elastomeric materials and the mechanical property that some of these materials exhibit. The term *resin* has also been used in reference to polymers. It generally refers to naturally occurring materials. Polymer resins are a main component in many plastics. Most polymers are synthetic materials that allow for a wide range of properties and applications. New polymeric materials are constantly being developed. This chapter provides an overview of the many types of polymers available, centering on the more common polymers used. The focus is on the basic nature, properties, and manufacturing processes related to these materials.

The modern polymer industry dates back to 1868, when John Wesley Hyatt replied to demands to find a replacement for the rare and expensive ivory used in the manufacture of billiard balls. This resulted in the development of cellulose nitrate, or celluloid. The first synthetic polymer was developed by Dr. Leo Bakeland, who was the first to successfully induce a controlled phenol-formaldehyde reaction in 1907. Since then, the quantity and popularity of polymers have increased tremendously.

5.2 POLYMERIC STRUCTURE

Polymers are a group of materials characterized by chains of molecules made up of smaller units called *monomers*, a majority of which are joined artificially. The term *polymer* comes from the Greek meaning "many parts," and the term *monomer* refers to a single large molecule that comprises the basic unit for the polymer chain. The *mer* is the repeated unit in larger molecules, such as the monomer and polymer. Most polymers are organic (carbon-based) materials that contain molecules composed of various combinations of hydrogen, oxygen, nitrogen, and carbon. These four elements are among the most common found in organic polymers. Carbon forms the "spine" of the polymer chain, and the other constituents attach themselves to the carbon. These polymer chains become entangled and form irregular coils, which give them added strength. A portion of this entanglement is natural and can be further induced by additives and controlled processes.

Most polymers are based on hydrocarbons, where the elements of carbon and hydrogen form predictable combinations based on the relationship, $C_nH_{2n}+2$. These petrochemical intermediates are chemicals manufactured from *paraffins* in petroleum and natural gas, which are further processed into polymer products. These intermediates are the basis for almost all rubber and polymer products. Intermediates are also produced from coal. Coal intermediates make up approximately one-third of the total produced in this country. The most important of these intermediates is ethylene. These are called *olefin intermediates* and include acetylene, propylene, butylene, isobutylene, and butadiene. Most of these are used in the production of rubbers. Nylon is produced from butadiene; polyvinyl chloride is produced from acetylene. The aromatic intermediates include benzene, toluene, and xylene. Another category of intermediates contains the naphthenic intermediates, of which cyclohexane is the most important, being used in the production of nylon. Aromatic, napthenic, and olefin intermediates are used to produce insecticides, detergents, rocket fuels, films, pharmaceuticals, explosives, alcohols, and other such products. Cellulosic plastics are produced from wood rather than petroleum, coal, or gas.

Single covalent bonds between atoms do not provide for additional atoms to be added, so they are said to be *saturated*. Saturated molecules have strong intramolecular bonds but weak intermolecular bonding. Methane and ethane are examples of saturated molecules. When carbon and hydrogen form *unsaturated* molecules, such as ethylene and acetylene, the molecules form double or triple covalent bonds. An unsaturated molecule doesn't have the necessary hydrogen atoms to satisfy the outer shell of the carbon atoms. Many molecules form many double bonds and are referred to as *polyunsaturated compounds*. These materials include common cooking oils and margarines. Unsaturated and polyunsaturated materials often give off smoke when burned; saturated compounds do not.

Polymerization, or the joining of large unit molecules called *monomers*, utilizes the valence of the partially filled outer shell of the carbon atom (carbon has a valence of 4) to join smaller units together to form larger chains of molecules. Oxygen, sulfur, silicon, or nitrogen can be used to replace the carbon atom. Notice where these elements occur in the periodic table of elements (Figure 1–1). Figure 5–1 illustrates the difference between the saturated ethane and the unsaturated ethylene.

Two conditions must be met for polymerization to occur. The first condition is that a molecule must have at least two locations that have unsatisfied bonds, which will easily join with other molecules. This requirement means starting with a molecule that has a double bond, such as carbon. Because the carbon molecule has a double bond, each bond is a pair of shared electrons. If one of the bonds between the carbon atoms opens up, a single bond exists, leaving the other two electrons to share with other atoms. If another carbon atom passes by that has opened up its double bond, the two can join to form a chain. This procedure continues, producing a polymer chain, and is called *polymerization*. The process continues as long as the second condition is met. The second condition necessary for polymerization is that after each reaction, at least two open locations must remain. Many chains form within polymers. These chains form threads that entwine themselves around each other for strength.

Figure 5–1
Ethane and ethylene.

(a) Ethane   (b) Ethylene

Figure 5–2
Propyl and isopropyl alcohols.

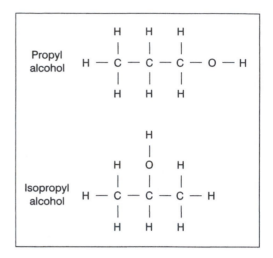

Polymers can be strengthened by *cross-linking*. Cross-linking occurs when the double bonds between atoms within a chain are broken and these atoms or molecules form, or link up, with neighboring atoms. This linkage provides additional strength to the chain and reduces the *slippage* that occurs between molecules. Slippage occurs when the polymer threads slide past each other when subjected to a load.

The properties of polymers also depend on the structure as well as the composition of the molecule. Two molecules with the same composition may form two different configurations having different properties, such as propyl (1-propanol) and isopropyl (2-propanol) alcohol. These variations are called *isomers*. Figure 5–2 shows how these two polymers are formed. Notice that the two materials have the same components, arranged differently. The two isomers are referred to by a standard designation system for organic compounds set by the International Union of Pure and Applied Chemists (IUPAC) to distinguish between isomers.

5.3 MECHANISMS OF POLYMERIZATION

Polymerization takes place through *addition polymerization, copolymerization,* or *condensation polymerization*. In the polymerization process, a large unit molecule, the *monomer*, is added to another monomer to form a larger chain, the polymer (referring to many parts), which has a number of repeated units, *mers*. Mers are the smallest units recognizable in the chain. The degree of polymerization is the number of repeating units that have identical structures within the chain formed by the polymer. Addition polymerization involves only one type of mer. Figure 5–3 shows polymerization by addition.

In copolymerization, more than one molecule makes up the mer. Acrylonitrile-butadiene-styrene (ABS) is an example of a copolymer. Figure 5–4 shows the copolymerization process for ABS polymers.

Condensation polymerization involves the chemical reaction of two or more chemicals to form a new molecule. This chemical reaction produces a condensate or

Figure 5–3
Addition polymerization.

Figure 5–4
Copolymerization.

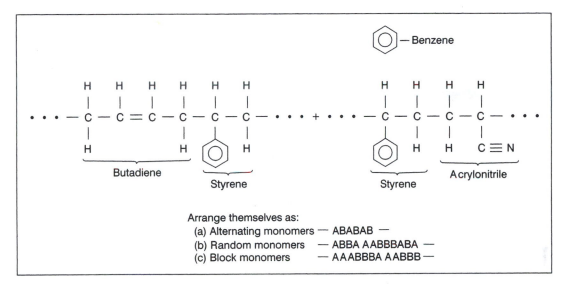

Figure 5–5
Condensation polymerization.

nonpolymerizable by-product, usually water. A catalyst is often required to start and maintain the reaction. It can also be used to control the reaction rate. The process of condensation polymerization is shown in Figure 5–5.

5.4 PROPERTIES OF POLYMERS

The terms *thermoplastic* and *thermoset* refer to the properties of polymers. Polymers are separated into these two general classifications.

Thermoplastic polymers often exhibit plastic, ductile properties. They can be formed at elevated temperatures, cooled, and remelted and reformed into different shapes without changing the properties of the polymer. However, the heat used to melt and remelt the thermoplastic must be carefully controlled, or the material will decompose. The properties of thermoplastic materials are determined by the bonding method between polymer chains; in thermoplastic materials, these bonds are weak, secondary bonds, as in van der Waals forces. Through the application of heat and pressure, these bonds can be weakened, and the material can be reshaped. Once the heat and pressure are removed, thermoplastic materials reharden in the new shape. Common thermoplastic polymers include acrylic, nylon (polyamide), cellulose, polystyrene, polyethylene, fluorocarbons, and vinyl.

Thermosetting polymers have strong primary bonds, often formed by condensation polymerization. Thermosetting polymers have these strong, primary bonds throughout, and their structure resembles one large molecule. Their properties are the result of chemical changes undergone during processing, under heat or through the application of a catalyst. Once hardened, thermosetting polymers cannot be softened or reshaped, due to the loss of part of the molecule (the by-product of the reaction) during the curing process. Once *cured*, if further heat is applied to a thermosetting material, it chars, burns, or decomposes. During curing, the thermosetting polymer becomes rigid and insoluble as the polymer chains become tangled and cross-linked. Common thermosetting polymers include phenolics, aminos, polyesters, epoxies, and alkyds.

Because the intramolecular forces of polymers are weaker than the intermolecular forces, deformation in these materials is the result of slippage between molecules rather than the breaking of the bonds. Through the application of heat and pressure, the polymer chains move apart and slide past each other. Once the heat and pressure are removed, the new formation is retained.

The principal mechanical properties of polymers are strength, stiffness, hardness, and melting point. Various materials can be added to polymers to improve certain properties, reduce the cost of the polymer, improve the formability of the material, and/or colorize the material. These additives are termed fillers, plasticizers, coloring agents, and lubricants. They are often divided into two groups: performance-related additives and processing-related additives, depending on their intended purpose.

The properties of polymers depend on the additives used, materials added to increase the polymer's strength; the amount and properties of the fillers used; and coloring agents used, and plasticizers, which are added as internal lubricants. Thermoplastic polymers are generally available in films, sheets, rods, tubing, and several molded or extruded shapes. The most common form of thermoplastic polymer is molding pellets. Thermosetting polymers are generally available as powders or liquids. These contain the basic polymer material, a filler, colorizing agent, plasticizer, a hardening agent (catalyst), which induces cross-linking, and an accelerator. To get the unsaturated mers to react with each other and join together, an initiator (usually a peroxide) is used. This initiator causes heat to be generated. The more heat, the faster the reaction. If too much heat is generated (by adding too much initiator), the material tends to bubble and foam up. This process is used to harden body-repair liquid resins. Generally, methyl ethyl ketone peroxide (MEKP) is added to the base resin to initiate the hardening process.

Fillers such as powders, fabrics, fibers, and so on, are used to alter the physical and mechanical properties of polymers. They are used in varying quantities, depending on the desired properties and cost of the product. If fillers are used in too high a quantity, they tend to clump up and produce weak points and voids, thus defeating their purpose. Their primary purpose is to restrict the movement of the polymer chain, thus increasing its strength but reducing its ductility. In addition, fillers are used to reduce the cost of the product. Fillers may also be used to control product shrinkage or increase the formability of the material. Table 5–1 displays some of the more common filler materials used.

The coloring agents added to polymers are generally either dyes or pigments. Dyes actually change the color of the product used, whereas pigments alter the color of the material through their presence. Most fillers don't produce attractive colors, so some form of coloring agent is used for aesthetics.

Plasticizers, in small amounts, are added to increase and control the flow of the material during processing. Plasticizers improve the wearability of the polymer by reducing the frictional wear effects and also increase the flexibility of the material by allowing easier chain movement. Plasticizers used in this fashion are employed as internal lubricants. These plasticizers are generally low-molecular-mass polymers that separate polymer chains and reduce *crystallinity*. The addition of plasticizers is kept to a minimum, because they can adversely affect the stability of the finished product during aging. Therefore, the Food and Drug Administration (FDA) has issued a warning and has banned the use of plasticizers in products manufactured for food or water containers.

Lubricants are also added in small amounts to increase the formability of the product and to facilitate removing the product from the mold after forming. Wax, stearates, and soaps are generally used as lubricants. Their proportion is also kept to a minimum due to their adverse effects on the material's properties.

Most polymers exist in an amorphous, or noncrystalline, state. These polymers exhibit two main types of structure: the amorphous, or isotropic, and the anisotropic.

Table 5–1
Common fillers and their uses.

Wood flour	General-purpose filler, low cost, fair strength, good formability
Cloth fibers	Improvement in impact strength, fair formability
Glass fibers	High strength, dimensional stability, translucence
Mica	Excellent electrical properties, low moisture absorption

Isotropic polymers exhibit the same properties throughout the material. The properties of anisotropic polymers depend on the direction in which measurements are taken. Some form and degree of crystallinity [formation of crystals (crystallization) and the ability to form crystals (crystallizability)] exist in polymers. Crystallization in polymers requires a regular structure and strong, evenly applied forces. Mechanical deformation and the addition of fillers are used to increase the crystallinity of polymers and to aid in the formation of crystals that improve the strength qualities of the material. The *degree of crystallinity* is expressed as a percentage of the material that has crystallized compared to its fully crystalline state. This is usually approximated by X-ray diffraction, although infrared spectroscopy and nuclear magnetic resonance techniques may also be employed.

Generally, polymers are lightweight, are good electrical insulators, are good thermal insulators, have a low modulus of elasticity, provide excellent corrosion resistance, provide abrasion resistance, and offer considerable resistance to chemical attack. They can be manufactured through many common processes, offer a good surface finish, and are available in a wide range of colors. However, without reinforcement, their principal drawback is their low strength as compared with other manufacturing materials such as metals and composites. Most polymers also exhibit poor dimensional stability due to their high creep rate and low fatigue strengths. *Creep* is the slow stretching of a plastic material subjected to a tensile load over time. *Fatigue* is the slow breakdown of a product subjected to repeated normal loads over extended periods of time.

Polymers may be used in the manufacture of lightweight materials requiring only low to medium strength; electrical insulation; thermal insulation; pliable foam used for packing or fillers; adhesives or bonding agents; or substitutes for more costly materials such as automobile fenders, doors, windows, and so forth. Polymers offer a wide range of applications and, with the addition of *reinforced polymers*, offer characteristics that, based on the combination of different polymers, offer properties not found in any other material.

Some common types of polymers, copolymers, and their characteristics are as follows:

ABS	Acrylonitrile-butadiene-styrene, light, good strength, excellent toughness
Acrylic	Excellent optical quality; trade names: Lucite and Plexiglas; high impact, flexural, tensile, and dielectric strengths
Cellulose acetate	Good insulator, easily formed, high moisture absorption, low chemical resistance
Cellulose acetate butyrate	Similar to cellulose acetate but will withstand more severe conditions
Epoxies	Good toughness, elasticity, chemical resistance, dimensional stability; used in coatings, cements, electrical components, and tooling
Ethyl cellulose	High electrical resistance, high impact strength, retains properties at low temperatures, low tear strength

Fluorocarbon	Inert to most chemicals, high temperature resistance, low friction coefficient (Teflon), used for self-lubricating bearings and nonstick coatings
Melamine	Excellent resistance to heat, water, and chemicals, excellent arc resistance; used in tableware and to treat paper and cloth for water resistance
Phenolic	Hard, relatively strong, inexpensive, easily formed, generally opaque; wide variety of available shapes
Polyamide	Good abrasion resistance and toughness, excellent dimensional stability; used in bearing materials that require little lubrication, textiles, fishing line, and rope
Polycarbonate	High strength and toughness; used in safety glasses
Polyethylene	High toughness, high electrical resistance; used in bottle caps, unbreakable utensils, and wire insulation
Polypropylene	Lightest weight, harder than polyethylene
Polystyrene	High-dimensional stability, low moisture absorption, excellent dielectric, burns readily, low resistance to certain chemicals
Silicone	Heat-resistant, low moisture absorption, high dielectric properties
Urea formaldehyde	Similar to phenolics; used indoors
Vinyl	Tear-resistant, good aging, good dimensional stability, good moisture resistance; used in wall and floor coverings, fabrics, and hoses

5.5 MANUFACTURING PROCESSES INVOLVING POLYMERS

Polymers are produced from a variety of naturally occurring materials, including wood, air, water, petroleum, natural gas, and salt. Chemists separate these naturally occurring materials into their basic elements and then reassemble these basic elements into different, distinct polymers using heat, pressure, and chemical action.

For example, polystyrene is produced from coal and petroleum or natural gas. The chemist begins by extracting benzene from the coal and ethylene gas from the petroleum or natural gas. The basic elements, benzene and ethylene, are combined to form ethylbenzene. Through the application of heat and pressure, the ethylbenzene is formed and ground to produce polystyrene beads. Many polymers are available in powder, pellet, or premixed form.

Many polymers are also available as liquid casting systems. These materials are typically two-part systems, where part A serves as a catalyst for part B, or vice versa. When mixed together in proper quantities, these two parts become a single, cured polymer. Epoxies are an example of a liquid casting system. Another type of casting polymer is *plastisol*. A plastisol is a liquid vinyl polymer that solidifies when it contacts a heated surface. Plastisols are commonly used in dipping processes for insulated coatings. An example is the insulated coating on tool handles.

Polymers are also available in expanded or expandable forms. Expanded polymers are low-density cellular polymers available in two varieties: open- and closed-cell polymers. Open-cell expanded polymers form one continuous-cell pattern and are interconnected. Closed-cell expanded polymers are formed into a series of bubbles, or air pockets. These expanded, or *foamed, polymers* come in two forms: rigid and flexible. An example of a rigid, foamed polymer is polystyrene, or Styrofoam, which is commonly used in packaging. One type of flexible foamed polymer is urethane, which is used in seat cushions and craft work.

Expandable polymers are preformed expanded polymers. They are ready to be foamed through the mixing of two liquids, through the mixing of gases with a liquid, or by applying heat to gas-filled bubbles. Once mixed or heated, depending on the raw state, these materials expand to several times their preformed state. An example of expandable polymers is foam insulation. The raw materials exist in separate containers, one liquid and one gas. These materials are mixed in the applicator nozzle and immediately begin to expand. They are used to fill voids and provide thermal and acoustic insulation.

Several methods are used to fashion products from polymers. Some of the more common production methods are casting, blow molding, compression molding, transfer molding, injection molding, extrusion, lamination, vacuum forming, cold forming, filament winding, calendaring, and foaming. Each of these methods has advantages and disadvantages, depending on the type of material used, the product design, and the cost involved.

Casting is the simplest of the manufacturing methods, because no fillers are used and no pressure is applied. However, a mold is required. There are four basic types of casting methods: simple casting, dip casting, slush casting, and rotational casting. In the simple casting process, a prototype, or model, of the material is made first. This model, usually steel, is typically dipped into a pot of molten lead, where it picks up a thin layer on the outside of the model, or mandrel. The mandrel is then removed from the thin sheath of lead, and the polymer is poured into the mold. The product is then cured and removed from the mold. The simple casting process is inexpensive, but it is limited to small, simple, solid shapes.

Dip casting requires a preheated mandrel, a dipping station, a curing oven, a cooling station, and a method to remove the finished casting. Dip casting can be performed by hand or through some automated means. In either case, some accurate means of timing is required. This method is commonly used to cast hollow objects. In this process, the preheated mandrel is lowered into a liquid polymer or plastisol. The mandrel is then withdrawn from the mold at a slow, predetermined rate using a timer. The slower the rate at which the mandrel is withdrawn from the mold, the thicker the coating remaining on the mandrel. The mandrel and the layer of plastisol are then cured in a curing oven. Once the plastisol has cured, the mandrel is withdrawn from the plastisol, leaving the finished part. Dip casting is used to produce parts such as overshoes, spark plug boots, and gloves.

Slush casting uses a split mold, which has an internal cavity formed in the shape of the finished part. A plastisol or powder is poured into the preheated slush mold, in which the plastisol solidifies where it comes into contact with the mold cavity walls. Leaving the plastisol in the mold for a longer period of time and pouring more polymer into the mold increases the thickness of the finished part. The excess liquid is poured out of the mold, and the mold is then placed in the curing oven. Once the part has cured, the mold is split, and the part is removed. Slush casting is used to produce hollow, open items, such as toys, storage tanks, and piggy banks.

Rotational casting or molding utilizes a predetermined amount of plastisol, which is placed inside a preheated split mold. The rotational mold is then revolved around two—or sometimes three—axes. The revolving of the mold helps to distribute the plastisol evenly throughout the mold. The mold and plastisol are then cured in an oven. After curing, the mold is split, and the part is removed. Rotational molding is used to produce hollow, seamless parts, such as floats, squeeze bulbs, and plastic ornaments.

Extrusion blow molding—or, more precisely, blow forming—is used to form hollow shapes, such as bottles and containers. A thermoplastic tube called a *parison* is fed through the opening formed by the two halves of a split mold. Once the necessary length of parison has entered the mold, the two halves close, sealing one end of the parison. Compressed air is then forced into the open end of the parison, forcing the parison to expand, following the contour of the inner mold cavity. The mold is kept cool during the blow molding process, so that the thermoplastic parison solidifies as it comes into contact with the mold. Once the material has cooled, the mold is opened and the part is removed. Two other forms of blow molding are injection blow molding and stretch blow molding. These differ in the way the raw material is introduced into the mold. The extrusion blow molding process is illustrated in Figure 5–6.

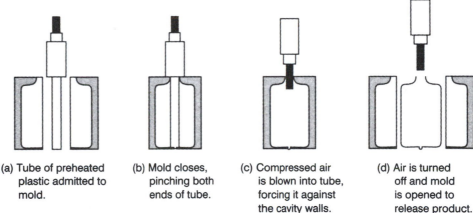

(a) Tube of preheated plastic admitted to mold.

(b) Mold closes, pinching both ends of tube.

(c) Compressed air is blown into tube, forcing it against the cavity walls.

(d) Air is turned off and mold is opened to release product.

Figure 5–6
Extrusion blow molding.

These compression-molded pistons are used in air spring assemblies for heavy-duty truck suspensions.
Courtesy of Goodyear.

A 50-lb, one-piece molded truck hood is removed from a compression mold. The hood will be fitted with structural sheet molded compound reinforcements and metal components.
Courtesy of Goodyear.

Compression molding is the most common method of shaping thermosetting plastic materials such as alkyds, melamine, urea, and phenolics. It may also be used for thermoplastic materials when the part is too large to be economically produced by injection molding or when extreme accuracy is required, as in the case of vinyl and styrene phonograph records. Compression molding requires that granules or preformed tablets of the raw material be placed in the cavity of an open, heated mold. The plunger, or male member of the mold, descends and closes the mold, which creates enough pressure to force the heated, fluid plastic material into the cavity. After the material has cured, the mold is opened, and the part is removed. In the compression molding process, several cavities are typically contained within a single mold. Compression molding is simple in operation but is generally restricted to thermosetting materials. For thermoplastic materials, alternate heating and cooling of the mold is required, which isn't always economical. Compression molding is often used to produce parts such as knobs, handles, buttons, and electrical components in multiple-cavity molds. One of the most common applications is in the auto industry for forming parts such as hoods and doors. Figure 5–7 shows three types of compression molding procedures.

In transfer molding, the raw material is placed in the plunger cavity and heated until it melts. The plunger then descends and forces the molten material into the die

Figure 5–7
Compression molding.

cavities. Transfer molding is used for delicate or intricate shapes, when a flash is not acceptable, for thin sections, for detailed parts, and to achieve good tolerances and finishes. It does this by transferring the material in the molten state, thus avoiding the high pressures and turbulent flow produced by compression molding. Transfer molding is often used in thermosetting polymers, where it allows a more even distribution of heat and finer detailed parts. Typical parts produced through the transfer molding process include distributor caps, drinking cups, and bottle caps. Figure 5–8 illustrates the transfer molding process.

The extrusion process is used to form materials that have long, uniform cross sections. It is typically involved in forming thermoplastic polymers, such as acrylics, cellulosics, fluorocarbons, nylon, styrene, polyethylene, and vinyls. In the extrusion

Step 1: Unpolymerized pellets or granules are loaded into hot die.

Step 2: Melted polymer is ready to be transferred.

Step 3: Transferred polymer begins to polymerize under induced heat and pressure.

Figure 5–8
Transfer molding.

A section of welding hose emerging from an extruder; the hose is next channeled through a trough of cool water.

Courtesy of Goodyear.

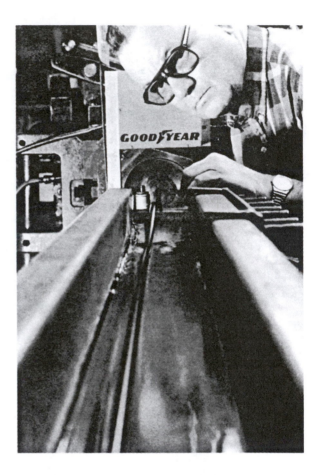

process, raw material is fed into a hopper, where it is gravity-fed into a screw chamber. In the screw chamber, a rotating screw forces the material into and through a preheated section. Here the material is compressed and then forced through a heated die onto a conveyor. As the material leaves the die and lands on the belt, it is cooled by an air or water spray to harden it sufficiently, preserving its shape during transportation. Usually, the material is then cut to length or coiled. Extrusion can be used

Figure 5–9
Extrusion.

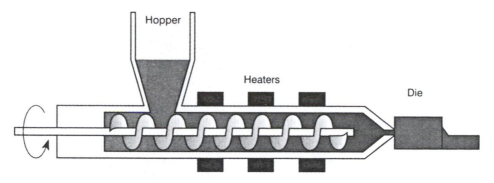

Figure 5–10
Injection molding.

to produce any material that has a uniform cross section, such as pipe, tubing, rods, and sheets. It is also used to form pellets for injection molding. Extrusion is very fast and allows for continuous production. The extrusion process is shown in Figure 5–9.

Injection molding is used to form thermoplastic materials. Commonly used thermoplastic polymers include acrylics, fluorocarbons, nylon, polyethylene, polystyrene, and vinyls. It is an extremely fast method of production. In injection molding, raw material

Injection-molded air ducts for automobiles.

Courtesy of Goodyear.

is fed into a hopper, which is screw-fed into the pressure chamber. As the screw turns, the material is fed into a heating chamber, where it is melted. The melted material continues through a nozzle that is seated against the mold. The nozzle allows the molten material to flow into the cavity through a system of gates and runners. In injection molding, the die remains relatively cool; thus, the product solidifies almost immediately upon filling the cavity. To prevent the premature cooling (and resultant solidifying) of the molten material, the material is forced through the barrel into the cavity under pressure. Premature cooling would result in defective parts. The mold is then opened, and the part is ejected. This process requires only a few seconds to complete and is shown in Figure 5–10. Injection molding is used to produce such items as toys, plastic models and parts, tile, food containers, battery cases, and radio cabinets.

Laminated plastics can be produced in the forms of sheets, tubing, and rods. Laminations are typically made from thermosetting polymers, such as phenolics, melamine, silicone, epoxy, and polyester. Laminated sheets are made of paper or cloth, which are constructed of glass or other types of fibers that have been impregnated with thermosetting polymers and stacked to produce the desired thickness. This sandwich-type structure is then cured under pressure and elevated temperatures. Laminates exhibit greater strength than single layers. The resulting properties are dependent on the type and quantity of sheet filler that is used. Phenolic laminates offer good properties at a reasonable cost. Melamine laminates are more expensive but offer enhanced electrical properties and greater flame resistance. Silicone laminates are used in high-temperature applications, where they offer better retention of their properties. Epoxy laminates offer greater water resistance, and polyesters are the lowest in cost. Epoxy and polyester laminates are used to produce gears, pulleys, countertops, paneling, and electrical panels and boards. Sheet laminates are manufactured through the process illustrated in Figure 5–11.

Figure 5–11
Lamination.

Figure 5–12
Lamination for rods and tubing.

Laminated rods and tubing are rolled over a mandrel to build up layers. Laminated sheet stock is wound over a mandrel to obtain the proper diameter. The product is then cured by heat and pressure in a molding press. The mandrel is removed and the material cut to length, if required. The difference between manufacturing rods and tubing is in the size of the mandrel used to wind them. This process is illustrated in Figure 5–12.

Some laminated products, such as boat hulls, auto bodies, and safety helmets, are manufactured by layering fabric or glass cloth that has been dipped in resin in or over a mold. The material is then allowed to cure. If additional layers are desired, they can be added after each subsequent layer has cured. This simple process has a low tooling cost but can produce only a limited number of parts economically.

Another common manufacturing process is vacuum forming. Vacuum forming is often used to form thermoplastic materials. In this process, a sheet of plastic material is placed over a die that contains the form of the finished product. The material is then

Figure 5–13
Vacuum forming.

Figure 5–14
Cold forming.

Raw material is loaded in press.

Cold-formed product is taken to oven, where it is cured.

heated until it becomes soft and pliable. After the material becomes pliable, a vacuum is introduced through the die, which draws the material into the die cavities. The material is then cooled, the vacuum is removed, and the product is removed from the mold. The entire process takes only a few minutes and is used economically to produce items such as panels and Braille pages for the visually impaired. Vacuum-forming processes are used in low-volume production. The vacuum tends to cause localized thinning of the material. The vacuum-forming process is shown in Figure 5–13.

The cold-molding process presses the raw material into final form while it is at ambient temperature. The product is then removed and cured in an oven. Because the cold-molding process does not require elevated temperatures or pressures, it has an economical advantage. However, the products manufactured using this process have poor surface finishes and don't exhibit good control over dimensional tolerances. Cold molding is primarily used to produce electrical parts such as switch plates, plugs, handles, knobs, and similar articles. The cold-forming process is illustrated in Figure 5–14.

The filament-winding process can be used to manufacture polymeric products such as storage tanks, which require high strength-to-weight ratios. This is possible due to the availability of plastic-coated, high-strength filaments composed of various materi-

Figure 5–15
Filament winding.

Mandrel

Resin bath

Filament

als, such as glass, graphite, and boron. These filaments are wound over a form, using some style or pattern, such as longitudinal, circumferential, helical, or a combination of any of these patterns. These materials and the patterns used to manufacture the product tend to generate higher strength in the direction required by the application. Based on the application of the finished product, different filament materials and winding patterns can be used to obtain desired results. A variety of polymers and resins can be used (epoxies are very common), making this a flexible and economical method of production. The filament-winding process is shown in Figure 5–15.

Calendaring has long been used in the rubber industry and has been adapted to the manufacture of such materials as paper, linoleum, sheet metal, and polymer films and sheets. In calendaring, the premixed thermoplastic is rolled, or calendared, through a series of heated, paired rollers. The distances between the rollers in the series are varied and are used to control the thickness of the sheet or film produced. After calendaring, the material is chilled and wound into rolls. In certain applications, the rollers are embossed with designs, logos, lettering, and so on, thus transferring the embossing to the finished film or sheet. Calendaring is a high-speed, high-volume process used to manufacture such items as shower curtains, raincoats, film liners, and seat covers. Figure 5–16 shows the calendaring process.

Foamed polymers are becoming more common. In foamed polymers, a foaming agent is added to the polymer, which releases a gas when heated. The foamed polymer product has a low density and can be either rigid or flexible, depending on the application. These products are used in packaging and shipping cartons, coatings, fillers for thin-walled components, thermal insulation, and acoustic insulation. This process is very flexible, and several thermoplastic and thermoset plastics are used for making foamed materials and products.

There are primarily two principles used in the foaming process: physical foaming and chemical foaming. Physical foaming requires the injection of a gas during the molding process. When this combination is heated, the gas expands and causes the combination to expand. Expanded polystyrene is an example of a polymer that has been physically foamed.

Chemical foaming is the result of the chemical reaction that occurs when two chemicals are mixed. One part of the reaction is the base polymer. The other part of the reaction is a catalyst, or initiator, and a chemical that reacts with the polymer to

Figure 5–16
Calendaring.

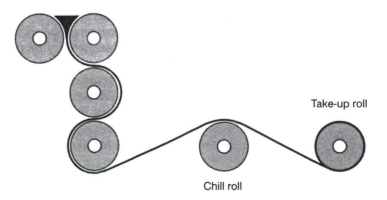

Take-up roll

Chill roll

Figure 5–17
Foaming processes.

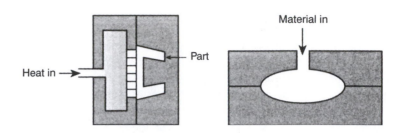

form a gas. When these two parts are combined, they produce a gas, which causes the material to expand. Polyurethane foam is a chemically foamed polymer.

There are two basic types of foaming processes: molding expandable polystyrene and casting polyurethane. When molding expandable polystyrene, styrene beads of the desired density are loaded into a closed mold cavity. The mold is then heated to approximately 275°F (135°C). The heat expands and fuses the styrene beads. The expanded beads take the form of the mold cavity. After the material has expanded, the heat is turned off, the part is cooled, the mold is separated, and the part is removed.

The second foaming process is used in casting polyurethane. The proper amount of polyurethane is mixed with a foaming agent. The mixture then expands to fill the mold cavity. Heat is sometimes used to speed the process. The two constituent materials can be sprayed through a nozzle, where they are mixed. The mixture can then be sprayed into recesses between walls, floors, or similar cavities, where it expands to fill the void.

Many polymers can be foamed. The most common are the styrenes and urethanes. However, vinyl, epoxy, polyethylene, silicone, cellulosic, and phenolic can also be expanded. Figure 5–17 shows the two basic foaming processes.

5.6 NATURAL RESINS

The term *resin* is generally applied to naturally formed materials such as saps and extracts. Natural resins are obtained either directly through the excretions of certain trees or from insects that feed on the sap juices. Natural resins are used in such products as paints, varnishes, enamels, soaps, inks, glues, and as plasticizers of other polymeric materials. These include shellacs, rosins, and copal resins.

Shellac is a thermoplastic material used as a base in a wide variety of chemical solvents, as a hard gloss coating, and as a bonding agent. Its use has declined in favor of less brittle materials; however, it is still used in the manufacture of abrasive grinding wheels where some elasticity is required. It is also used in polishes, waxes, inks, and lacquers. Rosin is a distillate of pine tree sap; turpentine is produced as a by-product. Rosins are used in paints, varnishes, soap, paper, and ink. Copal resins are used in paints, varnishes, and lacquers. When copal resins are mixed with celluloid, they form a hard film that is moisture and abrasion resistant.

Figure 5–18
Phenols.

5.7 THERMOSETTING MATERIALS

Thermosetting materials include phenols, aminoformaldehyde, melamine, urea, polyester, allyl, casein, epoxy, and urethanes. Diagrams and examples of the polymers are included with the discussion of each type.

Phenolic materials (Figure 5–18) were the first thermosetting polymers to be used in large quantities to produce consumer products. Many of the early plastic products were made of Bakelite, named after Dr. Leo Bakeland, who was the first person to produce the phenolformaldehyde polymer successfully. Phenolformaldehyde is produced through a series of condensation reactions, which form a rigid, cross-linked polymer. In its pure form, it is an opaque, milky white substance that discolors with age and exposure. It is, therefore, manufactured primarily in dark brown or black, although it can be produced in a variety of colors through the use of coloring agents. The trade name for phenolformaldehyde is Bakelite.

The properties of this material vary depending on the filler material used. Phenolics are medium weight and are among the hardest of all polymers. Their principal feature is their high compression strength. They show very little elongation before breaking. Phenolic products must typically be produced through compression molding and are primarily used for electrical applications and insulation. Phenolics are generally resistant to water, grease, alcohol, oil, and household chemicals. Because they are resistant to gasoline and oil, they have been widely used for molded engine parts in small engines, automobiles, and other similar applications. They can also be produced in laminate and foam forms or used as a brush- or spray-on preservative.

Amino polymers (Figure 5–19), such as amino-formaldehyde, are also formed through condensation reactions that involve aldehydes and an amino group. The most important of these materials are urea formaldehyde and melamine formaldehyde. The properties of urea formaldehyde are similar to phenolformaldehyde, but it has a lower resistance to moisture and heat. It is also produced by compression molding, but the molding powder is slightly more expensive. The principal advantages are that urea formaldehyde can be produced in a wide range of stable colors,

Figure 5–19
Urea and melamine.

and it doesn't leave disagreeable tastes or odors in foods that have been stored in it. It is resistant to detergents, cleaning fluids, oils, grease, gasoline, kerosene, and lacquer thinners. However, it tends to have a low moisture resistance.

Melamine formaldehyde is more expensive to produce but offers greater mechanical strength and increased moisture and heat resistance. Based on its high abrasion resistance, excellent surface finish, and resistance to attack by heat and chemicals, amino polymers are commonly used to produce countertops and shelving for household use. They are also used in kitchenware and dishes (under the name Melmac) for the same reasons. Ureas are used as adhesives in the manufacture of plywood and particle board. They can be used as a surface coating on appliances or in combination with cellulose to produce high-quality paper. The melamine-formaldehyde-mica polymer is used to produce *Formica*, which forms facing materials for countertops and cabinets.

Esters are produced as reaction products of acids and alcohols. Most of the polyesters are thermosetting, but Dacron (Figure 5–20) is a thermoplastic polymer commonly woven as a fabric. The most common application of the thermosetting polyesters is in combination with glass fiber fillers, those of low molecular mass, and a catalyst or hardening agent. The resultant mixture produces a rigid structure. The thermosetting polymer is used as the bonding agent for what is commonly referred to as *fiberglass,* which has been manufactured in the United States since the 1930s. Another group of thermosetting polyesters are the alkyds. Alkyd resins are commonly used in paints, enamels, lacquers, and related surface-coating applications. They can be formed by compression molding, transfer molding, or injection molding. Alkyd resins exhibit low moisture absorption, excellent dimensional stability,

Figure 5–20
Polyester (Dacron).

and good electrical insulation properties. When solvents and oils are added to the polyester, it can be used as a hard coating material. Unsaturated polyesters can be blended with other unsaturated monomers, such as styrenes.

Allyl polymers include diallyl phthalate (DAP) and diallyl isophthalate (DAIP). These phthalate esters of the allyl family can be used as monomers or converted to molding compounds. They are often reinforced by fibers and particulate fillers. They maintain their electrical properties at higher temperatures and higher humidity levels and possess excellent dimensional stability, along with resistance to moisture, most acids and alkalis, and solvents. They have the advantages of lower toxicity, lower losses during fabrication, and freedom from strong odors over the polyesters. Applications include electronic parts and components used under extreme environmental conditions, such as pump impellers and medical sterilizing equipment.

Casein is a thermosetting polymer that was used as a water-resistant bonding agent for plywood. Casein resins were developed in Europe and brought to the United States about the turn of the century. Casein resins have very limited properties, and their use has declined in favor of substitutes that have a wider range of properties and uses. Casein moldings offer good strength properties when first molded, but they suffer when subjected to sunlight and weather. They have average electrical resistance and thermal properties but are poor insulators due to their high water absorption rate. They also have a weak chemical resistance. Their primary uses include toys, knitting needles, buttons, and inexpensive buckles.

Epoxies (Figure 5–21) tend to be more expensive than other types. Their use in manufacturing is among the newest of the polymers. Their first application was as an adhesive to bond metals. Epoxies are medium weight, yet they exhibit high-strength properties. With proper reinforcement, their mechanical strength properties exceed those of structural steel. Epoxies are manufactured in a variety of solid and liquid forms, and a hardener (usually containing amino groups) is applied to the resin to produce the necessary cross-linking. The reaction of the hardener and the base doesn't produce by-products; therefore, less shrinkage occurs. Fiber reinforcements or filler powders are sometimes used to strengthen the material. Their principal outstanding characteristic is their ability to adhere to all types of surfaces. They cure at room temperature and have very low viscosity. They offer outstanding chemical

Figure 5–21
Epoxy.

Figure 5–22
Polyurethane.

(where R and R′ are any two radicals)

resistance and have a low water-absorption rate. Epoxies are used in various applications as high-strength adhesives. They are also used as surface coatings for appliances and floors and as coatings for severe service applications. As a reinforced product, they are excellent for use in manufacturing tooling, dies, and fixtures. They may also be foamed for use in sandwich constructions.

Research into the production and use of urethane polymers was first done in Germany by C. A. Wurtz in 1848. Urethane polymers are used as polymers, elastomers, and as adhesives, but their principal use is in the foam form. The term *urethane foam* has been adopted by the polymer industry to describe urethane foam and similar foam materials. Polyurethanes (Figure 5–22) are produced by the condensation reaction between isocyanate and an alcohol. They can produce either thermosetting materials or elastomers such as synthetic rubber. They have a high abrasion resistance and are, therefore, used as coatings. The properties of urethanes vary considerably according to the density of the material. Their flexibility and tear resistance make them excellent choices for cushion materials, mattresses, and chair pads. Their adhesion properties make them excellent insulation materials (both for thermal and acoustic insulation) in such applications as aircraft wings and aircraft bodies. They are also used to provide shock resistance in safety helmets for pilots, race car drivers, and motorcyclists. Their resilience allows them to be used as carpet padding and backing for floor coverings. They are also used in their foam form as insulation for applications such as refrigerators and over-the-road tractor trailers.

Silicone polymers (Figure 5–23) are often classified as inorganic materials. They are based on the silicon-oxygen mer. Silicon is immediately below carbon in the periodic table. Therefore, it also requires four bonds for each atom. Silicon is a very abundant material. Silicone polymers are more expensive to produce than organic polymers. Silicone-based polymers are lightweight, are produced in liquid or paste wax form, and are water repellant. They are also available in the form of silicone rubber, which requires a catalyst and cures into a thermosetting material. The principal advantage of silicone polymers is their wide temperature application. Silicone rubbers retain their flexibility at low temperatures, and polymers retain their properties at elevated temperatures. They are also used as antifoaming agents and to reduce the flammability of lubricants. Silicone polymers are used in scratch removers and in waxes for furniture and cars. Due to their high heat and chemical resistances, they are used as mold releases. They have excellent adhesion properties

Figure 5–23
Silicone.

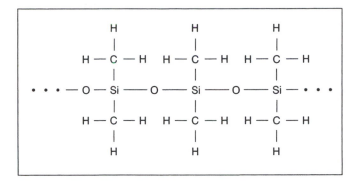

and are commonly used in pressure-sensitive adhesives and tapes. With the addition of glass fibers, silicone-glass laminates are excellent electrical insulators. Silicon polymers are also used in composites, exhibiting properties very similar to their carbon counterparts.

5.8 THERMOPLASTIC MATERIALS

Thermoplastic materials include polyethylene, polypropylene, polyvinyl chloride (PVC), polystyrene, acrylonitrile-butadiene-styrene (ABS), polytetrafluoroethene (PTFE, or Teflon), polyamides and polyesters, acrylics, cellulose, cellulose acetate, cellulose acetate butyrate, polyacetals, polycarbonates, polyetheretherketone (PEEK), and polyphenylene sulfides (PPS). One drawback to the polyethylene-based polymers is that they react readily with chlorine, bromine, and other chemicals. This drawback is overcome by replacing the hydrogen atoms with fluorine, a strongly negative element. This produces tetrafluoroethene. If the double bond between the carbon atoms is opened up, tetrafluoroethene can be polymerized into PTFE.

Polyolefin polymers include polyethylene and polypropylene (Figure 5–24). Polyethylene is one of the simplest polymers, chemically. It is one of the lighter polymers. It exhibits a high elongation—as much as 500% before breaking—and is, therefore, difficult to tear. Polyethylene, or just polythene, is a tough, flexible material that is water repellant and is used for electrical insulation, cold-water pipes, containers, film, and packing sheets. Polyethylene is used to wrap meat and fresh produce. It is best suited for these applications because it lets the material breathe. Oxygen can pass through the wrapping, and the carbon dioxide released by the material inside the wrapping can be released into the air. Because of its low temperature resistance, polyethylene has an additional advantage in film-wrapped, heat-sealed applications such as wrapping. Due to the lower mass of polyethylene, more film can be produced per pound. Based on their flexibility, toughness, and low water absorption, polyolefins are used in drinking glasses, squeeze bottles, and kitchen items such as ice-cube trays. The material's properties depend on the chain length. Classifications are generally in terms of high and low density.

Figure 5–24
Polyethylene and polypropylene.

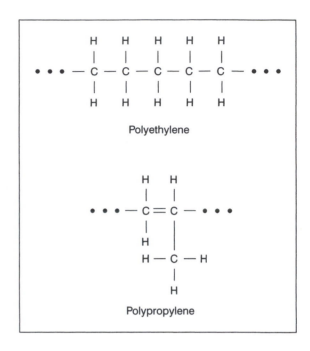

Polyethylene

Polypropylene

Early polyethylene plastics were produced using very high pressures and temperatures. They can be and, to a limited extent, are still produced in this fashion. These polyethylenes are termed *low-density polyethylene*. This label or grade refers to the normal density of the polymer. For example, low-density ranges from 0.910–0.925 g/cm^3, medium-density ranges from 0.926–0.940 g/cm^3, high density ranges from 0.941–0.960 g/cm^3, and very high-density ranges from 0.960 g/cm^3 and higher. Density influences cost, structure, ease of processing, and properties. Low-density polyethylene has a lower melting point, lower strength, and less elasticity than other grades. They are often used in packaging and coatings.

High-density polyethylene is produced under low-pressure and low-temperature conditions. The difference is the catalyst used to produce polymerization. Due to lower energy costs, high-density polyethylenes are more economical to produce. These are often used for injection-molded items, blow-molded bottles, and underground pipe.

Ultra-high-density polyethylene (UHMWPE) is a recent grade that has gained popularity due to its suitability in industrial applications requiring wear resistance. It is often used in manufacturing extruded shapes such as gears and bushings.

Polypropylene is a crystalline polymer that is stronger and more rigid than polyethylene, yet lighter weight and more expensive to produce. Its main uses are for sterilized chemical containers (based on its higher temperature resistance) and for high-fatigue-strength parts.

Vinyl polymers are among the oldest of the polymers. Pure vinyls tend to be hard and brittle. Therefore, their use is limited without the addition of plasticizers. The properties of vinyl polymers are wide ranging, depending on the particular type and the additives used. In general, vinyl polymers tend to be light- to

Figure 5–25
Polyvinyl chloride (PVC).

Figure 5–26
Polystyrene.

medium-weight, have low strength, and be tear-resistant. They have a low temperature resistance and are adversely affected by many common chemicals and solvents. One particular vinyl polymer—polyvinyl chloride, or PVC (Figure 5–25)—is used in the manufacture of pipe, fittings, and plumbing materials. PVC with a plasticizer, polyvinylidene chloride (trade name Saran), produces a soft, rubbery material used in the manufacture of upholstery, draperies, and window screens. Polyvinyl chloride acetate is used extensively in the manufacture of rain gear and weather-resistant clothing. PVC exhibits good chemical resistance and is very economical when filler materials are used. Other vinyl polymer applications are polyvinyl acetates, used in adhesives; polyvinyl acetals, used as interlayers in safety glass; polyvinyl alcohol polymers, used in hoses to transfer liquids; and vinyl foams, used in cushions and clothing materials.

Polystyrenes (Figure 5–26) are among the most common polymers used. Polystyrenes are classified into two groups: general-purpose polystyrenes and styrene alloys. General-purpose polystyrenes are lightweight, rigid, but brittle materials best used as electrical insulators and as expansion foams in thermal insulation. Polystyrenes are nontoxic materials that are odorless and tasteless, making them suitable for food and drink containers. They are used indoors and have a tendency to crack and discolor unless light stabilizers are added.

Styrene alloys are generally less expensive but more difficult to produce than general-purpose polystyrenes. Copolymers of styrene and acrylonitrile can be toughened through combination with butadiene acrylonitrile rubber to form the copolymer acrylonitrile-butadiene-styrene (ABS). Styrene alloys are used in plastic model parts, molded toys, frozen-food containers, and carrying cases and in their monofilament form can be used in brushes of various types. ABS is particularly important due to its chemical resistance and high impact strength. It is often used in molding car bodies, battery cases, and telephones. Expanded styrenes are used in flotation devices and as thermal insulation.

Figure 5–27
Fluorocarbons.

Fluorocarbons (Figure 5–27) can be classified under the ethylene family or separated into a distinct group. Fluorocarbons, as a group, exhibit the best overall properties of polymers. The oldest and most common member of the fluorocarbons is polytetrafluoroethylene (PTFE), commonly known as *Teflon*, and chlorotrifluoroethylene (CFE). The advantages of PTFE and CFE are chemical resistance, high impact strength, excellent electrical properties, and low coefficient of friction. They have zero water absorption. They are unaffected by weather, sunlight, and temperature. PTFE is opaque and naturally white, whereas CFE is transparent or translucent. They may be colored, often for aesthetics. They are often used as nonstick coatings, nonstick films, pump pistons, wire insulation, corrosion-resistant linings, fittings, coverings, bearing materials, and other similar applications. PTFE is hot-pressed or powder-sintered. All fluorocarbons are relatively expensive.

Polyamides (Figure 5–28) are the products of condensation reactions involving organic acids and amines. These are generally called *nylons*. DuPont was responsible for marketing nylon in 1938 as a replacement for the silk used in women's hosiery. Nylons are very strong, tough, and flexible, have high impact strengths, and are abrasion resistant. Some polyamides will withstand up to 300% elongation. They have extremely low coefficients of friction. They are best suited for indoor use and are often produced and used as fibers. They suffer a high shrinkage rate when molded and stain easily. Their main disadvantage is high moisture absorption, with accompanying dimensional changes and reduction in strength.

Nylons are expensive to produce and, due to their particular properties, are difficult to mold. However, they are well-suited for applications such as gears and gear coverings, food-processing equipment where no lubricant is used, household appliances, and textiles such as parachutes. Aromatic polyamides are used in composites and as reinforcement in tires. One product is polyaramid fibers (marketed under the trade name *Kevlar*) used in helmets and as reinforcing fibers in composites.

Figure 5–28
Polyamide.

Figure 5–29
Polymethyl methacrylate
(PMMA).

Polyesters can be either thermoplastic or thermosetting, depending on the acid and alcohol used in their production. One such polyester is polyethylene teraphthalate, which is produced in the condensation reaction between teraphthalic acid and ethylene glycol. It is mainly produced as extruded fiber, similar to the production of nylon fibers.

Acrylics are based on acrylic acid. The most common acrylic is polymethyl methacrylate, or PMMA (Figure 5–29), commonly known as *Plexiglas*. It is manufactured by the reaction of methylacrylic acid and an alcohol. An intermediate product is methacrylate. PMMA is a hard, rigid, transparent material that is easily formed by injection molding. Its uses include guards, glasses, lenses, and other clear optical applications. It is resistant to most chemicals but is attacked by gasoline, acetone, and most cleaning solvents. Included in the acrylics is acrylonitrile. Acrylonitrile is produced by addition polymerization and is typically manufactured as a fiber. Common trade names for acrylic polymers include *Acrilan, Orlon*, and *Lucite*. Acrylic polymers are relatively expensive, so they are limited to applications where their light weight, high transparency, and resistance to weather are used to the greatest advantage.

Cellulose polymers are based on the cellulose molecule (Figure 5–30). The discovery of cellulose polymers is attributed to the Hyatt brothers, in 1869. The Hyatt brothers were searching for a replacement for the ivory used in the manufacturing of billiard balls. Cellulose polymers are made from natural wood or cotton fibers. Early cellulose polymers were used to produce movie film (celluloid). The five basic cellu-

Figure 5–30
Cellulose.

lose polymers are cellulose acetate, cellulose nitrate, cellulose acetate butyrate, ethyl cellulose, and cellulose propionate.

Polymerization of cellulose is achieved through oxygen bonding. Celluloid (cellulose nitrate) is a tough, water-resistant material. It is the toughest of all thermoplastics. However, it is highly flammable and, therefore, cannot be molded. For this reason, it has been widely replaced by other less flammable materials. One application of cellulose nitrate is the manufacture of pen bodies and caps. Its flammability limits its applications.

One substitute for cellulose nitrate is cellulose acetate, which is inexpensive, is easily formed, is less flammable than nitrocellulose, and can retain its color. Cellulose acetate is tough, has a high impact strength, and can be flexible, transparent, and easily solvent-bonded. The excellent surface finish and formability of cellulose acetate make it a practical material for such applications as display racks and shelves, toothbrushes, toiletry items, and packaging and in the manufacture of shoes. Cellulose acetate monofilaments are very strong and are used as woven cloth fibers.

Cellulose acetate is not water-resistant, but the derivative cellulose acetate butyrate is. Cellulose acetate butyrate materials are used in outdoor applications. They often give off a foul odor when used indoors. These materials are found in outdoor advertisement bulletins, advertising signs, steering wheels, and eyeglass frames. Ethyl cellulose and benzyl cellulose are also common cellulose polymers. They are more commonly known as *rayon* and *cellophane.*

Polyacetal polymers (Figure 5–31) are based on carbon-oxygen bonds, which are actually formaldehyde monomers, with the exception of groups formed on the ends of the polymer chain. Acetal is the common name for this group of polymers. They are often found as replacements for nylon, because polyacetals have similar properties to nylon yet exhibit lower moisture absorption. Acetals have high tensile and compressive strengths. They exhibit no true yield point. They have excellent dimensional stability, have excellent abrasion resistance, and provide excellent electrical insulation. Their most outstanding quality is their resistance to chemicals, especially solvents.

Polycarbonates (Figure 5–32) are polyesters composed of carbonic acid and phenol. They are transparent and have low melting temperatures. They can be cut or scratched rather easily but have properties similar to the acrylics and have excellent impact strengths. One trade name is *Lexan.* It is used as a bullet-resistant shield over window glass.

Polyetheretherketone (PEEK) is a crystalline polymer that exhibits excellent thermal and combustion characteristics for a thermoplastic material. It also has a high resistance to chemical solvents and fluids. PEEK is used primarily as a coating and

Figure 5–31
Polyacetals (polyoxymethylene, or POM).

Figure 5–32
Polycarbonates (Lexan).

insulator material for high-performance wiring applications in the areas of aerospace, military, nuclear power plants, and oil drilling. PEEK materials are produced in reinforced and nonreinforced versions, using glass or carbon fibers for reinforcement.

The crystallinity of PEEK materials is controlled during the production process. If the material is hot-formed, it is highly crystalline and exhibits good dimensional stability. If the material is cold-formed, it maintains an optimum mechanical strength. These materials are used primarily for electronic and electrical applications, such as telecommunications and computer equipment, for pumps for the petroleum and chemical industries, and for automotive engine parts. In addition, PEEK materials are transparent to microwave radiation. Through proper production control, PEEK materials have been developed for kitchenware and kitchen appliance parts.

Polyphenylene sulfide (PPS) materials have high temperature stability and good chemical resistance. They are flame-retardant, have good dimensional stability, and retain their mechanical properties at elevated temperatures. PPS forms long polymer chains, which are used in a matrix with glass and graphite fibers to produce very strong composites.

5–9 LIQUID CRYSTAL POLYMERS

Liquid crystal polymers (LCPs) have proved their commercial success in various applications throughout the 1990s. They continue to gain in popularity as research and development continues to find new markets and applications.

These polymers may be made up of different mers. However, they share the characteristic of being crystalline in the liquid or melt state as well as the solid state. Kevlar, a polyamide, is a liquid crystal. Properties of the liquid crystal polymer depends on the component mers, but they generally exhibit high strength and temperature properties.

Cost of these LCPs continues to decrease, being used initially in many electronics such as liquid crystal displays (LCDs). As further development continues, LCPs may replace many conventional displays.

5.10 REINFORCED POLYMERS

Reinforced polymers are polymers whose properties have been altered through the use of a reinforcement material. Glass, mineral, and carbon fibers can be added to polymers to alter their properties. The additional costs involved are offset by the increase in their overall performance. Most polymers are available as glass fiber–reinforced products. The glass fibers, whose diameters range from 0.0002 to 0.001 in. in diameter, are coated with a resin and a coupling agent. Property increases of 200% or better can be obtained through glass fiber reinforcement.

The strength of these reinforced polymers depends on the type and quantity of filler used and the degree to which the resin has wetted the reinforcement material. One of the more common filler materials is glass fiber. Glass fiber–filled polyesters exhibit very high strength properties. They can be cured at room temperature, which makes their use practical and economical. Because polyesters cure at room temperature, are easily repaired, and are very resilient, they are more convenient and economical for use in the manufacture of such items as boat hulls, car bodies, small appliances, and similar applications. However, the material costs are higher and prohibit any large-scale production. Fiberglass car bodies are used in high-speed racing applications based on their reduced weight and easy repair. Polyester-reinforced materials exhibit high strength and flexibility, which makes them excellent material for fishing rods and similar sports equipment. They are also used to produce Mylar film. Mylar film is used as capacitor dielectrics, coil insulation, and photographic film.

Carbon fiber–reinforced materials are more expensive than glass fiber–reinforced materials, often between two and four times greater in cost. Carbon fiber–reinforced materials have high tensile strength, stiffness, and greater related mechanical properties. They also produce lower coefficients of expansion, improved creep resistance, better wear resistance, greater toughness, and higher strength-to-weight ratios than glass fiber–reinforced materials.

Glass and carbon are the most common fiber reinforcements. Other materials used in fiber reinforcement include graphite, boron, and cotton.

Reinforcement materials are available in several forms, including mat, fiber bundles, chopped fibers, and cloth. Mat fibers provide medium improvement in strength; however, they allow for a rapid buildup at economical costs. Fiber bundles are used to provide high directional strengths at low costs. Chopped fibers provide a medium increase in strength and are used in premix compounds, where material flow at reasonable costs is a factor. Of all reinforcement materials, cloth provides the greatest strength improvement. Cloth reinforcement is also the most expensive form of reinforcement.

Most of the reinforced thermosetting polymers (90% to 95%) are polyesters and epoxies, with the majority being polyesters. Phenols and silicones are also used in their reinforced forms, to a lesser degree. These reinforced polymers are used in boat hulls, car bodies, storage tanks, circuit boards, aircraft equipment, gears, and thermal and electrical insulation.

These reinforced polymers are manufactured through a variety of methods, including hand lay-up, spray-up, matched molding, premixed molding, and vacuum- or pressure-bag casting.

Hand lay-up involves a single mold, which has been precoated with a mold-release agent to allow for part removal. The liquid base is then applied in thin, even coats to the mold by brush, roller, or pressure spray. Reinforcement layers of cloth or mat are then soaked in the liquid and placed in the mold. The desired surface of the mold is smoothed to get the air bubbles out and to saturate the reinforcement material. The combination is then allowed to cure. Curing can take place at room temperature, or the combination can be oven-cured. The part is then removed from the mold. The mold can be made from a variety of materials, including wood, plaster, another polymer, or metal. Hand lay-up is the least expensive process used to manufacture reinforced polymers.

Spray-up processes require a two- or three-nozzle pressure spray gun, which is used to meter the required amounts of resin, catalyst, promoter, and reinforcement material. A single mold is used, which has been previously coated with a mold-release agent. The pressure spray gun is then used to spray the reinforced material onto the mold. Spraying continues until the required thickness has been achieved. This may require multiple coats. The material is then allowed to cure or is induced to cure in an oven, and the part is removed from the mold. The pressure spray gun used is similar to, but more intricate than, the spray gun commonly used for applying paints.

In matched molding, there are two mating halves of the mold. These halves are coated with a mold-release agent. The reinforcement material is cut to the approximate finished form. The preformed reinforcement material is laid over one-half of the mold and saturated with liquid resin. The two halves of the mold are then pressed together and heated. The increased heat and pressure induce the liquid to flow over the reinforcement material and saturate it. In addition, the heat and pressure cause the polymer to cure. After curing, the mold is then separated, and the part is removed. The equipment used in matched molding is similar to that used in compression-molding polymers.

Premixed molding requires a special set of matched molds. The molds are coated with a release agent to facilitate removal of the part after molding. A special batch is mixed. The batch consists of short glass fibers, which are mixed into the liquid to form a putty called the *premix*. A specified amount of the premix is then dropped into the mold cavity. The mold is closed and heated. Due to the heat and pressure, the material cures inside the mold cavity. After curing, the mold is separated, and the part is removed. Premixed polymers can be purchased or produced by the manufacturer.

Pressure-bag molding requires a mold and a rubber bag. The polymer and reinforcement materials are placed in the mold or are used to coat a partially filled rubber bag, which is then placed in the mold. The bag is inflated with air pressure, ranging from 20 to 50 lb/in^2. The internal pressure within the bag forces it outward. The bag and the uncured, reinforced polymer material are forced to follow the contour of the mold cavity. The mold may be heated. The material is allowed to cure within the mold cavity. The mold is then split, the bag is deflated, and the part is removed.

Vacuum-bag molding requires placing the polymer mix and reinforcement material over the desired mold, which has been pretreated with a mold-release agent. This combination is then covered with an airtight rubber bag. The air is evacuated from the space between the bag and the reinforced material. The air pressure outside the bag forces the material against the contour of the mold. The entire setup can be placed in an autoclave (a pressure tank that is used to maintain increased pressure on the outside of the bag). A vacuum is maintained within the bag during the curing process. After the part has cured, the vacuum, bag, and the part are removed.

Hand lay-up is used in low-volume applications due to its labor-intensive requirements. Consistency between parts suffers, but reduced equipment costs, portability, and flexibility are advantages in home and job shops. Spray-up processes require more expensive equipment than hand lay-up, and it is difficult to obtain complete impregnation. This process is commonly used for repair operations. If increased production rates are desired, matched molding provides increased rates and increased product finish. The high initial cost of the matched-mold process can quickly be recovered at medium- to high-volume production rates. Matched molding is typically used in producing safety helmets and sports car bodies. Premixed molding has features similar to matched molding; additionally, by using chopped fibers, premixed molding can be used to produce delicate or intricate parts, where the putty can flow into smaller areas. However, if strength is a concern, additional form thickness may be required. This requirement reduces the advantage of the premixed molding process. Premixed molding is typically used to produce parts such as propellers, impellers, and housings. Bag molding, whether pressure- or vacuum-applied, provides the highest production rate at reasonable costs. However, only one side of the part, the side covering the mold, will have an excellent surface finish; the other side will suffer from irregular surface-finish thickness. Because of the higher pressures involved, pressure-bag molding is used when better surface finishes are desired.

Reinforced polymers are replacing metals in a variety of applications. They offer similar or better strength characteristics at lower weight and, often, at lower cost. They are frequently easier to produce, are resistant to corrosion and chemicals, and offer a wide variety of manufacturing processes, from the very simple and inexpensive to the higher volume and more intricate processes.

5.11 ELASTOMERS

Elastomers have properties similar to polymers. Elastomers are materials that may be repeatedly stretched or elongated and will return to their original condition upon release of the force producing the elongation. Thus, they exhibit elastic behavior, as compared to polymers, which exhibit greater plastic properties. Included in this category are rubber and rubberlike materials. A distinction is made between rubber and elastomers. Rubber must withstand a 200% elongation and rapidly return to its original dimensions. This ability of a material to recover from elastic deformations is called *resilience*.

Natural rubber is produced from the gum resin of certain trees and plants that grow in southeast Asia, Ceylon, Liberia, and the Congo. The sap, which is an emulsion containing about 40% water, contains suspended rubber particles. The rubber particles are coagulated through the use of formic or acetic acid. These coagulated particles are then dried, rolled, and shipped to the manufacturer for further processing. Rubber in this form is often called *latex*. Raw rubber has few uses, because it has a low tensile strength and melts at elevated temperatures.

To increase the properties of natural rubber, the material is rigorously rolled, and additives are blended into the rubber. Additives include plasticizers, antioxidants,

A skid steer loader forges through a muddy quarry on endless rubber tracks.

Courtesy of Goodyear.

stabilizers, and inert fillers. Sulfur is added to the mix to aid in *vulcanization*. In 1836, Charles Goodyear vulcanized rubber by adding sulfur. The sulfur produced cross-linkages between the chains of the rubber molecules, which made the rubber stiffer and harder. These cross-linkages reduced the slippage between chains, increasing the elastic properties of the material. In the vulcanizing process, some of the double covalent bonds between molecules are broken, allowing the sulfur atoms

Coils of grade R acetylene welding hose.

Courtesy of Goodyear.

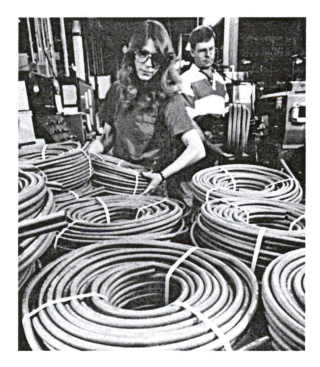

NAC III poly-V belts.

Courtesy of Goodyear.

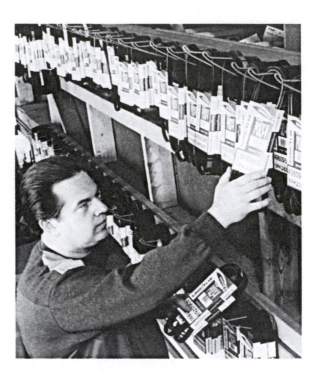

to attach themselves and form cross-linkages between molecules. Typical soft rubber contains approximately 4% sulfur by weight. This produces about 10% cross-linkage. Full vulcanization requires approximately 45% sulfur by weight and produces a hard rubber known as *ebonite*.

Rubber can be molded or extruded after mixing. It is then cured at an elevated temperature, about 300°F (150°C). Hard rubber has the appearance of phenol-formaldehyde (Bakelite). It provides excellent electrical insulation and is used in such applications as battery casings. Rubber strips and bushings are often used to provide vibration damping, such as in motor-mount applications.

The fillers used in the production of rubbers include silica and carbon black, which may comprise as much as half of their volume. Fillers have various purposes, including lowering the cost of the material, providing reinforcement, improving abrasion resistance, improving tear resistance, and increasing the wear properties of the material when exposed to varying conditions of heat and light. As an example, tires produced from latex contain approximately 30% carbon black filler by weight. Carbon black filler is used to provide the body and abrasion resistance in the tires. Therefore, the higher the percentage of carbon black, the harder and longer lasting (and more expensive) the tires will be.

Thermoplastic elastomers, sometimes referred to as elastoplastics, result from the copolymerization of two or more monomers. One monomer is used to provide the hard, crystalline features, whereas the other monomer produces the soft, amorphous features. When combined, these form a thermoplastic material that exhibits proper-

Figure 5–33
Isoprene.

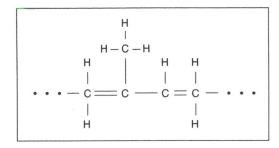

ties similar to the harder, vulcanized elastomers. Thermoplastic elastomers can be either molded or extruded into finished shapes.

Urethanes were the first thermoplastic elastomers to be produced. These products are used in such applications as gaskets, gears, and fuel lines. Other thermoplastic elastomers include copolyester, styrene, styrene-butadiene, and the olefins. Copolyester elastomers have more properties but are more expensive than urethanes. They are used in hydraulic hoses, couplings, and cable insulation. Styrene copolymers are

A fuel vapor recovery hose. The spiral design prevents kinking, stretching, puncturing, and tearing.

Courtesy of Goodyear.

Figure 5–34
Butadiene.

less expensive than either urethanes or copolyesters. They have a lower tensile strength but exhibit greater elongation. Styrene-butadiene rubber (SBR) is used in medical products, tubing, packaging materials, adhesives, and sealants. Thermoplastic olefins (TPO) are used in tubing, seals, gaskets, and electrical products.

Chemists had been working on discovering the basic structure of the rubber molecule for many years. It remained undiscovered until 1860, when the structure was determined to be that of isoprene. The structure of isoprene is illustrated in Figure 5–33.

Once the structure of the rubber molecule was discovered, chemists began trying to find ways to produce rubber artificially. World War I provided an impetus for developing synthetic rubber, yet the rubber produced was of a low quality and was subsequently open to attack by organic materials, such as fuels and lubricants. Not until 1930, when thiokol rubber was developed, were chemists able to produce a rubber suitable for widespread application. Thiokol rubber can withstand gasoline and other organic solvents and can therefore be used to replace natural rubber in such applications as fuel hoses, hydraulic hoses, and oil lines.

Originally, the demand for developing synthetic rubbers was small, because natural rubber was available in generous supply from Malaysia. However, with the advent of World War II, the Japanese occupied the Malaysian peninsula and cut off

Figure 5–35
Buna-S.

Styrene-butadiene rubber (SBR)

Figure 5–36
Buna-N.

Nitrile rubber (buna-N)
Copolymer of acrylonitrile and butadiene

other countries from their primary rubber supply. Rationing and severe cutbacks on the use of rubber were instilled, and great pressure was generated to develop higher-quality synthetic rubber products. Although these synthetic rubber products were of low quality during the war, further research stemmed from these initial efforts. Synthetic rubber production has evolved into a vital industry.

Synthetic rubber constitutes the majority of the world's consumption of rubber materials. Butadiene forms the basis for what is known as Buna rubber. When butadiene (Figure 5–34) is mixed with other unsaturated compounds, the resulting rubber exhibits excellent properties that are desirable in such items as tires, hoses, and gaskets.

Butadiene also forms the basis for styrene-butadiene rubber (SBR). SBR, also known as buna-S rubber (Figure 5–35), is an example of a synthetic rubber made through the copolymerization of styrene and butadiene.

Butadiene can also be mixed with acrylonitrile to form buna-N rubber. This nitrile rubber, shown in Figure 5–36, is a copolymerization product of butadiene and acrylonitrile. Nitrile rubber is more expensive than SBR.

Polychloroprene, or neoprene rubber, is reputed to be the synthetic rubber closest to natural rubber. Neoprene (Figure 5–37) is lower in tensile strength and elasticity than natural rubber but is higher in chemical resistance. Therefore, it is used in applications such as fuel lines, hoses, and gaskets. However, neoprene is also more expensive to produce.

Butyl rubber is an addition polymer of isobutylene. When mixed with isoprene, butyl rubber provides excellent resistance to oxidation and many solvents. Therefore, butyl rubber is a very good material for applications such as tubing, hoses, and tires. Figure 5–38 illustrates butyl rubber.

Figure 5–37
Neoprene.

Neoprene or polychloroprene rubber

Figure 5–38
Butyl.

Butyl rubber
Copolymer of isobutylene and butadiene, or isoprene

Thiokol rubber
Produced from ethylene dichloride and
sodium polysulfide

Figure 5–39
Thiokol.

Thiokol (Figure 5–39) is produced by polymerizing ethylene dichloride with sodium polysulfide. The sulfur makes thiokol rubber self-vulcanizing.

Elastomer molecules form chains, similar to polymers. These chains may take one of two forms, either the *cis-* form or the *trans-* form. The chains of the elastomer rotate and spin endlessly around single carbon-to-carbon bonds, as is the case with isoprene. When these single bonds become double bonds, they are no longer free to rotate. Looking at the molecules that make up the elastomer, if the two end groups are on the same side of the enclosed double bond, the molecule is in the cis- form. If the end groups are on opposite sides of the double bond, it is in the trans- form. The two forms are illustrated in Figure 5–40.

Gutta percha is the trans- form of the isoprene molecule. It forms a hard rubber, which is used in the production of such items as bowling balls, billiard balls, and

Figure 5–40
Cis- and trans- forms.

Cis-butadiene   Trans-butadiene

canes. The cis- form of the isoprene molecule tends to coil. These coils give the elastomer its elasticity.

Some rubber products, such as those products made from isoprene and butadiene, tend to oxidize, get brittle, and crack with age. This is due, in part, to exposure to oxygen in the environment and to other organic mers, which, due to their affinity, break one of the double bonds and attach themselves to the molecule. In order to help protect these products from cracking and becoming brittle, these rubbers are

Table 5–2
Properties of thermoset polymers.

Polymer	Density	Tensile Strength lb/in.² (MPa)		Compressive Strength lb/in.² (MPa)	
Aminos	1.5	5,000–13,000	(34–90)	25,000–45,000	(172–310)
Casein	1.35	10,000	(69)	27,000–53,000	(183–365)
Epoxies	1.1–1.8	4,000–13,000	(28–90)	13,000–28,000	(90–193)
Phenolics	1.25–1.55	4,000–9,000	(28–62)	15,000–50,000	(103–345)
Polyesters	1.3	4,000–25,000	(28–172)	10,000–35,000	(69–241)
Silicones	1.5–2.0	5,000–35,000	(34–241)	10,000–15,000	(69–103)
Urethanes	1.2–2.0	5,000	(34)	Varies	

	Impact Strength	Electrical Resistance	Water Absorption	Chemical Resistance
Aminos	Good	Good	Low	Good with the exception of strong acids and alkalies
Casein	Fair	Fair	High	Poor
Epoxies	Excellent	Excellent	Low	Good
Phenolics	Good	Good	Low	Good with the exception of strong acids and alkalies
Polyesters	Excellent	Good	Low	Poor
Silicones	Good	Excellent	Very Low	Good
Urethanes	Varies	Good	Very Low	Excellent

Table 5–3
Properties of thermoplastic polymers.

Polymer	Density	Tensile Strength lb/in.^{2} (MPa)		Compressive Strength lb/in.^{2} (MPa)	
Acetals	1.4	10,000	(69)	18,000	(124)
Acrylics	1.16–1.2	7,500–15,000	(52–103)	12,000–18,000	(83–124)
Cellulosics	1.15–1.4	2,000–8,500	(14–59)	12,500–35,000	(86–241)
Fluorocarbons	2.1–2.2	6,500–10,000	(45–69)	1,600–80,000	(11–552)
Polyamides	1.09–1.14	7,500–11,000	(52–76)	7,500–13,000	(52–90)
Polyolefins	0.9–0.97	1,500–5,500	(10–38)	2,500–10,000	(17–69)
Polystyrenes	0.98–1.1	3,500–12,500	(24–86)	5,000–16,000	(34–110)
Vinyls	1.2–1.6	1,500–9,000	(10–62)	1,000–13,000	(7–90)

	Impact Strength	Electrical Resistance	Water Absorption	Chemical Resistance
Acetals	Excellent	Excellent	Low	Fair
Acrylics	Excellent	Good	Low	Fair
Cellulosics	Good	Good	High	Fair with the exception of strong acids and alkalies
Fluorocarbons	Excellent	Excellent	None	Excellent
Polyamides	Excellent	Excellent	None	Good with the exception of strong acids
Polyolefins	Excellent	Excellent	Low	Fair
Polystyrenes	Good	Good	None	Good with the exception of strong acids
Vinyls	Good	Good	Low	Good with the exception of strong acids

vulcanized. Vulcanization removes some of these double bonds, lessening the degree to which the material may be attacked.

Tables 5–2 and 5–3 list the major characteristics of the various elastomers discussed.

5.12 SUMMARY

This chapter deals with the theory of polymerization, the different types of polymers and elastomers, and various polymer-strengthening mechanisms. In addition, various common manufacturing methods are described for different polymers and elastomers. This is a comprehensive, yet not exhaustive, list of the common varieties of polymers and elastomers used today.

QUESTIONS AND PROBLEMS

1. What are meant by the terms *saturated* and *unsaturated* in relation to polymers?
2. Define the term *isomer*.
3. Explain the three types of polymerization.
4. What are the basic differences between thermosetting and thermoplastic polymers?
5. Name three commonly used thermoplastic polymers and their applications.
6. Name three commonly used thermosetting polymers and their applications.
7. List some common characteristics for thermosetting and thermoplastic polymers.
8. List and describe the six major production techniques that are used in the plastics industry.
9. Explain the lamination process and the purpose for laminating plastics.
10. Explain the difference between the terms *plastic* and *polymer*.
11. Explain the types of bonding that exist in polymers.
12. How does the bonding type affect the polymer's properties?
13. How can the properties of polymers be improved?
14. Define the term *amorphous* as it relates to polymers.
15. What is the difference between isotropic and anisotropic polymers?
16. Explain the difference between crystallization and crystallinity.
17. List the basic thermoplastic polymers and their distinguishing characteristics.
18. List the basic thermosetting polymers and their distinguishing characteristics.
19. Define the term *rheology*. How does this relate to polymers?
20. Discuss viscoelasticity as it applies to polymers.
21. What two basic methods are used to foam polymers?
22. What two principles cause polymers to foam?
23. List the two principal polymers used in foaming applications.
24. List at least one advantage and one disadvantage for each of the thermosetting and thermoplastic polymers discussed.
25. What is an elastomer?
26. Explain the purpose and procedure of vulcanization.
27. List the elastomers discussed in the chapter and an application of each.
28. Define and discuss the cis- and trans- forms of elastomers.
29. In what ways do elastomers differ from polymers?
30. Why are elastomers and not polymers used in applications such as tires?
31. Do polymers have a crystalline structure? If so, how does this type of structure influence the plastics properties? What applications are there for crystalline polymers?
32. Why are tires made of rubber? Are there substitutes that could be used? Why or why not?
33. List five common applications of plastic products.
34. What factors influenced the selection of the polymers used in these applications.
35. Take five common plastic products and describe the processes used to make these products.
36. The substitution of plastics for metals in traditional applications continues to grow. Name five recent applications where plastics have replaced metals.
37. What properties do plastics exhibit that are advantages and disadvantages to metals?

CHAPTER 6

Wood and Wood Products

Objectives
- Distinguish between the types of woods.
- Recognize and describe pertinent features of woods and wood products.
- Describe various applications and grading systems of lumber and plywood.
- Describe how plywood is made and its applications.
- Define the physical and mechanical properties of wood.
- List various preservatives used for wood and explain why these properties are important.

6.1 INTRODUCTION

Wood is a natural, renewable, organic substance with a wide variety of uses. Of particular interest are trees that grow to sufficient height with diameters large enough to be useful in the production of lumber and wood products. Other plants, such as bushes, vines, and shrubs, are also wood, but these plants have little use in structural application. This chapter focuses on logs, lumber, laminations, plywood, particle board, and the related products of trees.

Wood is widely available and easy to work, although it burns easily, has a variety of defects, and is subject to attack by insects and fungi. Therefore, special precautions must be taken to preserve it. This chapter deals with the types and properties of wood and wood products and some of the processes connected with wood.

6.2 TYPES AND PROPERTIES OF WOOD AND LUMBER

Wood comes from trees. When wood has been prepared for subsequent use in building and construction, it is called *lumber*. Wood is a product of nature and is renewable through natural processes and reforestation programs. Without conservation practices and reforestation programs, the total supply of trees would soon be depleted. The particular properties of woods are dependent on the type of wood and the region from which it came. Figure 6–1 illustrates the regions of forests and the major classes of trees included in each region.

6.2.a Features of Wood

Trees grow in three directions: (1) They grow upward as trunks and limbs, (2) they grow downward through roots, and (3) they grow in diameter. Trees grow in diameter by adding material to their growth layer, or *cambium layer*, which lies just inside the bark. The cambium layer is not normally visible in a tree's cross section, but as the bark is pulled away, the cambium layer can be seen as a sticky film that exists on the outer portion of the sapwood. Trees grow radially through division of these cells. Cell division on the inner side of the cambium layer forms new cells, whereas cell division on the outer side of the cambium layer gives rise to new bark. The cambium layer itself is only one cell thick.

From the outside of the tree toward the core, a tree is composed of bark (inner and outer), wood, and pith. The bark is a flaky layer of corky material that protects the tree. The *inner bark*, nearest the cambium, is alive and provides protection around the tree. The *outer bark* is not alive and gradually becomes dry and brittle. The bark protects the tree from attack by various insects and diseases.

Trees grow outward from a central core, or *pith*, which is the soft center of the tree. This portion is often decayed and darker than the rest of the cross section. The pith may be thought of as the original tree sprout, around which the rest of the tree grew. The pith has no particular use in manufacturing or construction. It remains intact throughout the life of the tree.

Northeast and North Central

Black ash
Green ash
White ash
Bigtooth aspen
Quaking aspen
American basswood
White basswood
American beech
Paper birch
Sweet birch
Yellow birch
Ohio buckeye
Yellow buckeye
Butternut
Eastern redcedar
Atlantic white-cedar
Northern white-cedar
Black cherry
Eastern cottonwood
Swamp cottonwood
Plains cottonwood
Flowering dogwood
American elm
Rock elm
Slippery elm
Winged elm
Balsam fir
Hackberry
Eastern hemlock
Bitternut hickory
Mockernut hickory
Pignut hickory
Shagbark hickory
Shellbark hickory
Honeylocust
Black locust
Black maple
Red maple

Silver maple
Sugar maple
Black oak
Bur oak
Chestnut oak
Chinkapin oak
Northern red oak
Nuttall oak
Overcup oak
Pin oak
Post oak
Scarlet oak
Shumard oak
Southern red oak
Swamp chestnut oak
Swamp white oak
White oak
Willow oak
Pecan
Common persimmon
Eastern white pine
Jack pine
Pitch pine
Red pine
Virginia pine
Balsam poplar
Yellow poplar
Sassafras
Black spruce
Red spruce
White spruce
Sweetgum
American sycamore
Tamarack
Black tupelo
Black walnut
Black willow

Southeast and South Central

Green ash
White ash
Baldcypress
White basswood
American beech
Sweet birch
Yellow birch
Ohio buckeye
Yellow buckeye
Butternut
Black cherry
Eastern redcedar
Atlantic white-cedar
Eastern cottonwood
Swamp cottonwood
Flowering dogwood
American elm
Rock elm
Slippery elm
Winged elm
Hackberry
Eastern hemlock
Bitternut hickory
Mockernut hickory
Nutmeg hickory
Pignut hickory
Shagbark hickory
Shellbark hickory
Water hickory
Honeylocust
Black locust
Southern magnolia
Black maple
Red maple
Silver maple
Sugar maple
Black oak
Bur oak

Cherrybark oak
Chestnut oak
Chinkapin oak
Laurel oak
Live oak
Northern red oak
Nuttall oak
Overcup oak
Post oak
Scarlet oak
Shumard oak
Southern red oak
Swamp chestnut oak
Water oak
White oak
Willow oak
Pecan
Common persimmon
Eastern white pine
Loblolly pine
Longleaf pine
Pitch pine
Pond pine
Sand pine
Shortleaf pine
Slash pine
Virginia pine
Yellow-poplar
Sassafras
Sugarberry
Sweetgum
American sycamore
Black tupelo
Ogeechee tupelo
Water tupelo
Black walnut
Black willow

Rocky Mountains and Great Plains

Green ash
Quaking aspen
Paper birch
Eastern redcedar
Western redcedar
Black cherry
Eastern cottonwood
Plains cottonwood
Slippery elm
Douglas-fir
Grand fir
Subalpine fir

White fir
Hackberry
Western hemlock
Bitternut hickory
Rocky Mountain juniper
Western larch
Lodgepole pine
Ponderosa pine
Western white pine
Pinyon pine
Engelmann spruce

Pacific Northwest and Southwest

Alaska-cedar
Red alder
Quaking aspen
Paper birch
Incense-cedar
Port-Orford-cedar
Western redcedar
Black cottonwood
Douglas-fir
California red fir
Grand fir
Noble fir
Pacific silver fir
Subalpine fir
White fir
Mountain hemlock
Western hemlock
Western juniper
Western larch

California-laurel
Pacific madrone
Bigleaf maple
Oregon white oak
Jeffrey pine
Lodgepole pine
Monterey pine
Ponderosa pine
Sugar pine
Western white pine
Balsam poplar
Redwood
Giant sequoia
Black spruce
Engelmann spruce
Sitka spruce
White spruce
Tanoak

Figure 6–1

Tree species by regions.

Source: Silvics of Forest Trees of the U.S., Agriculture Handbook No. 271, USDA Forest Service, April 1975.

Wood is generally divided into two easily recognizable areas: heartwood and sapwood. The woody layer just beneath the cambium layer is used to transport moisture through the tree and is called *sapwood*. The region of sapwood cells, which are often light in color, lies outside the heartwood cells. Although sapwood cells are mature cells, most of which are dead, the sapwood conducts food to other parts of the tree. *Heartwood* comprises the darker cells surrounding the pith. Heartwood is also called *truewood*. Heartwood cells are dead and no longer contribute to the life processes of the tree. Sapwood is less dense than heartwood and is less resistant to decay and insect attack than heartwood, which is more durable (not stronger) in use.

Very young trees are all sapwood, which serves the dual purpose of supporting the tree while conducting minerals, or *sap*, from the roots to the leaves. Photosynthesis occurs between the available sunlight and the chlorophyll present in the leaves. These sapwood cells are made of cellulose. As growth continues, the cell walls become thicker, and the process of lignification occurs. These growth processes are known collectively as *secondary thickening*. This thickening marks the change of sapwood into heartwood. Along with secondary thickening, gums and resins are deposited in these cells, which renders the cells unable to conduct any longer, and they die. Heartwood is, therefore, dead wood, and it contributes very little to further tree growth, except as storage.

Wood typically consists of tiny, hollow, fibrous cells that are composed of *cellulose* and *lignin*. The typical composition of wood is 60% cellulose, 28% lignin, and 12% other materials. Lignin acts primarily as a cement to hold the wood cells together, thus supporting the tree. Cellulose is made up primarily of carbohydrates produced by photosynthesis. The sugars produced by photosynthesis in the leaves are dissolved into the sap and converted to carbohydrates, which are known as *cellulosics*. Wood cellulose is used as a raw material in the production of synthetic silk, celluloid, lacquers, plastics, and explosives. Lignin has very little commercial value and is usually stripped from the wood with concentrated acid solutions before the wood is used. For example, wood used to produce paper must be completely delignified prior to its use. In addition to cellulose and lignin, wood contains the following materials in small quantities: starch, resins, fats and waxes, tannins, oils, and mineral matter.

Due to climate trends, a tree grows more rapidly in the spring than in the summer. Cells that grow in the spring tend to be longer, with thinner walls, whereas the cells that grow in the summer are short, with thicker walls. Because of this, *summerwood* tends to be denser, stronger, and usually darker in color than *springwood*. This can be seen in the distinct annular rings, or *growth rings*, when a tree is cut. The age of a tree can be determined from these rings. However, trees grown in warmer climates with little variation in seasonal temperatures, such as the tropical rain forests, have less distinction in their growth rings.

When discussing the structure of wood, there are three primary planes of reference used. These planes of reference are shown in Figure 6–2. One reason for referencing a particular plane when discussing properties is due to the variation in properties across different planes. This variation is called *anisotropy*. For example, wood exhibits greater strength along its length than it does from the side. We can observe this by walking across a plank supported on both ends. The plank readily bends up

Figure 6–2
Planes of reference.

and down, possibly breaking. If the plank is supported so that it is vertical or horizontal (on edge), it exhibits much greater strength.

The transverse is the most often cited, because cell arrangement is best viewed from this direction. The transverse view is what we see when looking at the end of a log or piece of lumber. Tangential views are common to plain- or slab-sawn lumber, and radial surfaces are common to quarter-sawn lumber.

Wood fibers are arranged longitudinally in the tree, which gives the tree strength in that direction. This arrangement produces the wood *grain*. The wood's strength depends on the direction of loading. Wood tends to be strongest longitudinally, or along the grain. In other directions, strength tends to be lower. These other directions are often referred to as parallel or perpendicular to the grain. The inner construction of a tree is shown in Figure 6–3.

6.2.b Hardwoods and Softwoods

Trees are often classified as *hardwoods* or *softwoods*. This distinction is based on their cell structure, not on their mechanical properties. However, most hardwoods

Figure 6–3
Inner construction of a tree.

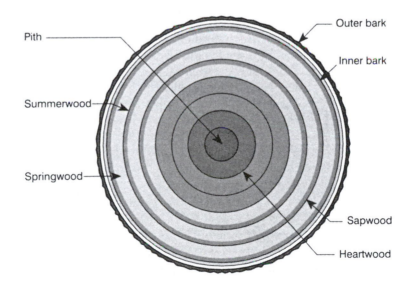

Table 6–1
Principal types of wood.

Softwoods	Uses
Cedar	Clothes closets, pencils
Cypress	Tanks, silos, storage containers
Douglas fir	Plywood
Sitka spruce	Aircraft applications
Southern pine	Railroad ties, boxes, trim
Western white pine	Matches, boxes, crating

Hardwoods	Uses
Ash	Tool handles, baseball bats
Basswood	Boxes, crating
Birch	Butcher block, study desks
Hickory	Tool handles
Maple	Flooring, boxes, crating
Oak	Posts, flooring, pallets, trim

are harder and more durable than softwoods, but this relationship is not an absolute. For example, balsa is classified as a hardwood, but it is the lightest and softest of all woods. In general, hardwoods are deciduous, or broad-leaved, trees, which drop their leaves in the fall. Softwoods are conifers, which bear needles instead of leaves and produce seed cones. The softest and hardest of all woods are found in the tropics; the lightest is balsa, and the heaviest is ironwood, or lignum vitae.

Softwoods are generally found in the cooler climates at higher altitudes. Hardwoods favor warmer climates at lower altitudes. In all, there are approximately 30 softwoods and 50 hardwoods commercially harvested in the United States and Canada. The individual characteristics of these woods depend on the extracts in the wood, principally in the heartwood. These extracts generally preserve the wood and make it more resistant to attack by fungi and insects. Darker colors generally indicate more durable woods. These natural dyes and extracts are destroyed by exposure to ultraviolet light or through evaporation.

The principal distinction between the two types of wood is that hardwoods contain very large cells, known as *vessels*, or *pores*, whereas softwoods do not. Vessels are large tubular cells that run vertically up the tree. Vessels provide the primary conducting channel within the tree. They are often large enough to be seen in a hardwood tree's cross section. Along with vessels, hardwoods have wood fibers that run vertically up the tree. These fibers are smaller and shorter than vessels but are thick-walled. Their only function is to support the tree during its lifetime.

In addition, hardwoods have wood parenchyma and ray cells. Wood parenchyma are thin, soft-walled cells that often surround the vessels and contribute to the grain pattern of the wood. *Medullary rays* are cell zones that radiate outward from the inner heartwood to the sapwood. They provide channels from the heartwood to the sapwood.

Table 6–2
Woods used for construction.

Use	Wood
Structural	Yellow pine, Douglas fir, hemlock, redwood, cedar
Light framing	Spruce, hemlock, yellow pine, fir, cedar
Siding	Yellow pine, cypress, redwood, spruce, cedar, hemlock, fir
Exterior finish	White pine, cypress, redwood, spruce, cedar, hemlock, fir
Interior finish	Painted: White and yellow pine, birch, gum, redwood, poplar, spruce, cedar, fir
	Natural or stained: oak, white pine, birch, redwood, yellow pine, walnut, ash, cherry, gum, maple
Sashes, doors, frames	White pine, fir, birch, redwood, yellow pine
Sheathing, roofing	
Flooring	Fir, hemlock, pine, redwood, spruce
	Yellow pine, maple, oak, hemlock, birch, beech, spruce, gum, fir

Softwoods are simpler in structure. *Tracheids* serve the functions of both the vessels and wood fibers. They provide support for the tree while conducting food for the tree. Softwoods may have lighter, less pronounced rays, but parenchyma is virtually nonexistent. Softwoods are dominated in composition by tracheids.

Softwoods are primarily coniferous, having needlelike leaves and bearing cones, whereas hardwoods tend to be broad-leaved and mostly deciduous, although there are exceptions to these generalities. Softwoods tend to have many regularly spaced branches, whereas hardwoods have fewer, randomly spaced branches. Softwoods suffer less shrinkage than hardwoods. Hardwoods tend to be very strong and durable, the better hardwoods being used for furniture, flooring, and marine work. The principal woods used are listed in Table 6–1.

6.2.c Applications of Wood and Lumber

Wood, for construction purposes, is primarily used as structural timber; light framing; siding; exterior finish; interior finish; sashes, doors, and frames; sheathing, roofing, and subflooring; and flooring. The principal woods used for each of these purposes are shown in Table 6–2.

6.2.d Lumber Production

Historically, trees were generally cut in the fall and winter seasons, because drying is slower in the fall and winter than in the spring and summer. As the wood dries, it

Figure 6–4
Typical cuts for lumber.

(a) Slash-cut

(b) Quarter-sawn

shrinks. Because it dries more rapidly around the circumference than it does radially, the tree tends to crack. Today, logs are immediately transported to sawmills or log ponds, where they remain wet. Drying presents little or no problem based on modern logging and transportation methods.

The round logs that result from cutting a fallen tree have a few uses as pilings in bridge construction, as poles for farm buildings, and as fence posts. These logs tend to be irregular in shape and construction, and their use is, therefore, limited. Round logs are cut into regular, rectangular pieces, which offer greater flexibility of use.

A round log is cut along its length into rectangular pieces of varying sizes. Typically, a log will be cut into *boards* less than 1.5 in. (38 mm) thick; *strips*, or *laths*, which are narrow lengths less than 1.5 in. (38 mm) thick; *dimension lumber*, such as 2 × 4 in. (50 × 100 mm) up to 3 × 16 in. (75 × 400 mm), sometimes called *studs* (depending on the intended purpose); and larger pieces known as *timbers, balks*, or *flitches*. Typical cuts from a round log are shown in Figure 6–4.

Logs are cut into lumber using two principal methods: slash-cut and rift-cut. Slash-cut lumber is produced by sawing the log into parallel strips, beginning on the outside of the log and working across, as in Figure 6–4(a). Slash-cut hardwoods are called *plain-sawn*, whereas slash-cut softwoods are known as *flat-grain*. In the second method, Figure 6–4(b), lumber is sawn radially and is called *quarter-sawn* (hardwoods) or *edge-grain* (softwoods). Slash-cut lumber is less costly because it involves less waste and less labor. Quarter-sawn lumber exhibits less shrink and swell and is preferred for its appearance in certain applications. Other methods are used to produce desirable effects in the lumber.

The orientation of the grain in lumber that has been cut from a round log differs according to the method used. The grain direction also affects the shrinkage effect of the lumber. Structural lumber is preferably quarter-sawn (with the growth rings forming a close, uniform, parallel pattern along the length of the piece). Decorative boards are preferably *back-sawn* (cut with the growth rings roughly parallel to the length and converging at various points along the piece to form more interesting and decorative patterns). Quarter-sawn pieces of lumber shrink with little distortion, whereas back-sawn pieces tend to cup.

In addition to the previous cuts, logs can be cut into *veneers* for use in dressing plywood or pulped for use in particleboard and hardboard. Veneers and plywoods are discussed later in this chapter.

After cutting, lumber is seasoned. Seasoning is the conditioning of wood through drying. Lumber is dried to reduce the shrinkage while in use; to reduce checking (cracks that develop in the lumber due to uneven drying) and warping; to increase its mechanical properties; to increase its resistance to decay; to prepare it for further treatment by preservatives, and to reduce its weight for transportation. The sapwood in a living tree may contain over 30% moisture, consisting of the moisture in the fibers and the free sap present in the tree. Once the tree is cut, the log gives up this moisture to the surrounding air until it reaches an equilibrium with the ambient moisture conditions. Once dried, the lumber may reabsorb from 5% to 15% water, depending on environmental conditions. This fact is primarily due to the hygroscopic nature of wood. *Hygroscopy* is the property of a material in which the material absorbs water from the air when it is present in the surrounding environment and gives up water when the surrounding air is less moist than the material. Lumber is typically either air dried or kiln dried.

Air drying takes place in the open air. The lumber is covered to prevent direct contact with rain, snow, and so on, and allowed to dry hygroscopically. The lumber is stacked and separated so that air can circulate through the stacks, thus allowing it to dry completely and uniformly. Air drying may take anywhere from 6 months to 2 years for softwoods and up to 3 or 4 years for hardwoods.

Kiln drying is conducted using moist air or steam. Temperatures may range from 70° to 200°F (21° to 94°C) and commonly vary from 160° to 180°F (71° to 82°C). The advantage of kiln drying is that less drying time is required. Kiln drying takes from a few days to several months, depending on the type of wood.

Structural timber, dimension lumber, railroad ties, and lower grades of lumber are typically air dried. Some mills may partially kiln dry these types of lumber, depending on market demand and available supply. Lumber designated for interior work and flooring is kiln dried. Most hardwood lumber is air dried. Subflooring and flooring lumber and interior finish pieces may be partially air dried before being kiln dried.

Although air and kiln drying are the most common methods of drying lumber, chemical seasoning is sometimes used. Chemical seasoning is accomplished with chemicals that are hygroscopic. The chemicals keep the surface of the lumber moist while replacing the lost water with chemicals as drying progresses. Thus, shrinkage and checking are minimized. Various chemicals may be used, but urea is the most common, based on its properties and economy.

Lumber should be chemically treated as soon after cutting as possible. Chemical treating can be accomplished through steeping the lumber in large tanks for long periods of time, by dipping, by spraying, or by spreading the chemical over the lumber. Following the chemical treatment, the lumber is then dried in the solution at high temperatures, kiln dried, air dried, or dried in use. Chemical treatment is faster and is preferred for wider or thicker pieces to avoid excessive checking.

6.3 PHYSICAL AND MECHANICAL PROPERTIES OF WOOD

The physical properties of wood, including the grain pattern, physical characteristics, weight, and moisture content, relate to the structure of the wood. Mechanical properties, such as compressive strength, flexural strength, and modulus of elasticity, are associated with the ability of the wood to withstand applied forces. (Additional information about the physical and mechanical properties of selected wood types can be found in the appendices.)

The terms *grain* and *figure* are sometimes used synonymously. However, the grain can be thought of as the direction, size, arrangement, appearance, and/or quality of the wood fibers. The figure of a piece is the pattern produced on the surface of the piece by uneven coloring, growth rings, medullary rays, knots, and other visual characteristics.

Physical characteristics determine the quality and, therefore, the price of the lumber. The number and extent of these characteristics affect the lumber's use. These characteristics are not necessarily defects, although they may affect the useful properties of the lumber. A *defect* in a piece of lumber is any irregularity or abnormal condition that lowers the lumber's strength, durability, and/or utility. A *blemish* is something that mars the visual appearance of the lumber. Blemishes can make the piece unsuitable for interior finish yet not affect the strength or durability of the piece. For example, knots and burls are unacceptable in structural lumber, yet these features may produce a desirable effect in panels, trim, and some furniture pieces.

Physical characteristics are the result of the structure of the wood, shrinkage effects, fungi attack, and manufacturing techniques. Structural characteristics include grain irregularities, knots, bark pockets, pitch, shakes, and wane.

These grain irregularities are commonly characterized as *cross-grain*, which refers to fiber direction in a piece of lumber that is not parallel to the axis of the piece. Cross-grain irregularities with a slope of less than 1:10 do not seriously affect the mechanical properties of the lumber. Irregularities with slopes greater than 1:10 seriously affect the following mechanical properties: the modulus of elasticity, the modulus of rupture, and the shock resistance of the lumber.

A *knot* is the visual evidence of a branch or limb that was cut off during the manufacture of the piece of lumber. Knots may be round, spike, or encased knots. When the branch or limb is cut squarely, a roughly round knot appears in the lumber. When the branch or limb is cut lengthwise, the knot appears bullet-shaped and is called a *spike knot.* Encased knots result from further growth that has encased or enclosed a dead branch or limb. Knots affect the strength of the lumber when the grain of the piece has been forced to go around the knot or when checking occurs in or around the knot.

A *bark pocket* is a piece of bark around which the tree has grown. Also, *pitch* can occur in lumber as streaks or pockets that have collected in the tree. Pitch is a resin product of the tree. *Shakes* are separations that occur in the grain. Most of them occur between growth rings and may be caused by bending of the tree during its life due to high winds, tornadoes, and such high-force situations. *Wane* is characterized by bark, or lack of wood, on the corner of a piece of lumber.

Uneven shrinkage causes checking, warping, loosening of knots, and collapse. *Checking* is a separation of the wood that typically occurs across the growth rings. Checking can occur on the surface of the piece or can occur completely through a piece of lumber; the latter is termed *splitting*. *Warping* is the deviation of a piece of lumber from a flat, plane surface. Warp is further identified as *cup, bow*, or *crook*, or any combination of these. A bow is when a piece of lumber deviates from a straight line along the length of its flat surface. Cup is the curvature of the piece of lumber across the width of the piece from edge to edge. Crook is the deviation that occurs along the edgewise length of the piece. Loosening of knots produces voids in the lumber when the knot cannot be held in place based on growth or location. *Collapse* is also a common defect in wood. It occurs when the cells of the heartwood collapse during drying. Collapse is characterized by the corrugated effect that it produces in wood. Hardwoods are more susceptible to collapse than are softwoods. Figure 6–5 illustrates the various types of warping.

Various molds, stains, and decay in wood are produced by fungi. Molds and staining occur primarily in the sapwood and can be brushed off or removed by surfacing

Figure 6–5
Types of warping in lumber.

the lumber. Decay, however, breaks down the cell structure of the wood and weakens the wood. Some woods have greater decay resistance than others, depending on the type of fungi, character of the wood, and the type and extent of exposure to which the lumber is subjected.

Defects and blemishes caused by manufacturing techniques are primarily due to chipped grain, loosened grain, raised grain, torn grain, skips in surfacing, sizing variations, saw burns, gouging, and various other imperfections that can be traced to manufacturing.

The density of wood is affected by moisture content, gums, resin, and similar wood extracts. Density is an excellent indication of the strength of the wood in a dry condition. High density is desirable in wood that will be subjected to high stresses. Low-density wood is often used in situations where weight, thermal transmission, and ease of manufacturing are control factors.

Density can be classified as basic density, air-dry density, or green density. *Basic density* is the mass of a completely dry piece of wood divided by its volume determined by displacement. *Air-dry density* is determined at 12% moisture content, which is the moisture content after air drying. *Green density* is the density of fresh-cut wood and varies with the seasons of the year and geographic region. Most woods have densities less than 1 and will, therefore, float on water. However, due to the absorption of water into the tree's cells, the wood eventually becomes saturated and sinks.

Moisture content is commonly determined by oven-drying or electrical-resistance methods. Oven drying requires that the test piece be weighed in the moist condition and placed in an oven at 215°F (103°C) for 1 hour. The piece is then removed and reweighed. It is put back into the oven for another hour and then reweighed. This continues until the weight stabilizes. The following formula is then used to compute the percent moisture content:

$$\text{percent moisture content} = [(W_o - W_f)/W_f] \cdot 100$$

Electrical-resistance meters have sharp prongs about 1 in. (25 mm) apart. These prongs are driven into the surface of the specimen, and the moisture content is read directly from the meter, based on the resistance of the specimen. The resistance per length depends on the species.

6.4 CLASSIFICATION AND GRADING OF LUMBER

Lumber grading is performed to allow the user to select the best quality product suited to the application. In almost all grades, for all types of lumber, grading is based on the number, character, and location of various features and defects that may lower the strength, durability, or visual value of the product. Common features include knots, checks, splits, pitch pockets, and stains. The better grades of lumber do not contain as many or as severe defects as the lower grades do. These features often do not prevent the lumber product from being satisfactorily applied.

Lumber grades are based on different principles for hardwoods than for softwoods. Hardwood grades are based on the amount of clear areas in the board and

Softwoods are graded according to use, manufacture, and size:

Use

1. Yard lumber: Generally intended for ordinary construction and general building purposes.
2. Structural lumber: Two or more inches in nominal thickness and width; for use where higher working stresses are required.
3. Factory and shop lumber: Selected for remanufacturing uses.

Manufacturing

1. Rough lumber: Lumber that has not been surfaced but has been sawn, edged, and trimmed so that it shows on four edges of its length.
2. Surfaced lumber: Lumber that has been surfaced or dressed by planing on one side (S1S), two sides (S2S), one edge (S1E), two edges (S2E), or a combination of these (S2S1E).
3. Worked lumber: Lumber that has been further worked beyond surfacing.
   a. Matched lumber: Tongue-and grooved lumber that is machined to form a close-lapped joint when fitted together.
   b. Shiplapped lumber: Rabbeted on both edges to form a close-lapped joint when fitted together.
   c. Patterned lumber: Shaped to a pattern or moulded form beyond being surfaced, matched, or shiplapped, or a combination of these.

Size

1. Nominal Size
   a. Boards: Less than 2 in. in nominal thickness and 2 in. or more in nominal width. Boards less than 6 in. in nominal width may be classified as strips.
   b. Dimension: From 2 in. to 5 in. in nominal thickness and 2 or more inches in nominal width. Includes the common classifications as framing, joists, planks, rafters, studs, and other specific terms.
   c. Timbers: 5 or more inches in the smallest dimension. These include beams, posts, girders, and other large pieces.
2. Roughdry size: Minimum roughdry size thickness for finish, common boards and dimensions of sizes of one or more inches of nominal thickness cannot be less than ⅛ in. thicker than the correponding minimum finished dry thickness, except that 20% of a shipment may be not less than ³⁄₃₂ in. thicker than the minimum dry thickness for that dimension. The minimum widths of finish, common strip, boards, and dimension lumber shall not be less than ⅛ in. wider than the corresponding minimum finished dry width for the same dimension.
3. Dressed size: Dressed sizes of lumber should equal or exceed the minimum size shown for standard lumber.

Note: This is a general procedure for softwood lumber grading. Other criteria and procedures may be used in specific applications and by other manufacturers.

Figure 6–6
Lumber classification.

the size of those areas. Softwood grades are based primarily on the size of defects and the characteristics of these defects that relate to the mechanical properties of strength for that piece of lumber. Figure 6–6 shows one system for classifying softwood lumber. Hardwood-lumber grading is generally based on one side; the other side of the board may exhibit different features and characteristics.

Hardwood grading is based on the percentage of the piece of lumber that can be cut into smaller pieces of standard size and maintain one face clear of defects and the other face sound in composition. Based on this system, the standard grades for hardwood lumber are firsts, seconds, selects, no. 1 common, no. 2 common, sound wormy, no. 3A common, and no. 3B common. In general, firsts and seconds can be combined into one category; selects and no. 1 common into another, and nos. 3A and B common into another. This system allows each descending grade to have 8.33% less recoverable material than the previous grade. For example, firsts would have 100% to 91.67% recoverable wood, seconds would contain 83.33% to 91.67% recoverable wood, and so on.

Hardwood-lumber applications include trim and molding, stair treads and risers, paneling, flooring, dimension lumber, and structural lumber.

Softwood lumber is graded according to use, size, and method of manufacture. Lumber classified according to use includes yard lumber, structural lumber or timber, and factory, or shop, lumber. Lumber classified according to size includes strips, boards, dimension lumber, and timbers in increasing size. *Dimension lumber* includes planks, scantlings, and joists. *Timbers* include structural joists, beams, posts, and timbers. Lumber classified according to method of manufacture includes rough lumber, surfaced lumber, and worked lumber. *Rough lumber* is the form that comes straight from the saw. *Surfaced lumber* has been run through a planer and may be dressed on one or more sides. *Worked lumber* includes matched lumber, shiplapped lumber, and patterned lumber. Matched lumber has been worked to close tolerances to provide close fits between tongue-and-groove joints at the edges or at the ends of end-matched lumber. Shiplapped lumber is worked to provide close fits in rabbeted, or lapped, joints at the edges. Patterned lumber has been shaped into a patterned or molded form.

Softwood lumber comes in three grades: select, standard, and common. *Select lumber* is free of all major defects and has at least 75% of the strength of a clear test specimen. *Standard lumber* may contain minor defects but must maintain at least 60% of the strength of a clear test specimen. *Common lumber* may contain defects and is used where strength is not a primary concern. Common lumber must have at least 30% of the strength of a clear test specimen.

6.5 PLYWOOD, RECONSTITUTED WOOD PANELS, MODIFIED WOODS, AND THEIR APPLICATIONS

Plywood design has changed very little since its introduction in 1905. Softwood-veneer plywood is the most widely used panel for industrial and construction applications. Hardwood-veneer plywood is a high-quality interior panel used in applications such as furniture and cabinets. Softwood veneers often include pine, fir, spruce, and hemlock; hardwood veneers for plywood include birch, oak, lauan, walnut, and ash. The upper limit for thickness for veneers is ⅜ in., and solid wood thicker than this is termed *lumber.* Some veneers are as thin as 1/200 in. Standard furniture veneers are 1/28 in. thick.

Standard thicknesses for plywood are 1/4, 3/8, 1/2, 5/8, and 3/4 in. Plywood is graded according to the quality of the veneer on the front and back surfaces and the type of adhesive used in the construction of the plywood. Several adhesives are used to produce plywood, particleboard, and hardboard and to assemble veneers. These adhesives are primarily animal glues, blood-albumin glues, fish glues, glues made from starch, casein, and synthetic resin glues. Chapter 10 discusses these adhesives in further detail.

Animal glues are made from the hides, hooves, and other products of animals. These glues are generally available in a dry form and are mixed with water and melted before application. The glue solution is kept warm during the application, and the assembly is then pressed and cooled. Animal glues set quickly, exhibit great strength, stain the wood very little, and have less of a dulling effect on tooling. However, animal glues have a low moisture resistance. They are used in the making of furniture, cabinets, and products intended for interior use.

Blood-albumin glues are mixtures of blood serum and other products introduced at the time of application. They are used at ambient temperature, require hot pressing, and set quickly. They stain the wood very little and dull tooling only slightly. They offer a high moisture resistance and high durability in wet conditions.

Fish glues are made from fish and other animal products. They are typically available in liquid form ready for application. They may be applied and pressed at ambient temperature, set quickly, and stain little. However, fish glues have low moisture resistance and low durability under wet conditions. They are also more expensive than other types of glues.

Starch pastes are used primarily in the assembly of veneers. They are relatively inexpensive, but stain the wood slightly, moderately dull tooling, and offer low durability and moisture resistance under wet conditions. Casein, a derivative of skim milk, is generally available as a dry powder. It is mixed with water, applied, and pressed at ambient temperatures. Casein glues set quickly, stain some woods, and dull tooling badly, yet they possess high strength, good moisture resistance, and durability under wet conditions. They are used primarily in the assembly of veneers and plywood.

Synthetic resin glues are made from phenolic and urea resins. Phenolic resin glues offer high strength and are waterproof. They must be pressed hot and are relatively expensive. Urea resin glues offer good moisture resistance and may be pressed hot or cold. Both of these are widely used in the manufacture of plywood.

The term *veneer* refers to a thin sheet of material or a form of protective facing. In the wood industry, veneers are used as decorative or protective facings and are rotary cut, sliced, or sawed from a log, or flitch. When these veneer pieces are assembled in panels, each sheet is referred to as a *ply*. Plywoods, therefore, come from stacking or layering these plies for strength, durability, and other useful properties.

Veneers are thin pieces of wood ranging from 0.010 to 0.250 in. (0.25 to 6.4 mm) thick. Veneer may be sawn, sliced, or rotary-cut. Even though there are six cutting

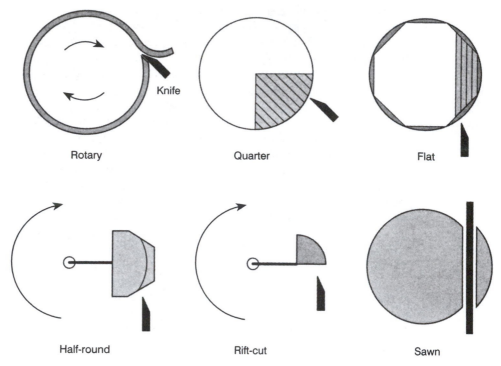

Figure 6–7
Types of veneer cuts.

methods (see Figure 6–7), the majority (more than 75%) of veneers are cut using the rotary method.

Sawn veneer is produced by running the wood through a circular saw, which is thicker in the middle than on the outside. The saw blade has one straight face and one convex face. Sawn veneer tends to be thicker and made of less costly woods, due to the waste involved in sawing. However, both sides of a sawn veneer may be exposed. Sliced veneers are produced by running a block of wood known as a *flitch*

Figure 6–8
Checks developed during cutting.

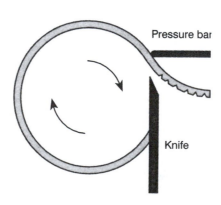

across a heavy knife in a diagonal direction. The veneer face that lays closest to the knife comes out checked and cracked due to the forces exerted by the knife blade in separating it from the flitch. Figure 6–8 illustrates the checks developed during cutting. The damaged side is used as the glue side. Sliced veneer is used primarily in paneling, where such woods as oak and mahogany produce a handsome figure.

Rotary-cut veneer is produced in a continuous sheet by rotating the log against a stationary knife, which is fed into the log at a preset speed to obtain the desired thickness. The log is soaked in water before cutting to make it less brittle. Most veneers are produced by rotary cutting, because larger pieces and high production rates can be achieved. Most veneer for use in plywood is rotary-cut.

Plywoods are made from several pieces of veneer that have been glued together. The grains of the plies run at angles to each other, typically 90° between adjacent

(a)

(c)

(b)

Figure 6–9
Common plywood construction types.

Photographs courtesy of American Plywood Association.

plies. The outside plies are called *faces* or, when one face is to be exposed, the face and the back. The central ply is called the *core* of the piece of plywood. The remaining plies are known as *crossbands,* or *crossplies.* The core may be made from different pieces of wood. It may be veneer, lumber, or a combination of these. Plywood rarely exceeds 3 in. in thickness, the more common being less than 1 in. The number of layers of the plywood varies according to the desired thickness of the piece and the wood used. Figure 6–9 illustrates common plywood construction.

Plywood is produced in a balanced construction fashion, where an odd number of plies are arranged in pairs on opposite sides of the core. The grain runs parallel for the pair. Balanced construction is used to increase the strength and to reduce the shrinkage and swelling effects on the wood. This result is particularly important for thinner pieces.

NOTE: All-veneer APA Rated Sheathing Exposure 1, commonly called "CDX" in the trade, is frequently mistaken as an Exterior panel and erroneously used in applications for which it does not possess the required resistance to weather. "CDX" should only be used for applications as outlined under Exposure 1 above. For sheathing grade panels that will be exposed permanently to the weather, specify APA Rated Sheathing Exterior (C-C Exterior under PS 1.)

See Treated Plywood (page 18) for recommended plywood grades and exposure durability classifications in applications requiring fire-retardant-treated or preservative-treated plywood.

(1) Some manufacturers also produce a premium "N" grade (natural finish) veneer, available only on special order.

(2) Although glue bond durability is described by exposure durability classification, panel surface may become uneven and irregular under prolonged moisture exposure. Panels should be allowed to dry, and panel joints and surfaces may need to be sanded before applying some finish materials.

(3) Exposure 1 panels may also be used when exposure to the outdoors is on the underside only, such as at roof overhangs.

Figure 6–10
Grading of softwood plywood veneers.
Courtesy of American Plywood Association.

Several types of wood are used in plywood construction. Among these are oak, walnut, mahogany, pine, cedar, birch, maple, poplar, and fir. The type of plywood used and the thickness desired for an application depends on the strength required, the use of the plywood (whether interior or exterior, finish or general construction, and so on), and the cost.

Plywood is assembled in layers and then glued together in a heated press. Plywood is made in sheets, the most common being a 4×8 ft (122×244 cm) sheet. One important advantage of using plywood in construction is that it provides strength in more than one direction. Plywood provides strength parallel to the face and at right angles to it. This allows plywood to be used in a variety of situations, including sheathing, paneling, subflooring, cabinets, and siding.

Plywood generally comes in two types: interior and exterior. Interior plywood has a moisture-resistant glue line, whereas exterior-grade plywood has a waterproof glue line, enabling it to be used in permanent outdoor conditions. Plywood is available in grades N, A, B, C, and D. Grade N is a special-order natural-finish veneer that is free from open defects. All heartwood or all sapwood may be specified. Grade A, or sound, plywood is firm, is smooth cut, and can be painted. Neat repairs are permissible in grade A. Grade B, or solid, plywood is also paintable and must be free of any open defects. In grade B, tight knots and repair plugs are permitted. Grade C, or exterior

Grade	Face	Inner Plies	Back	Uses
A-A interior	A	D	A	Interior cabinet, furniture and places where both sides show
A-B interior	A	D	B	Same as A-A except back is less quality
A-D interior	A	D	D	Furniture and paneling where one side is not seen
A-A exterior	A	C	A	Highest quality for buildings, furniture and boat interiors
A-C exterior	A	C	C	Cabinets and construction where one side is not seen
B-C exterior	B	C	C	Rough work panel for shelves and buildings
Marine	A or B	B	A or B	Marine applications

Figure 6–10, *continued*

back, plywood may have defects that do not detract from its serviceability. Knotholes up to 1 in. in diameter, splits, and visual defects that do not detract from its serviceability are permitted. Grade C is the lowest exterior-grade plywood. Grade D, or utility back, plywood is an interior-type plywood, which may contain several defects that do not affect the performance in strength or serviceability of the plywood. Knots and knotholes up to 2 1/2 in. in diameter and limited splits are permitted. Figure 6–10 is a summary of the grading procedure for softwood plywood veneers.

General uses for softwood plywoods include sheathing for walls, floors, and roofs; concrete forms and retainers; and paneling and siding. Further information on the grading and selection of plywoods can be obtained from various plywood manufacturers' associations, such as the American Plywood Association, the Department of Commerce, or your local lumberyard.

Hardwood and decorative plywoods are used for decorative purposes where appearance is important, such as for furniture, cabinets, paneling, and marine applications. They are often used for chests, desks, beds, and chairs. They may also be used in skis, sleds, golf clubs, bowls, tools, and guitars.

Hardwood plywood is typically graded by the species and grade of the face veneer and the type. The species of the face is an important characteristic in the grading of hardwood plywood. The species is categorized by the specific gravity of the hardwood. There are three basic categories for hardwoods and five groups for softwoods, as shown in Figure 6–11.

Grades for veneers used in hardwood plywoods include premium grade (A), good (1), second (2), utility (3), backing (4), specialty (SP), and certain softwood veneers. Softwood veneers have face grades specified, whereas others are graded under the same system as hardwoods.

Premium-grade veneers are smooth and full length. Edges of veneer pieces used for the face are matched for best appearance in color and grain. Allowance is made for certain small defects. Good grade is the same as premium, although edges do not need to be matched as closely. Second-grade veneers are free from open defects. Edge matching is not important.

Utility-grade veneers allow open defects, with knotholes up to 1 in. in diameter and open splits or joints up to 3/16 in. that run half the length of the panel. Backing grade is basically the same as utility, except larger defects are allowed. Backing grade allows knotholes up to 3 in. in diameter, and some smaller splits and joints may run the length of the panel. Specialty-grade veneers are agreed upon by buyer and seller prior to sale. These have special characteristics important for the buyer, such as a particular species (wormy chestnut, birdseye maple, or similar woods that contain desirable features).

Lumbercore plywood is the highest-quality panel. A core of solid hardwood strips is covered by hardwood veneers on the face and back. It is used for higher-quality furniture, cabinets, desks, and tabletops.

Medium-density-overlay (MDO) exterior plywood has veneers of an opaque resin-treated fiber overlay. This finish produces a smooth, tough surface suitable for painting. MDO panels are generally used for signs, soffits, and cabinets.

High-density-overlay (HDO) exterior plywood has an abrasion-resistant, resin-impregnated overlay that provides a hard, smooth surface similar to MDO, only

Figure 6–11
Species for plywood construction.

Courtesy of American Plywood Association.

CLASSIFICATION OF SPECIES

Group 1	Group 2	Group 3	Group 4	Group 5
Apitong	Cedar, Port	Alder, Red	Aspen	Basswood
Beech,	Orford	Birch, Paper	Bigtooth	Poplar,
American	Cypress	Cedar, Alaska	Quaking	Balsam
Birch	Douglas-	Fir,	Cativo	
Sweet	Fir 2(a)	Subalpine	Cedar	
Yellow	Fir	Hemlock,	Incense	
Douglas-	Balsam	Eastern	Western	
Fir 1(a)	California	Maple	Red	
Kapur	Red	Bigleaf	Cottonwood	
Keruing	Grand	Pine	Eastern	
Larch,	Noble	Jack	Black	
Western	Pacific	Lodgepole	(Western	
Maple, Sugar	Silver	Ponderosa	Poplar)	
Pine	White	Spruce	Pine	
Caribbean	Hemlock,	Redwood	Eastern	
Ocote	Western	Spruce	White	
Pine, South.	Lauan	Engelmann	Sugar	
Loblolly	Almon	White		
Longleaf	Bagtikan			
Shortleaf	Mayapis			
Slash	Red			
Tanoak	Tangile			
	White			
	Maple, Black			
	Mengkulang			
	Meranti,			
	Red(b)			
	Mersawa			
	Pine			
	Pond			
	Red			
	Virginia			
	Western			
	White			
	Spruce			
	Black			
	Red			
	Sitka			
	Sweetgum			
	Tamarack			
	Yellow-			
	Poplar			

(a) Douglas-Fir from trees grown in the states of Washington, Oregon, California, Idaho, Montana, Wyoming, and the Canadian Provinces of Alberta and British Columbia shall be classed as Douglas-Fir No. 1. Douglas-Fir from trees grown in the states of Nevada, Utah, Colorado, Arizona and New Mexico shall be classed as Douglas-Fir No. 2.

(b) Red Meranti shall be limited to species having a specific gravity of 0.41 or more based on green volume and oven dry weight.

tougher. HDO is commonly used for concrete forms, signs, countertops, and workbenches.

Laminated lumber has a similar construction to plywood, with the exception that the grain of all the lumber layers runs in the same direction. Laminated lumber allows larger pieces, in thickness and cross section, than standard lumber. It is stronger and less expensive than solid lumber of the same size. One application for laminated beams is in home construction, where they are used for their appearance and strength in longer spans.

Particleboard and waferboard are often called reconstituted wood panels. Particleboard is the most widely used reconstituted wood product. It is nonveneered and is produced by bonding wood chips under heat and pressure and adding an adhesive such as urea formaldehyde. The total percentage of adhesive ranges from 6% to

10% and, therefore, must be inexpensive. The two most common types of particle-board are industrial-grade, or medium-density, fiberboard (MDF) and underlay-ment. MDF grade particleboard is a high-quality, smooth-finish panel used for fur-niture, cabinetry and countertops under a plastic laminate. It is made up of finer particles than underlayment particleboard. Underlayment particleboard is used over the subflooring and under carpet or tile. Particleboard is suitable for interior use but quickly disintegrates when wet. Particleboard also suffers from high creep when exposed to sustained loads such as shelving. Particleboard is generally less expensive than other similar materials, but it is difficult to join to other pieces except by gluing.

Waferboard is a nonveneered panel produced from wood wafers. It is commonly used for wall and roof sheathing, subflooring, and underlayment. Due to its wafer pattern, it is sometimes used for cabinets, paneling, and craft projects.

Hardboard is produced by interlocking wood fibers into a mat and pressing them together under heavy rollers. Its strength is based on the interlocking of the wood fibers and the natural adhesiveness of the lignin in the pulp. This process produces one smooth side and one side with a fine mesh pattern on it. Masonite is a common brand of hardboard. Hardboard is produced in service, standard, and tempered types. Service hardboard is an interior panel used where weight is important and high strength is a secondary consideration, such as for cabinet backing. Standard hardboard has higher strength and a better finish. It is often used in furniture, pan-eling, and door skins. Tempered hardboard is standard hardboard that has been chemically and heat-treated to improve its strength properties. Tempered hardboard is often used for exterior finishing applications. Perforated hardboard panels are also available. These are often called *pegboards*.

Modified woods include impreg and compreg. *Impreg* is a term applied to phenol-treated wood in which the resin has been cured without the use of pressure. The wood is soaked in a phenolformaldehyde solution. Once soaked with the resin, the piece is cured through heating. Treated to the 30% to 35% level, by weight, the shrinking and swelling of the product are reduced. Improvements in decay resis-tance, termite and borer resistance, and strength properties make impreg suitable for many industrial and commercial applications. Impreg is widely used for pattern and die models.

Compreg is another resin-treated wood product, although it is cured with pressure. Phenolformaldehyde has been found to be the most effective resin for soaking wood pieces. Varying percentages of resin produce various results. Compreg has properties similar to impreg. It offers increased strength, which makes it useful for jigs and dies, pulleys, electrical insulators, and decorative items such as utensil handles.

6.6 PRESERVATIVES

Wood preservatives are used to repel attacks by organisms and natural elements such as wind, water, sunlight, and fire. Different preservatives offer different proper-ties and features for protection. In general, these preservatives should be perma-nent, penetrate the wood, be safe to apply and handle, be economical, and not destroy the wood. No one preservative offers protection in all situations, but several offer a wide range of protection.

Insects that present a danger to wood come in two categories: termites and borers. Termites are similar in shape and social structure to ants. They feed on wood for food. They are more prevalent in warmer climates and are particularly destructive in the tropical regions. The most common of these are the subterranean termites, which build nests underground similar to ant colonies. A less common type of termite is the drywood termite, which builds its nest in wood above ground. Termites avoid the light and therefore eat the wood from the inside, which makes their detection very difficult until it is too late.

Woodworms or borers include the anobium, or furniture beetle, the lyctus, or powder post beetle, the ambrosia, and the bostrychid. The anobium attacks primarily the sapwood in softwoods. The anobium larvae live inside the tree for up to 6 years. This makes it difficult to detect their presence until considerable damage has been done.

The lyctus feeds on the sapwood of hardwoods that contain a high starch content. Its larvae live about a year before appearing from the wood. The lyctus digs tunnels parallel to the grain and may completely destroy the sapwood.

The ambrosia and bostrychid attack unseasoned wood, digging holes through exposed wood. They do not present any structural danger; they just dig frustrating holes.

Various fungi exist in the air at any time. Several of these fungi use wood with a high moisture content (about 20%) as a growth medium. The most common of these is the merulius lacrymans, commonly referred to as *dry rot*. If allowed to progress unhindered, it eventually reduces the wood to a spongy mass.

In order to protect the wood, some mechanical protection or preservative treatment action should be taken. Mechanical protection involves erecting barriers to attack, such as painting, ventilation, and water drainage. Soil poisoning presents a chemical barrier to insect attack.

Preservatives can be used to deter insects or fungi. Liquid chemical preservatives are applied by brushing or spraying to protect the wood's surface. Dip diffusion involves dipping the wood into a tank of preservative and allowing the wood to soak up the preservative, thus permeating the wood. Pressure-impregnation consists of placing the wood in a pressure vessel and forcing the preservative into the pores of the wood.

Standard preservatives are grouped into three categories: creosote, oil-based preservatives, and water-based preservatives. Creosote and creosote solutions are used for protection against decay and organisms. They are not used where the wood needs to be painted or where their odor would be objectionable. There are three basic types of creosote and creosote solutions: creosote, creosote-coal tar mixtures, and creosote-petroleum mixtures.

Creosote is probably the most widely used of all wood preservatives. It is the most effective of materials for general protection against organisms. Creosote is a distillate of coal tar. Creosote–coal tar mixtures are widely used for marine and saltwater applications. They are effective against water organisms. Mixtures ranging from 50% creosote and 50% percent coal tar to 80%-20% mixtures are used for applications. Creosote-petroleum mixtures are used where economical considerations are most important. They are generally mixed 50%-50%. The petroleum is generally used to reduce the cost of the mixture. Creosote-petroleum treatments leave an oily film on the treated material.

Oil-based preservatives are used in any application other than saltwater situations. These preservatives are chemical compounds that vary in component strength and composition according to application. The most common of the oil-based preservatives are pentachlorophenol and copper naphthenate. Pentachlorophenol is the most widely used oil-based preservative. It is used in concentrations of approximately 5%. It is generally applied through the dip process, soaking, and brushing. It widely used for treating posts, timbers, poles, plywood and siding. Copper naphthenate is a salt manufactured from the combination of copper and naphthenic acids. This combination produces a green color in the treated wood.

Water-based preservatives include chemical salt combinations of arsenic, chromium, copper, and zinc, such as zinc chloride, chromated zinc chloride, copper-chromate-zinc chloride, and sodium fluoride. The percentages used in these combinations vary, depending on the properties required in the preservative and the cost involved. These preservatives are used in above-ground applications and situations where there is limited contact with water. Because these preservatives are water-soluble, they tend to leach out when exposed to moisture over a period of time. Their primary advantages are these: They are low cost, they are paintable, and they are toxic to organisms but not to humans. The prime disadvantage is that water causes the wood to swell. Therefore, treated wood must be redried to achieve the proper moisture content.

Preservatives are applied through pressure treatments and nonpressure treatments. Pressure treatments involve placing the wood to be treated in a closed container, introducing the preservative into the container, and applying high pressure to force the preservative into the grain of the wood. Nonpressure treatments include brushing, spraying, dipping, and diffusion. Brushing and spraying are direct applications of the preservative, where penetration of the preservative relies on capillary action or the wood soaking up the preservative. Dipping involves immersing the wood in a bath of preservative for a specific length of time. The term *dipping* is often limited to immersions up to 15 min, whereas the term *soaking* is reserved for immersion times greater than 15 min. Posts, pilings, and timbers may be soaked for days, whereas common treatments of general-purpose materials take approximately 5 min. Diffusion involves allowing the preservative to penetrate into the wood through the free water in the wood. It is also used in treating posts and similar wood products.

6.7 PAPER AND CARDBOARD

The term *paper* brings to mind the newspaper we read or the ruled paper on which we write every day. Paper and paper products such as cardboard have a wide variety of industrial uses. Paper is often overlooked as an engineering material.

Paper products are divided into two groups: paper and paperboard. Paperboard is heavier and more rigid than ordinary paper. Approximately 50 million tons of paper are consumed annually in the United States. This paper is used in newsprint, books and magazines, sanitary paper, tissue paper, wrappings, writing pads, structural materials, and other related applications. Overall, roughly one-third of paper production ends up in industrial or packaging applications.

Paper is made of wood pulp and other ingredients, such as wastepaper and used cloth. Wastepaper or recycled paper is hard to use because it often contains foreign items such as adhesives, staples, plastics, and other trash. Wastepaper also has to be de-inked and bleached. Low-grade rag paper is commonly used in roofing felt paper, whereas higher-grade rag stock is commonly used in legal documents, bonds and bank notes, drafting paper, stationery, and cigarette papers.

Figure 6–12

Paper recycling operation. In the drawing, bales of old newspapers and shredded magazines and catalogs (A) are moved onto a weigh station equipped with computerized controls that ensure an optimum mix of newsprint and coated paper. The dewired bales travel up twin automated conveyors (B) into one of two pulpers (C).

A mixture of chemicals and water is added to create a slurry, which then proceeds through a series of cleaning and screening stages (D). The contaminant-free slurry is pumped into the first of several de-inking cells (E). The addition of soap creates bubbles that attract ink particles as they rise to the surface, where the foam is removed. The de-inked pulp proceeds through thickening stages (F and G), which remove water. (All water is recycled and clarified to improve pulp brightness.)

The clean pulp is stored in a high-density chest (H) to await transport to the #10 paper machine (I), which produces rolls of 100-percent-recycled paper (J). This de-inking process achieves a much higher fiber yield, compared to standard newsprint de-inking.

Illustration and photographs courtesy of International Paper.

Figure 6–12, *continued*

The two main chemicals in wood are lignin and cellulose, as previously mentioned. Papermaking relies on the fact that cellulose fibers adhere to each other to form sheets. Lignin is present in low-grade paper and newsprint. It is chemically removed in higher-grade paper, stationery, and permanent paper.

The majority of common paper is produced using one of three methods: the sulfite process, the kraft process, and the groundwood process. Special processes may be required to produce nonstandard paper and paper products.

Wood pulp is produced by dissolving the lignin in chemicals to free the cellulose fibers. The *sulfite process* uses sulfurous acid and bisulfites to free the fibers. It was the most popular process until the 1960s, when the *kraft process,* or sulfate, became the most popular.

The kraft process (kraft, from the German word for *strength*) produces a stronger pulp by using caustic soda and sodium sulfide to remove the lignin. The kraft process is able to pulp woods, such as many hardwoods and resinous woods not suited for the sulfite process. Because the majority of kraft pulp mills are in the southern regions of the United States, they are able to harvest both the resinous pines and hardwoods of the area. The reactions of the sulfite and sulfate (kraft) processes take place in large tanks called *digesters*. These tanks are approximately 16 ft in diameter and roughly 50 ft tall (see Figure 6–12).

In the photographs above, a conveyor (top left) takes bales of old newspapers, magazines, and catalogs to a pulping tank. The paper is converted into pulp for cleaning and screening. The pulp is moved into one of several de-inking cells (top right).

Figure 6–13
Fourdrinier paper machine.
Courtesy of American Forest & Paper Association.

The cleansed pulp produces rolls of 100-percent-recycled paper (bottom left). They are cut to size and packed for use in copiers, laser printers, plain paper fax machines, and offset presses. They are also available in rolls for conversion to envelopes, forms and tablets.

The sulfite and kraft processes produce chemical pulp. The third process, the *groundwood process,* grinds the wood while it is being washed with water. Groundwood mechanical pulp is used in newsprint, tissue, and towels and for packaging materials.

Raw pulp produced by these methods may require bleaching and must be run through a beating or brushing process prior to papermaking. This process improves the strength and other properties of the finished paper. Additives, such as dyes, clays, and sizing, may be added to improve the characteristics and properties of the finished paper.

Once the pulp is ready, paper is produced on a long woven wire belt that forms a screen of roughly 60 to 85 mesh. The belt is made of bronze or synthetic fiber. Coarser mesh is used to produce paperboard, and finer meshes are used for more delicate papers, such as cigarette papers. The pulp is pumped onto the belt from a preformed nozzle. Pressure rollers are used to drive moisture out and to press the paper into preliminary shape. Dryers are also used to evaporate any remaining moisture. Once the paper is dry, it is calendared over a series of paired vertical rollers, as shown in Figure 6–13. Calendaring controls the paper's thickness and smoothness. Sizing, such as starch, restricts the penetration of inks into the paper and produces a smooth, hard finish. Sizing and other liquid coatings are applied by running the paper through a bath and wiper system extending over the width of the paper.

Paper may be converted into any number of finished products or used as is in packaging, wrapping, or as an overlayment, such as roofing felt paper. Converting operations for paper include laminating, coating, corrugating, and being formed into bags.

When paper is laminated, a combination of properties produces a heavier or stronger paper or paperboard. Paper and aluminum foil are laminated to produce wrappers, for instance. Paper coatings are often required to help the paper retain ink, and coatings are often required to help the paper retain ink and to help reproduce sharper, clearer images. In addition, coatings are applied for purposes such as asphalt-coated roofing felt. Corrugated paper is formed in three layers. The outside layers are normal, whereas the middle layer is corrugated on rolls. An adhesive is applied to the two outer layers, and the three are sandwiched together. The sandwich construction provides strength with low weight. This type of construction, using a variety of layers and materials, can be found in applications such as aerospace, semitrailers, housing construction, and other applications. Another use of paper is the common, ordinary paper bag in which we carry many items, from groceries to lunches. When we think of paper bags, we think of the brown kraft paper grocery bag or the multi-ply shipping bags used to carry sand and cement. These bags are made from long tubes of paper, which are then formed, folded, and glued by a bottoming machine.

Paper and paperboard are important materials in decorative wrappings as well as protective coverings. The use of paper for its recyclability and widespread application continues to grow.

6.8 SUMMARY

Wood is a natural, renewable, organic substance with a wide variety of uses. This chapter presents the uses and products of trees with sufficient height and sufficient diameters, such as the production of lumber and plywood. The focus of this chapter is on logs, lumber, laminations, plywood, particleboard, and related products.

Wood is widely available and easy to work, although it burns easily, has a variety of defects, and is subject to attack by insects and fungi. Therefore, special precautions must be taken to preserve it. Wood also has a tendency to absorb moisture in the air. For this reason, lumber and wood products are dried prior to use. In addition, there are grading systems used for lumber and veneer plywoods to aid in selecting the best product for an application. These grading systems are based, in part, on the physical and mechanical properties of the species of wood used in the product's manufacture. There are also several different production methods used to make lumber and plywood, such as different saw cuts and different angles of cuts. Wood can be reconstituted into particleboard, waferboard, and similar products. Modified wood products include impreg and compreg, which are resin-treated woods that offer increased properties.

Preservatives are used to protect wood against various forms of attack. Preservatives include creosote, oil-based preservatives, and water-based preservatives. Each of these offers particular protection in specific applications. Wood is still a prime factor in construction and structural applications and retains an important position in the study of materials.

QUESTIONS AND PROBLEMS

1. List the two basic types of wood and five species of each.
2. Differentiate between sapwood and heartwood.
3. Differentiate between springwood and summerwood.
4. List and describe two principal methods of cutting lumber.
5. Define *hygroscopy*.
6. Define and illustrate the terms *warp, bow, crook, cup, wane, shake, check,* and *split*.
7. Describe the process for determining moisture content in lumber.
8. Find the moisture content of a piece of wood if its wet weight is 1500 g and its dry weight is 1383 g.
9. List five common defects in lumber and wood products.
10. List the grades of hardwood lumber.
11. List the grades of softwood lumber.
12. List five adhesives used to construct plywoods.
13. Describe the three processes used to cut veneer.
14. Describe particleboards and hardboards.
15. List three materials used to preserve wood.
16. Describe the pressure treatment process for preserving wood.
17. Describe three nonpressure treatment processes for wood.
18. List three distinct applications of dimension lumber.
19. List three distinct applications of plywoods.
20. List three distinct applications of modified wood panels.
21. Describe *impreg* and *compreg*.
22. Describe the papermaking process.

CHAPTER 7

Ceramics and Glass

Objectives • Understand the compositions and applications of various ceramic materials and products, including glass, stone, clay, cermet, brick, and refractory materials.
• Describe the development of these materials as well as the various products made from them.
• Explain the major types and uses of modern synthetic abrasives.

7.1 INTRODUCTION

Ceramic materials are complex compounds and solutions that contain both metallic and nonmetallic elements and are typically heated at least to incandescence during processing or application. They are typically hard and brittle, and they exhibit high strength, high melting points, and low thermal and electrical conductivity.

They have a wide variety of applications, from pottery to brick, tile to glass, ovenware to magnets, and refractories to cutting tools. Due to their high resistance to heat, ceramics are used in furnace linings and tiles for the space shuttle. Ceramics also are used in superconductivity applications. This chapter deals with the structure, processing, properties, and applications of various ceramic materials. Ceramic materials included in this chapter are glass, stone, clay, brick, and refractory materials. Each of these materials is unique in composition and application. Each is examined in terms of types, components, properties, and applications. Typical ceramic materials are given in Table 7–1.

7.2 CERAMICS

Traditional ceramics are made from three basic components: clay, silica, and feldspar. Structural clay products such as bricks, sewer pipes, drain tiles, and floor tiles are made from natural clays, which contain all three of the basic components. China, dental porcelain, and sanitaryware are all examples of traditional ceramic products.

In contrast to traditional ceramics, which are based primarily on clays, are the *technical ceramics*. Technical ceramics are mainly pure compounds or nearly pure compounds of primarily oxides, carbides, or nitrides. Alumina (Al_2O_3), silicon nitride (Si_3N_4), silicon carbide (SiC), beryllia (BeO), and barium titanate ($BaTiO_3$) are all examples of technical ceramics. These are normally hot-pressed in dry powder form into useful products.

Ceramics are a diverse group of nonmetallic, inorganic solid compounds with a wide variety of compositions and properties. Ceramics are crystalline compounds consisting of metallic and nonmetallic elements whose properties differ from the constituents. In other words, the product of the combination of the metallic and nonmetallic elements has differing properties than the metallic and nonmetallic elements from which it was made.

Ceramics, in the form of pottery, are among the oldest products manufactured by humans. Clay is a relatively inexpensive material found almost everywhere on earth. Early ceramic products were sun dried and not fired, as are today's ceramics. Firing, as used in pottery, dates back to around 2000 to 3000 B.C.

The term *ceramics* comes from the Greek word *keramos*, which means *potter's clay.* Ceramics are crystalline, like steel, and have few free electrons at room temperature. Because they have few free electrons, ceramics are generally electrically nonconducting and exhibit low thermal conductivity. Ceramics, in general, are hard, brittle, and stiff. They are generally higher in compressive strength than tensile.

Table 7–1
Common ceramic materials and their uses.

Category	Common Ceramics	Applications
Carbides	Boron carbide	Abrasives
	Chromium carbide	Wear-resistant coatings
	Silicon carbide	Abrasives
	Tantalum carbide	Wear-resistant materials
	Titanium carbide	Wear-resistant materials
	Tungsten carbide	Cutting tools
	Vanadium carbide	Wear-resistant materials
Intermetallics	Nickel aluminide	Wear-resistant coatings
Metalloids	Germanium	Electronic devices
	Silicon	Electronic devices
Nitrides	Boron nitride	Insulator
	Silicon nitride	Wear-resistant materials
Oxides	Alumina	Electrical insulators
	Chromium oxide	Wear-resistant coatings
	Kaolinite	Clay products
	Magnesium oxide	Wear-resistant materials
	Silica	Abrasives, glass
	Titanium oxide	Pigment
	Zirconia	Thermal insulation
Sulfides	Molybdenum sulfide	Lubricant
	Tungsten disulfide	Lubricant

Metalloids are ceramic elements which exhibit both metallic and nonmetallic behavior. Intermetallic ceramics are compounds produced from the combination of two metallic elements.

They are totally elastic, meaning they exhibit no plasticity when a load is applied. There is little or no deformation prior to fracture.

In general, ceramics have the highest melting points of any materials. Their melting points range from 3500°F (1930°C) to as high as 7000°F (3850°C), except for clay, which softens at 2000°F (1090°C).

Typically, ceramics manufacture involves blending fine starting materials with water to form a plastic mass that can be formed into final shape. Although water is present, clay is formable and exhibits high plasticity. Formation processes generally include extrusion, pressing, and casting. With the invention of the potter's wheel, clay cups, bowls, saucers, and other round or cylindrical objects were produced. Clay can also be extruded through a die to form continuous shapes in bar form. It can also be made into a slurry called *slip* and poured into a mold.

After the material is formed, the product is dried to remove excess water. Dried clay products are porous and not very strong. The material must be fired at elevated temperatures to provide fusion and cause chemical reactions in the material that produce the desired properties. The fusion process is called *sintering*. As a result of sintering, the edges of the individual particles fuse together to form bonds. This bonding gives ceramics their strong, rigid, brittle nature.

After firing, a ceramic coating or glaze may be applied to produce a smooth, impermeable surface on the product. Glass and refractory materials differ in that forming of the finished shape is performed after the starting materials are fused.

Properties of ceramic materials vary greatly, depending on composition, microstructure, and forming method used. Most ceramics are highly resistant to compressive loads but fail readily under tensile loads. Resistance to abrasion, heat, and staining as well as chemical stability, rigidity, and weatherability contribute to the wide application of ceramic materials. Ceramics are susceptible to spalling. *Spalling* is defined as the thermal cracking of ceramic materials. Most ceramics will crack if rapidly heated or cooled. Ceramics may contain other compounds to help reduce the chance of spalling. For example, Pyrex glass may contain boric oxide to provide better thermal expansion properties.

Most of the cost of a ceramic product is contained in the manufacture of the product. The raw materials represent a very small percentage of the total cost. Many ceramic products, such as building bricks, are produced locally using readily available materials.

Ceramic materials are commonly used as dielectric materials for capacitors, disk capacitors (mainly barium titanate) being the most common application for ceramic capacitors. Some ceramic materials, such as sintered oxides, exhibit semiconducting properties. One application for semiconducting ceramics is in thermistors, which are thermally sensitive resistors used in temperature measurement and control. Barium titanate and many other ceramic materials exhibit the *piezoelectric effect*. The piezoelectric effect is the electrical response that a material exhibits when subjected to a mechanical force or the mechanical effect that an electrical force creates. Piezoelectric ceramics are used in accelerometers and speakers. These *transducers* can convert the input sound energy into electrical response or can convert electrical inputs into sound energy.

7.3 GLASS

7.3.a Introduction

Glasses are traditionally described as supercooled liquids. The reason for this description is that glasses do not behave like metals or polymers when they cool from a molten condition. Metals, as an example, exhibit a definite quantity of heat given off when they cool from the molten state. This quantity of heat is called the *heat of solidification.* Polymers exhibit changes in volume when heated, and the point of inflection on a graph of specific volume versus temperature is known as the *transition tempera-*

ture. In general, glass is a transparent silica product, which may form an amorphous or crystalline structure, depending on the heat treatment during production.

Most glasses do not exhibit any indication of transition or a clear point of inflection when cooling from the molten state. Therefore, glass is described as a viscose liquid when it has cooled from a molten condition. Glass has the appearance and feel of a solid material but will flow like a liquid, given time. One example of the flow of glass can be seen by measuring the window panes in an old house. The pane will be thicker at the bottom than the top. The reason that the pane is thicker at the bottom is that flow has occurred. This supports the theory that glass is a supercooled liquid. Figure 7–1 illustrates the liquid property of glass in old houses.

To determine the degree of crystallinity, glass is examined with X rays. X-ray diffraction can be used to determine the crystallinity of glass and glass products. The higher the degree of crystallinity in the glass, the more the rays are diffracted. Normally, glass does not crystallize, but technology has developed that allows special types of glass to be made crystalline through heat treatment.

So, is glass a solid or a liquid? Glass is a solid material from the standpoint that it is rigid and hard, like most solid bodies we encounter. However, glass lacks the continuous lattice associated with solid materials such as metals. Any crystal lattice in glass is at the submicroscopic level; therefore, glass is a liquid. Glass in the molten state is readily identifiable as a liquid, but when cooled, it tends to flow and exhibits high viscosity. Its high viscosity impedes crystal formation and, as a result, glass is

Figure 7–1
Glass panes in old houses.

Glass flows slowly downward.

Glass flow over time.

"frozen" in the liquid state. The unique properties of glass are a result of the unique noncrystalline nature of glass, and these properties are adversely affected by crystal formation.

7.3.b History

Glass is also one of the oldest manufactured materials known. No one knows exactly when, where, or how glass was discovered. Nature began producing different forms of glass long before humans. One style of natural glass is called *obsidian*. Obsidian is a common natural glass product formed by intense volcanic heat. Obsidian is usually black, but, depending on the minerals and coloring agents present during the time it is formed, it may be red, brown, or green. Early people used obsidian to manufacture weapons, tools, and ornamental jewelry.

The Roman historian Pliny wrote that Phoenician sailors had discovered glass around their campfires on the sandy beaches around the seashore approximately 5000 years before the birth of Christ. Historians place the first attempts at glass manufacture by humans at around 5000 B.C., in the region of Syria. Glass beads and bottles have been found in the Mesopotamian region that date around 4000 B.C. Wherever and whenever glass was discovered, early people produced ceramic pottery, jewelry, and ornamental products as well as colorful glazes to adorn their products. The Egyptians used glass products as adornments for the rich and powerful. Because glass production was very expensive at that time, only the rich could afford to lavish themselves with glass. Glass was an expensive and sought-after commercial product.

The first craftspeople who worked with glass were artisans and potters. Potters produced glass jars, bowls, dishes, cups, and hollow vessels, and artisans produced jewelry, fine threads, and luxurious glass products for display. Glass manufacture centered around Alexandria, where Phoenician traders carried the process to the rest of the Mediterranean.

As the glass-manufacturing industry spread throughout the Mediterranean, the Egyptian dominance declined. Other cultures began producing and improving the production of glass and glass products. The Babylonians invented the process of glass blowing around 200 B.C., which brought an entirely new dimension to glass production. Products that had required hours for manufacture could now be done in just minutes with the blowpipe.

The glass blowpipe is just a hollow iron tube from 4 to 5 ft long, with a knob at one end and a mouthpiece at the other. The glass-blowing process, which is still in use today, involves dipping the knob end of the pipe into the melted glass, where a glob of glass sticks to the end of the pipe. The worker then blows gently but firmly into the mouthpiece while rotating the blowpipe. This produces a hollow bulb of glass, where the thickness of the bulb depends on the size of the bulb. This process made possible the manufacture of almost any symmetrical shape. With the added use of molds, the glass can be blown into shape quickly and with repeated accuracy. Products produced through this method include bottles, dishes, lamps, and jars. These glass products replaced many of the clay and wooden products of the time.

The Romans were the first to produce a relatively clear glass product. With the decolorization of glass products, the quality and quantity of glass products increased. Glassmakers began to roll hot glass over stone slabs. The slabs were covered with sand, which left pockmarks in the glass, but this was the earliest recorded attempt at producing flat glass products. Due to the rough surface finish, the flat glass produced during this period severely distorted the light passing through it, making it unsuitable for window glass. Flat window glass came much later.

Glassmaking swiftly spread throughout the Roman Empire. Countries such as Syria, Egypt, Greece, Italy, and the western provinces of Gaul and Brittany all engaged in glassmaking. When the barbarians invaded Rome and the Roman Empire fell, glassmaking shifted to the east to Constantinople and the Byzantine Empire.

The decline and Dark Ages of western Europe were a period of progress for the artisans of the Islamic world. Around A.D. 1000, artisans began producing stained glass windows for churches. Slowly, over the years, the use of stained glass in churches grew, until the Middle Ages, when no church would be without a glorious stained glass window. The Byzantine artisans became known and admired for their use of color and artistic taste in the production of these pieces.

Meanwhile, the Syrians discovered an efficient way to manufacture sheet glass in the seventh century. This method involved blowing a bubble of glass, where a solid rod, called a *pontil*, or *punty*, was stuck to the bubble at the opposite end. The blowpipe was broken off the bubble, and the bubble was spun. The forces produced by spinning the glass flattened the bubble, leaving a "crown" in the center where the rod was pulled away. Sheet glass produced by this method became known as *crown glass*. Most of the window glass produced until the nineteenth century was spun using the crown method.

After the Crusades, the focus of glass production shifted to Venice. During the four centuries of Venetian influence, glassmakers were elevated to a noble status. Craft guilds guaranteed glassmakers trade status, and glassmakers were treated with the utmost respect. The glassmaking industry was moved to the island of Murano to protect the closely guarded secret of glass production. One reason for taking extra security precautions was because the Venetians produced mirrors of a quality that could not be reproduced anywhere else in the world. Reportedly, the glasshouses of Murano stretched for an unbroken mile. These craftspeople were closely guarded and were prohibited from emigration under penalty of death.

In the sixteenth century, the Venetians perfected the first colorless and transparent glass in the world. They called this product "cristallo" glass. Significantly, cristallo had fallen into the hands of two scientists, Zacharius and Johann Jansen. The Jansen brothers were lensmakers, who stumbled on the principle of the microscope and the telescope around 1600. Other scientists who worked with cristallo were Hans Lippershey, credited with the earliest telescope in 1608; Galileo, who in 1654 invented the air thermometer, and Della Porta who, in combination with others, used a combination of lenses and mirrors in the camera obscura, which is a distant relative of the modern camera.

The sixteenth and seventeenth centuries are marked as the beginning of the modern period of glass. Since this time, glass has been the tool of science and has played a significant role in many of our greatest scientific discoveries. Also around this time, people began experimenting with different compositions for glass. Newton, in 1666, discovered the multichromatic nature of light by using a prism to diffuse white light into its constituent colors. Additionally, Newton is credited with the invention of the first reflecting telescope in 1668.

George Ravenscroft developed flint glass in 1675. Flint glass is produced with lead oxide, which gives it a distinctive brilliance and a comparative softness, making it easier to form and decorate. Lead oxide had been used as a flux by the Romans and Venetians. The name *flint* glass came from the use of a very pure form of silica, in the form of flint, in the production of this early glass.

Another invention came in 1688, when the French perfected a technique for producing quality flat glass. Molten glass was poured over flat iron tables and rolled out with rolling pins, much like bread dough. This produced a rough, uneven surface to the glass which was then polished out by hand. Similar techniques are used today, with machines doing the rolling and polishing. This "new" technique established French dominance in the manufacture of large, polished plate glass products during this period.

One problem that existed up to this time was the inability to produce optical glass that was free of imperfections, chemically homogenous, and available in a variety of compositions that would satisfy the needs for refractivity and dispersion. Pierre-Louis Guinand is credited with discovering a process through which he could achieve these requirements. In 1790, Guinand found that he could achieve chemical homogeneity by stirring the molten glass with a ceramic rod and then annealing the glass. This is the type of glass that was commonly used in microscopes and telescopes.

Glassmaking in America first started in the Jamestown, Virginia, colony. Two glass plants were started, one in 1608 and the second in 1620. These two plants operated for several years and then failed. Glass manufacture in the United States then shifted to the Dutch settlement of New Amsterdam, on Manhattan Island, where it continued from 1645 to 1767. Early U.S. glass manufacture focused around three individuals: Caspar Wistar, Henry William Stiegel, and Deming Jarves, of the Boston and Sandwich Company.

Caspar Wistar, a button manufacturer of German origin, imported Belgian glassworkers to Salem County, New Jersey, in 1739 to set up a glass business. For 42 years, Wistar was the most consistent producer of high-quality glassware in the United States. Another German, Henry William Stiegel, established a glassworks in Manheim, Pennsylvania, in 1765. The Manheim plant produced some of the best blown glass made in the United States and was probably the first flint glass manufacturer in the United States. The Stiegel plant folded in 1774; 7 years later, the Wistar plant closed. The third of the early U.S. glass plants was the Boston and Sandwich Company, which was founded in 1825 by Deming Jarves. This was the first U.S. glass plant that operated on a large-scale basis. The Boston and Sandwich Company is credited with inventing a method by which glass could be pressed into iron molds. This invention shifted plant production from expensive artware to relatively cheap pressed ware and allowed greater access to glass products by the public. In 1887, the Boston and Sandwich Company closed its doors due to a labor dispute.

Since 1887, many U.S. glass companies have been founded, and, unfortunately, many have failed. Early manufacturers had to compete with European glassmakers, which were well founded and could produce glass products in greater quantities and at reduced costs. However, people such as Owens, Colburn, Brooke, and Corning continued to develop and expand the production of glass and glass products. This expansion included new processes of production, such as bottling machines, light bulb manufacture, and continuous production methods, and new types of glass, such as Pyrex, safety glass, fiberglass, and foam glass. Each of these developments has influenced the shape of the glass manufacturing industry of today and, indeed, the products we buy and use everyday.

7.3.c Types of Glass and Glass Products

Glass is a superviscous liquid that contains mostly silica and one or more metallic oxides in an amorphous (literally, *without body*) or noncrystalline state. As a liquid, glass tends to "flow" over time. However, glass generally has a softening point of around 1300°F (700°C). The amorphous nature of glass accounts for its transparency and its brittleness. Glass can be strengthened by inserting a binding material such as a fine wire mesh, by laminating plastic sheets with glass, or by tempering or pre-stressing the outer layers of the glass through heat treatment.

Glass is generally either transparent (you can see through it, as in eyeglasses) or translucent (the image is blurred, as in frosted glass). As light passes through glass, it is bent, or *refracted*, by the particles in the glass. The type of glass determines the degree of refraction. Each type of glass has a unique *index of refraction*, or *refractive index.*

The angle that light enters the glass is called the *angle of incidence*, q_i. The angle that the light's path makes with the perpendicular is called its *angle of refraction*, q_r. The index of refraction for a particular glass is calculated by dividing the sine of the angle of incidence by the sine of the angle of refraction:

$$n = \sin \theta_i / \sin \theta_r$$

The angle of incidence and the angle of refraction are illustrated in Figure 7–2.

Figure 7–2
Angle of incidence and angle of refraction.

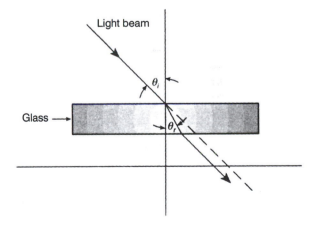

Example

If the path of light entered the glass at 20° from the perpendicular and was deflected to an angle of 10° from the perpendicular, what would the index of refraction be?

$$n = \sin 20°/\sin 10° = 0^{3420}\%.1735 = 1.9695$$

As the light passes through the glass, the light is bent away by the glass from the perpendicular surface it leaves. This principle is applied to lenses. By shaping the surface of the glass, light can be bent in different ways to meet the needed conditions. All light frequencies are not refracted the same. Newton discovered this in his experiments with prisms. Figure 7–3 illustrates the action of a prism in separating the various frequencies of light as they pass through.

Glass is made up primarily of sand, or silicon dioxide. Silica or quartz glass is pure silicon dioxide, a very stable material that gives glass its stable and durable nature. Window glass, such as the type that is ordinarily used in houses, is made from a mixture of sand (SiO_2), limestone ($CaCO_3$), and soda ash (Na_2CO_3). A typical glass batch might include 74% silica, 12% soda, 8% lime, and 6% other substances, such as colorizing agents, stabilizers, and other additives that enhance the characteristics of the glass. Early glass workers had a difficult time creating the 2930°F (1611°C) required to melt silica. They found that the addition of soda ash or lime reduced the temperature needed to melt the product to more attainable temperatures. In addition to providing colorization, metallic oxides helped reduce the tendency of the silica to form solids in the glass.

Various types of glasses are made by varying the amount of silica and adding other acidic oxides. Colored glasses are made by adding metallic oxides: Cobalt oxide is used to produce blue glass, iron oxide produces green-colored glass, and selenium is used to produce red-colored glass.

The ingredients for a glass batch are mixed in dry form and then melted in a special furnace made of heat-resistant brick. The temperature is raised to around 2930°F (1611°C), where the material becomes a viscous liquid. It is allowed to cool into an orange-hot mass that can be handled and shaped into products.

Figure 7–3
Prismatic action.

As a liquid approaches the point at which it changes state from a liquid to a solid, the volume of the material starts to reduce because solid crystals occupy less volume than liquid crystals of the same material. Glass does not undergo this change. It goes right through the change-of-state temperature, or freezing point, and becomes a supercooled liquid. *Devitrification*, which is partial or complete crystallization, may occur if the glass is left at the freezing point for a long period of time. Devitrification can be accelerated by adding a glass crystal to act as a "seed." This seeded glass, or *polycrystalline glass*, can be formed through normal means and possesses greater impact strength, hardness, and thermal shock resistance compared to noncrystalline glass. One application of polycrystalline glass is in kitchenware that is advertised as being safe to use from refrigerator to oven.

As the glass cools past the freezing point, its viscosity increases. This reduces the tendency of the atoms to migrate through the material and form crystals. This process depends on the rapid cooling of the material past the critical temperature. This property preserves the transparency of the glass.

Glass can be thought of as containing one large molecule. Thus, glass—like one large crystal or a liquid—doesn't contain any obstructions to the passing of visible light, such as holes, inclusions, or solid objects. However, different types of glass allow different wavelengths of light to pass and reflect others. The *greenhouse effect* is based on this fact. Common soda-lime glass will allow infrared radiation from the sun to pass through it but reflects the radiation reemitted by structures.

Glass is highly resistant to corrosion because all its constituents are oxides. Glass is also a brittle material, because once a fracture is started, such as the force of an impact that breaks the bonds between glass molecules, it travels unhindered through the material. The grain boundaries in metals resist propagation of the fracture. The same holds true for polymers, where chain links oppose fracture.

There are many different varieties of silica glass, but they can be divided into seven general categories. Each of these categories has many variations. Only the more common varieties are presented here.

The general categories of silica glass include soda-lime glass, borosilicate glass, lead oxide glass, aluminosilicate glass, high-content silica glass, fused silica glass, and pyrocerams. Soda-lime glass is the oldest category of glass. It contains approximately

Use	SiO_2	Al_2O_3	CaO	Na_2O	MgO	Properties
Containers	74	1	5	15	4	Good workability High durability
Plate	73	1	13	13		High durability
Window	72	1	10	14	2	High durability
Lamps	74	1	5	16	4	Good workability

Figure 7–4
Soda-lime glass composition and properties.

70% silica, 15% soda, or sodium oxide, and 10% lime, or calcium oxide. It has a tensile strength of approximately 10,000 lb/in.^{2}, a compressive strength of 50,000 lb/in.^{2}, and a modulus of elasticity of 10^7 lb/in^2. Soda-lime glass is still the most predominant glass in use and is the cheapest. It is used in the manufacture of windowpanes, bottles, light bulbs, and other similar items. Soda-lime glass is commonly available in sheets up to 6 ft wide in single strength (1/16 in. thick) and double strength (1/8 in. thick). It is produced in many thicknesses and grades, depending on the intended use. Figure 7–4 is an example of an application of soda-lime glass.

Borosilicate glass is also commonplace. In it, much of the sodium has been replaced with boron. It is produced from roughly 80% silica, 13% boron oxide (B_2O_3), 2% alumina, 4% sodium oxide, and 1% potassium oxide. Borosilicate glass (trade names include Pyrex, Kimax, and others) exhibits excellent thermal qualities and is, therefore, used in the production of ovenware and heat-resistant glass commonly used in scientific and technical products. Figure 7–5 is an example of borosilicate glass in application.

The lead glasses or lead alkali glasses come in a number of varieties. Optical lead glass is typically made of 58% lead oxide, 35% silica, and 7% potassium oxide (K_2O). This glass has a high refractive index, and it has one of the lowest softening temperatures of all glasses. One type of lead glass was originally made from flints, which were considered a pure form of silica; hence, it is often called *flint glass*. It is

Use	SiO₂	Al₂O₃	Na₂O	B₂O₃	Properties
Pyrex	81	2	4	12	Low thermal expansion High chemical stability Good electrical resistivity

Used extensively for piping, gage glasses, laboratory products, electrical insulation, and kitchenware products.

Figure 7–5
Borosilicate (Pyrex) glass composition and properties.

Use	SiO₂	Na₂O	PbO	K₂O	Properties
Tableware	67	6	17	10	High index of refraction

Lead glasses or flint glasses are used for high-quality tableware, optical uses, neon sign tubing, and in artistic applications.

Figure 7–6
Lead glass composition and properties.

primarily used in the production of lenses and prisms. Another lead glass formula has 57% silica, 29% lead oxide, 8% potassium oxide, 5% sodium oxide, and 1% aluminum oxide. This low-lead glass is used in applications where high electrical resistance is a design factor. It is typically used in the making of decorative cut crystal, due to the sparkling quality it radiates after it has been cut. Due to its high density, it is also commonly used as protection against X rays and gamma rays in laboratories. An example of lead glass is given in Figure 7–6.

In aluminosilicate glass (as shown in Figure 7–7), some of the silica has been replaced by alumina, Al_2O_3. It averages 59% silica, 20% alumina, 9% magnesia (MgO), 6% lime, 5% boron oxide, and 1% sodium oxide. Aluminosilicate glass has a high softening temperature, which enables it to withstand high temperatures. It is often used where chemical resistance is important to the application. It is expensive to produce and difficult to fabricate, which are its main disadvantages.

Use	SiO_2	Al_2O_3	Na_2O	CaO	B_2O_3	MgO	Properties
Aluminosilicate	57	20	1	6	4	12	High thermal shock resistance; high heat resistance

Figure 7–7
Aluminosilicate glass composition and properties.

Use	SiO_2	Al_2O_3	B_2O_3	Properties
Fused silica	99+			Very low thermal expansion; very high viscosity
	Transparent or fused quartz glass is highly transparent to ultraviolet, visible, and infrared radiation and is mechanically stronger than the translucent form. Translucent or vitreous silica glass is lower cost and is used for chemicalware and electrical insulation.			
96% silica	96	3.5	0.5	Very low thermal expansion; high viscosity
	96% silica glasses are expensive and are used chiefly for their extreme thermal shock resistance and high heat resistance.			

Figure 7–8
High-content silica glass composition and properties.

High-content and fused silica glass are also called *fused silicates,* or *quartz glass.* One hundred percent fused silica has approximately the same properties as quartz. It has a very high optical transparency and a very high softening point, and it will transmit ultraviolet waves (unlike most glasses). Fused silica glass consists entirely of silica in a noncrystalline state. It is the most expensive glass to produce, is the most difficult to fabricate, and can withstand greater temperatures. It is, therefore, used in severe conditions where there is no plausible substitute, such as telescopes and melting crucibles. Fused silicates are very expensive to produce due to the high temperatures and high purity required in their manufacture. High-content silica glass contains approximately 96% silica. Typically, it is made from 96.5% silica, 3% boron oxide, and 0.5% aluminum oxide. Their place is between the borosilicates and the fused silicates. This makes the glass highly heat resistant and easier to fabricate. Applications of high-content silica glass include space-shuttle windows, missile nose cones, laboratory glassware, and heat-resistant coatings. Figure 7–8 shows the composition and properties of high-content silica glass.

Pyroceram is a crystalline glass invented by Corning Glass Works. Pyrocerams are polycrystalline glasses that soften above 2000°F and exhibit excellent thermal properties with about the same rigidity as aluminum. Figure 7–9 lists the composition and properties of polyceram.

Two other glasses worthy of note are rare-earth and phosphate glasses. Rare-earth glass contains no silicates at all. It contains approximately 28% lanthanum oxide (La_2O_3), 26% thorium oxide (ThO_2), 21% boron oxide, 20% tantalum pentoxide (Ta_2O_5), 3% barium oxide, and 2% barium tungstate ($BaWO_4$). It has the highest refractive index of any glass. Because of its clarity and high refractive index, it is used almost exclusively in the manufacture of lenses and optical applications. Phosphate glass, like rare-earth glass, contains no silicates. It contains 72% percent phosphorous pentoxide (P_2O_5), 18% alumina, and 10% zinc oxide. It offers a high transparency to infrared wavelengths and is, therefore, used in the nose cones of heat-seeking missiles.

Typical compositions used for crystalline glasses:

Al_2O_3—Li_2O—SiO_2

Al_2O_3—MgO—SiO_2

Li_2O—MgO—SiO_2

Li_2—SiO_2—ZnO

Controlled crystallization is achieved through the addition of nucleating agents such as CaF_2, TiO_2, or ZrO_2 or metallic particles such as gold, silver, platinum, or copper. These glasses are subjected to special heat-treatment processes that convert the glass into a microcrystalline ceramic that may be translucent or transparent, depending on the crystal size.

Figure 7–9
Polycrystalline or Pyroceram product composition and properties.

Table 7–2
Glass composition.

Glass	Composition
Aluminosilicate	59% silica, 20% alumina, 9% magnesia (MgO), 6% lime, 5% boron oxide, 1% sodium oxide
Borosilicate	80% silica, 13% boron oxide (B_2O_3), 2% alumina, 4% sodium oxide, 1% potassium oxide
Flint glass	58% lead oxide, 35% silica, 7% potassium oxide
Fused silica	99%+ pure silica
High-content silica	96.5% silica, 3% boron oxide, 0.5% aluminum oxide
Lead glass	67% silica, 16% lead oxide, 10% potassium oxide, 7% sodium oxide
Low-lead glass	57% silica, 29% lead oxide, 8% potassium oxide, 5% sodium oxide, 1% aluminum oxide
Phosphate	72% phosphorus pentoxide (P_2O_5), 18% alumina, 10% zinc oxide
Pyrex	81% silica, 12% boron oxide, 3% aluminum oxide, 4% sodium oxide
Rare-earth	28% lanthanum oxide (La_2O_3), 26% thorium oxide (ThO_2), 21% boron oxide, 20% tantalum pentoxide (Ta_2O_5), 3% barium oxide, 2% barium tungstate ($BaWO_4$)
Soda lime	70% silica, 15% soda or sodium oxide, 10% lime or calcium oxide

Table 7–2 summarizes the composition of the different types of glass discussed in this section.

7.3.d Manufacture of Glass

Glass production originated in the Middle East, where it was used in the manufacturing of bottles, glasses, and jewelry. The first production method was casting. Blown glass was found in Sidon, today called Lebanon. Glass blowing (as shown in Figure 7–10) dates back to the third century B.C. Glass is the only known material that can be formed through the blowing process; because it is a supercooled liquid, it doesn't exhibit a marked yield point or proof stress, as in other material—especially metals. Based on this fact, glass doesn't neck down, as do metals and polymers in tension. Glass elongates uniformly. This principle forms the basis for modern glass-forming techniques.

The ancient Romans were the first to use glass as a construction material for windows, beginning in the first century A.D. Early windows were tinted, either green or blue, and ranged from 1/8 to 1/4 in. (3 to 6 mm) in thickness. Glass production improved up to the fourth century A.D., by which time glass had become semitransparent. Glass production then suffered a setback with the fall of the Roman Empire. Glass production thrived in the eastern regions and was reintroduced into Europe during the Crusades.

Figure 7–10
Glass blowing.

Glass production became increasingly dependent on blowing rather than casting, due to the fact that blown glass has a smoother surface than cast glass.

Plate glass was produced by the French in the early eighteenth century by casting and polishing the cast plate. Lucas Chance perfected a better method of producing sheet glass. Chance's method used a blown cylinder, which was split and flattened over a sheet of smooth glass, thus giving the rolled out cylinder sheet a smooth finish. This process was cheaper than casting and polishing.

Libbey-Owens patented a glass-production process in 1905 where glass ribbons or strips could be drawn directly from the furnace. Once drawn from the furnace, these strips were immediately cooled by water sprays. Later, these strips were reheated and straightened.

Float glass was developed in 1955 by the Pilkington brothers in England. The float glass process involved taking the glass as it was drawn from the furnace and floating it over a bed of molten tin, where heat was applied above and below the product. The glass was then cooled, giving it a "fire-polished" appearance on both sides.

Clear glass allows the entire spectrum of visible light to pass through it. Colored glass absorbs all other wavelengths except for its color. For example, yellow glass transmits yellow light and absorbs all other colors of the spectrum. This fact can be demonstrated by observing the colors emitted by a stained glass window.

Most early glass production resulted in colored glass, based in part on the decorative purpose of the glass. However, as little as 0.1% of a metallic oxide will tint the glass. A small amount of iron oxide, in the forms of ferrous oxide (FeO) and ferric oxide (Fe_2O_3), is present in most silica sand. Glass made from sand containing these metallic oxides is tinted; ferrous oxide produces a blue-green tint and ferric oxide produces a yellow to green tint, depending on its concentration.

Stained glass manufactured during the Middle Ages consisted of colored glass pieces that were joined using lead *cames*. Cames are H-shaped bars, or strips, that are slotted on the sides so that the glass pieces fit snugly into the slots. Later, during this same period, stained glass products were made of clear glass that had a thin layer of colored glass poured over it or by pouring a thin layer of colored glass over the original colored piece of glass. Figure 7–11 is a typical stained glass product. In either case, the colorizing agents in these pieces were metals or metallic ores. For example, the addition of manganese results in purple, cobalt results in blue, copper adds a green tint, silver results in yellow, and gold gives a red color. The colors derived from the addition of these substances varies according to the composition of the glass. In other words, the colorizing agents that produce one set of colors in soda-lime glass may not produce the same effect for Pyrex.

Glass may also contain a pattern. This pattern can be imprinted through mechanical means by engraving the rollers through which the soft glass passes. As the rollers contact the glass, an impression of the pattern is transferred to the glass. *Figured glass* is one type of patterned glass. It contains a repeating pattern, such as in shower stalls. *Cathedral glass* contains a regular, wavy pattern, which impedes visibility. This type of pattern is sometimes found in reception areas, such as in a hospital. *Opal glass* is produced through the addition of fine crystals in the clear glass matrix. These crystals provide a "frosted" effect. The greater the percentage of crystals added to the matrix, the greater the opacity of the glass. Patterns may also be introduced in glass by *etching*. Mechanical etching can be achieved through sandblasting the glass over a specified pattern. This produces a frosted surface where the sand particles hit the glass. Chemical etching involves the use of acids that eat into the surface of the glass, removing part of the surface. The result of chemical etching is similar to sandblasting. Figure 7–12 shows etched glass.

Mirrors are made by coating the glass on one side with a thin layer of metal. You can observe a mirrorlike effect by holding a piece of polished metal or foil behind a piece of clear glass. Originally, the Venetians coated the mirror with an amalgam, or blend, of tin and mercury. Of course, mercury is poisonous. Mirrors are usually

Figure 7–11
Stained glass.

Figure 7–12
Etched glass.

coated with metals. The metallic coating may be sprayed on or electrolytically deposited. Different-colored mirrors are achieved through the use of the appropriate metal and colored glass.

One-way mirrors are a special type of mirror. They can be manufactured by applying a thin mirror surface to one side of the glass. This allows a person sitting on the darker side to view through the glass. The person on the lighter side of the mirror sees only his or her reflection. Another type of one-way mirror is produced by applying thin [approximately 3/4 in. (20 mm)] strips of mirrored surface, spaced approximately 1/16 in. (1.5 mm) apart. The width of the strips and the spacing produce varying effects in the mirror's quality. A person looking into the mirror sees only his or her reflection, whereas a person behind the mirror can look through the spaces.

Safety glass has been produced in various forms, including *wired glass, tempered glass*, and *laminated glass*. Wired glass contains a fine wire mesh. The wire mesh does not add to the strength of the glass. When broken, the glass tends to be held together by the mesh and doesn't shatter as easily. Most wired glass has been replaced by either tempered or laminated glass products. Some wired glass doors may be found on gymnasiums and older public buildings. Figure 7–13 shows wired and tempered safety glass.

What happens when glass cracks? Glass should, theoretically, have an incredibly high tensile strength and be highly resilient to breakage. In actuality, the surface of the glass has minute cracks, which develop during the cooling period. These cracks concentrate the stress. When an object hits the glass, for instance, the cracks open up and break the bonds that hold the glass together. Bond after bond breaks, allowing the crack to propagate through the glass. If you have ever seen a star-shaped crack form on a car windshield and, over time, spread out through the windshield, you have visually confirmed what was just described. To help prevent this from happening, the glass may be tempered, filled with another material, or laminated to prevent the stress from building up and reducing the tendency to propagate cracks.

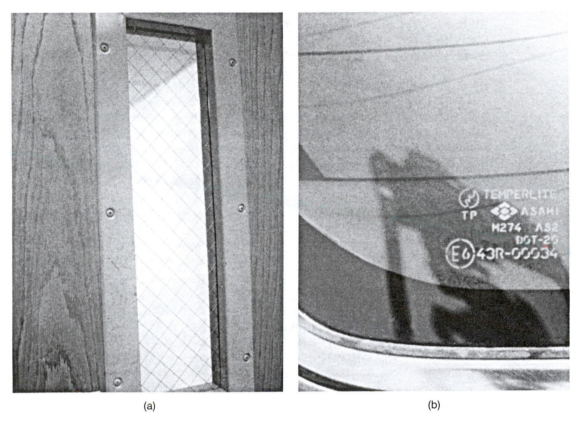

(a)   (b)

Figure 7–13
Safety glass: A. Wired. B. Tempered.

The tempering, or prestressing, of glass involves heating the glass until it is soft but doesn't flow or lose its shape. It is then sprayed with jets of cold air or quenched in oil. This cools the outside layers of the glass rapidly, putting them in compression. The cooling of the glass stops before penetrating the glass thoroughly. The interior of the glass is allowed to cool slowly. This process is called *thermal tempering*.

Another tempering process involves chemically tempering the glass. In chemical tempering, the glass is toughened by immersing it in a bath of molten potassium chloride. The potassium replaces the sodium atoms on the surface of the glass. The potassium atoms are larger than the sodium atoms, so the surface atoms tend to be squeezed together more tightly. Eyeglasses are strengthened using the chemical tempering process and can withstand considerable impact.

Tempered glass is tougher than normal glass due to the compression forces on the external layers of the glass. Annealing is also commonly used to reduce the internal stresses in the glass. One problem with prestressed glass is that if the surface of the glass is nicked or scratched, the internal stresses of the glass are released, and the glass tends to explode into small pieces. Therefore, tempered glasses are drilled and cut to size before tempering. The compressive forces on the outer surfaces must be exceeded before the glass will break. When they are exceeded, the glass tends to break into smaller pieces, which are smaller and less sharp than ordinary glass when it breaks. Tempered glass is often used in such applications as swinging doors, eyeglasses, sliding glass doors (such as patio doors), and safety glass.

Laminated glass is produced by adding a polymer layer, typically polyvinyl butyryl, between two layers of glass. This is done in a "clean" room, which is free from dust and moisture. The assembly is then passed through a set of heated rollers. Laminated glass can withstand tremendous impact without shattering due to the sheet of polymer, which tends to hold the assembly together. Several laminated layers are required in such applications as bulletproof glass. One of the most common applications of laminated glass is in front windshields on automobiles. If you look in the corner of your car's windshield, you should see a label stating that the windshield is made of laminated safety glass.

There are three special types of glass: photochromatic, photosensitive, and polychromatic. *Photochromatic glass* darkens when it is exposed to ultraviolet radiation, such as sunlight or heat, and lightens when the radiation is removed. It is primarily used in light-sensitive eyeglasses, which are made to darken when outdoors and lighten when indoors. It typically has a composition of 60% silica, 20% boron oxide, 10% soda, 10% alumina, 0.9% iodide ion, 0.6% silver, and 0.3% chloride ion. When struck by light, the silver ions in the glass become silver metal. This result causes the glass to darken and reflect more of the light. When the light is removed, the silver becomes silver ions, and the glass lightens again.

Photosensitive glass changes from clear to opaque when it is exposed to ultraviolet radiation or heat. The change is permanent. It acts like photographic film in that a pattern is produced on the glass when it is "exposed." It has a typical composition of 72% silica, 17% sodium oxide, 11% lime, 0.04% selenium, and 0.02% gold. Cerium and copper may replace the gold as sensitizing agents. When photosensitive glass is used, a pattern is embedded in the glass, which can then be chemically etched into the glass. Electronic circuit boards, microscopic in size, are manufactured in this fashion.

Polychromatic glass is a full-color photosensitive glass used in windows and as a data-storage medium.

7.3.e Applications and Properties of Glass

Glass has many applications, some of which have already been introduced. In addition, glass fibers have been used in insulation, and fiberglass (or glass fibers) has been used in construction materials as a lightweight, high-strength alternative to metal, in acoustic insulation, as fillers in polymers, as reinforcement in plastic laminates, and in fiber optic communications. Glass can also be used to shield or transmit radiation, such as in thermal pane windows. Glass-lined tanks are used in many industries to handle aggressive chemicals or to protect the product from contamination, such as in the brewing industry, where almost all beer is processed in glass-lined tanks. This practice protects the product from picking up unwanted tastes. One large use of glass is in building construction. Many of today's buildings rely heavily on glass in their construction. Through technology, the glass used in these buildings is able to withstand high wind loads, withstand the environment, and filter out sunlight. The applications of glass are numerous and continue to increase. As glass technology increases, it will continue to find more and more applications and replace more traditional materials in everyday applications.

Some general statements may be made concerning the properties of glass and glass products.

- Glass is harder than many metals (400 to 600 kg/mm^2, or 6 to 7 on Mohs' scale).
- Tensile strength ranges from 4000 to 10,000 lb/in.^{2} (10,000 lb/in.^{2} in a 1/2 in.–diameter rod, but as high as 3,500,000 lb/in.^{2} on fibers 5×10^{-5} in. in diameter).
- Glass is a brittle substance with low ductility when cooled.
- Glasses have low coefficients of thermal expansion compared with many metals and polymers (from 0.3 to 5×10^{-6} in./in.^{2} °F or 0.5 to 9×10^{-6} cm/cm °C).
- Glasses have low thermal conductivity compared with many metals: (0.5 Btu/h/ft^2/°F/ft or 9×10^{-3} cal/s/cm^2/°C/cm) compared to steel (25 Btu/h/ft^2/°F/ft or 0.1 cal/s/cm^2/°C/cm).
- The modulus of elasticity for amorphous glasses is about 1×10^7 lb/in.^{2} (7×10^4 MPa).
- Glasses have a compressive strength of about 140,000 lb/in.^{2} (945 MPa).
- Glasses are good electrical insulators.
- Glasses can be used at elevated temperatures. Glass fiber insulation can be used to temperatures up to 900°F (480°C).
- Glasses are resistant to most acids, solvents, and harsh chemicals. However, they are slowly fogged by water and attacked by some alkaline solutions.
- Glasses offer the finest optical properties for windows, lenses, and sight glasses.

7.4 TYPES AND PROPERTIES OF STONE

7.4.a Introduction

Stone is one of the oldest building materials known. In the strictest technical definition, stone is not a ceramic material, because it is not heated to incandescence during processing or application. However, it is related to ceramic materials and is included because of its nature and use. Stone is commonly available in the forms of limestone, sandstone, shale, slate, granite, and marble, to name a few common varieties. These formations are contained in three types of rock: igneous, sedimentary, and metamorphic.

7.4.b Types of Rock and Stone

Rocks and stone are solid materials formed by the molten material that has been brought to the surface from within the earth by volcanic eruptions and cracks in the earth's crust. Rock formed in this manner is called *igneous rock*. Igneous stones may be dense and glassy or very porous. Granite, lava, and basalt have igneous origins.

Igneous rock that has been broken down and, as a result, has moved into new layers and reformed is known as *sedimentary rock*. Sedimentary stones are composed of small particles cemented together. These particles are the eroded effects of wind and/or water that have been deposited in layers to form new rock formations. Sedimentary stones can often be split easily along these formation layers. Limestone, sandstone, and shale are examples of stones with sedimentary origin.

Metamorphic rocks are formations that have been subjected to heat and/or pressure after formation. They are composed of the broken remains of igneous and sedimentary rocks. Slate and marble are common forms of metamorphic rocks.

7.4.c Properties of Stone

The properties of stone vary according to the chemical and physical structure of the material. In addition, samples of various stones taken from different quarries will vary in chemical and physical characteristics. Three principal components affect the properties of stone: (1) silicon dioxide (SiO_2); (2) silicates of alumina; and (3) calcium carbonate ($CaCO_3$). Silicon dioxide, or silica, is a very durable material, which is acid-resistant except for hydrofluoric acid, which dissolves it. It is a major constituent of granite, sandstone, and other stones. Silicates of alumina, associated with other minerals, are the basic components of shale, slate, and feldspar. Silicates of alumina are less durable than silica. Calcium carbonate and, to some extent, magnesium carbonate are the principal ingredients of limestone, marble, and similar stones. These are soluble in dilute acid solutions and will, therefore, disintegrate in chemically polluted environments.

Limestone is a sedimentary stone composed of calcium carbonate and magnesium carbonate. When calcinated, limestone produces lime (CaO). A large percentage (roughly 75%) of all stone used in construction is limestone. Limestone has a specific weight of about 153 lb/ft^3 and an ultimate compressive strength of roughly 10,000 lb/in.^{2} at a porosity between 0.3% and 20%. Its low porosity and high workability make it very suitable in masonry work and building construction. It is also used in the production of acetylene gas. When limestone is burned with coke, it forms calcium carbide. Calcium carbide, when mixed with water, forms acetylene gas.

Marble is a metamorphic stone that has a hard, crystalline structure capable of being polished, such as in sculptured art forms and table tops. It is commonly used for decorative purposes, for monuments, and to some extent in building construction for steps, walls, and floors. Marble has the same chemical structure as limestone, but the presence of mica or iron pyrites causes imperfections, which disfigure the finished surfaces and disintegrate rapidly. Marble has a specific weight of approximately 165 lb/ft^3 and an ultimate compressive strength of roughly 18,000 lb/in.^{2} at a porosity between 0.2% and 0.6%.

Granite is an igneous rock composed of quartz and feldspar. Mica may also be found in granite and is harmless in small quantities. Large quantities of mica form stress concentrations that cause the granite to disintegrate. Granite varies widely in texture but is considered a dense, hard, strong material. It offers good abrasion and water resistance and is available in a variety of colors. Granite has a specific weight of approximately 165 lb/ft^3 and an ultimate compressive strength of roughly 25,000 lb/in.^{2} at a porosity of 15% to 30%.

Sandstone is a sedimentary rock composed of grains of silica cemented together with hydrated silica, iron oxide, calcium carbonate, or clay. The properties of sandstone are dependent on the cement, silica producing the strongest and most durable material and clay, the weakest and least durable material. Individual components produce various colors in sandstone. Pure sandstone has a milky white color. Iron oxide produces colors from yellow to red, depending on the amount of the oxide present. Glauconite produces a green tint. Magnesium oxide produces a black color. Many sandstones contain a combination of the various tints. Sandstones are used for general building purposes. Sandstone has a specific weight of approximately 145 lb/ft^3 and an ultimate compressive strength of roughly 13,000 lb/in.^{2} at a porosity between 7% and 20%.

Shale is composed of fine grains of silica and alumina. It is formed under pressure into a usable form. Shale is not porous like sandstone but is hard and has properties much like slate. One form of shale, oil shale, is impregnated with a black, oily substance, which can be heated and broken down into oil, gas, and coke. In the British Isles, oil shale can yield between 15 and 100 gal of oil per ton of shale. Oil shale refined in the United States typically yields from 15 to 30 gal of oil per ton of shale.

Slate is a metamorphic material formed from mud and clay deposits. It is a form of shale that forms into layers or very flat planes. This makes it possible to cut and grind flat slabs for use as blackboards, roofing shingles, and flagstones. It is a low-porosity material with high strength and weather resistance. Slate has a specific weight of approximately 175 lb/ft^3 and an ultimate compressive strength of roughly 17,000 lb/in^2. Slate is also ground into fine particles for use in caulking compounds and linoleum products.

Because rock is basically an aggregate of various minerals, hardness is an indication of the strength of the rock based on the combination of these mineral particles. The strength of the rock depends on the particle size, where a smaller grain size gives a harder rock. Other factors that influence the strength of the rock are porosity, crack formations, included flaws, and weak particles. To determine the hardness of rock, Mohs' scale for hardness has traditionally been used. It is a scale of 10 minerals, each of which will scratch the ones below it in the scale. Figure 7–14 gives Mohs' scale.

Figure 7–14
Mohs' hardness scale.

Mohs' Scale			
10	diamond	5	apatite
9	corundum	4	fluorspar
8	topaz	3	calcite
7	quartz	2	gypsum
6	feldspar	1	talc

To get a feel for what Mohs' scale means in terms of hardness, tungsten carbide inserts used in rock drills typically range from 8 to 9 on Mohs' scale. Hardened tool steels are between 7 and 8. Glass is typically 5.

Hardness in commercial stone is a function of the silica content of the material. The higher the silica content, the harder the material. Limestone, for example, contains little silica and is, therefore, referred to in the mining industry as *soft rock*. Materials such as quartz and quartzite are almost pure silica and are referred to as *hard rock*.

Stone may also be produced synthetically. *Synthetic stone* is produced by combining finely crushed stone powder with a polyester plastic binder. This material is cheaper to produce than natural stone and is widely used as a veneer on many building faces. It is also used as a brick veneer, in counter tops, in sinks, and in other related products. A typical batch of synthetic granite contains approximately 15% polyester resin, 60% finely crushed granite, and 25% fine silica sand. Synthetic marble, which actually contains no marble at all, might contain the following: 15% polyester resin, 2.5% pigment, and 82.5% fine silica sand. The fine swirls and designs that make marble aesthetically pleasing are a result of the pigment and mixing procedure.

7.5 TYPES AND PROPERTIES OF CLAY

7.5.a Introduction

Clay is a mixture composed of fine particles, which result from the disintegration of igneous rocks. When these fine particles are mixed with water, they become a plastic mass that can be molded or formed into shapes. Once fired, these shapes become hard and brittle.

7.5.b Types of Clay

Clay is composed of silica and alumina, usually in chemical combination as amorphous silicate of alumina. Other ingredients or impurities may be present, such as ferric oxide, lime, magnesia, or carbon dioxide, in quantities up to 20%. One exception is firebrick, which may have a silica content up to 98%. Various combinations of silica (SiO_2), alumina (Al_2O_3), impurities, and water produce different properties and clay products.

Four general types of sedimentary clays are fireclays, glacial clays, residual clays, and sedimentary clays.

Fireclays contain approximately 7% to 8% impurities and are very suitable for the manufacture of clay products. Fireclays are separated into three general categories: low-, standard-, and super-duty clays. Low-duty clays have an upper working limit of 1600°F (878°C). Low-duty clays are low alumina and silica clays. Standard-duty clays are used up to a limit of 2500°F (1382°C). Super-duty clays have a working limit up to 3000°F (1662°C).

Glacial clays have been deposited over a long period by the movement of glaciers. As the glaciers moved, they picked up sand, stone, and gravel. Glacial clays are not typically used in the manufacture of clay products.

Residual clays are formed by the disintegration of rocks due to environmental conditions. Residual clays contain many impurities, such as iron oxide and lime.

Sedimentary clays are transported by water flow and deposited by sedimentation. Sedimentary clay deposits may be found near oceans, lakes, and swamps. They are the best kind of clay for manufacturing clay products.

The following clays are used in general engineering applications: aluminous clay, ball clay, corundum, cristobalite, feldspar, flint clay, kaolin, mullite, and vermiculite.

Aluminous clay is composed of bauxite ($Al_2O_3 \cdot 3\ H_2O$) and diaspore ($Al_2O_3 \cdot H_2O$). *Ball clay* is a very plastic clay, which is used in the manufacture of electrical porcelain and whiteware products.

Corundum (Al_2O_3) is an aluminum oxide mined as a gemstone in its natural state. When corundum contains chromic acid, it has a ruby color and is known as ruby; iron oxide and titanic oxide produce sapphire; and the presence of ferric oxide produces yellow topaz. Corundum may also be produced synthetically for use as bearings in watches and precision instruments, among other uses. Aluminum oxide, after fusion, can be crushed and used as an abrasive material or coating on sandpaper, emery cloth, grinding wheels, and abrasive powders. In its brick form it is used as a refractory for furnace linings.

Cristobalite is a quartz material that differs from quartz in crystal structure. Above 2478°F (1370°C), pure silicon dioxide (SiO_2) is called cristobalite; below 2478°F it is called tridymite. As in steel production, the temperature greatly affects the properties of the resultant material.

Feldspar is found in clay and is classified as a rock. It is often used as a flux material in the production of ceramics and ceramic products. *Flint clays* are somewhat rigid clays that are often used to reduce the plasticity of other clay batches. *Kaolin* ($Al_2O_3 \cdot 2\ SiO_2 \cdot 2\ H_2O$) is a common clay, which appears white in its pure form. Kaolin is used in the manufacture of porcelain, china, paper, some rubber, and firebrick and as a pigment in paint.

Mullite ($3\ Al_2O_3 \cdot 2\ SiO_2$) is a refractory material that may be retrieved from the mineral buckite or produced by heating and fusing silica sand and bauxite. Mullite is used in crucibles, extruding dies, and spark plug insulators. It may also be polished for a gemstone. Vermiculite is heated and ground to form an insulating material that may withstand up to 2000°F (1102°C).

7.5.c Processing of Clay Products

Few clay products are made from one kind of clay. Brick and tile are pressed or extruded into final shape, dried, and fired. Higher firing temperatures or finer particle sizes produce more vitrification, less porosity, and higher product density. These qualities improve the mechanical properties of the material but reduce the insulating properties of the brick or tile. Vitrification involves using a glassy substance to hold ceramic substances together.

Earthenware products are fired at relatively low temperatures, where little vitrification occurs and the porosity is high. Therefore, earthenware products are prone to leakage and must be covered with an impermeable glaze prior to use.

At higher firing temperatures, greater vitrification and less porosity occurs. This result is called *stoneware*. Stoneware is used for drainage and sewer pipes. China and porcelain require even higher firing temperatures to reduce the porosity even further and aid in complete vitrification.

In order to process clay mixtures into final products, the raw materials must be tempered with water. There are four basic methods associated with clay manufacturing: slip casting, soft-mud process, stiff-mud process, and dry-press process. These four processes vary in the amount of water used to temper the clay mixture.

Slip casting uses between 12% and 50% water in relation to the weight of the clay mixture. In this process, the mixture is called a *slip* and is poured into a plaster of Paris mold, which absorbs the water and causes a solid shell to form. The slip enclosed in the center of the mold is poured out, and the outer mold is removed after drying.

Soft-mud processing is generally used in the manufacture of bricks, and the *stiff-mud process* is used specifically for firebrick. In both processes, the bricks are extruded through a die in a continuous length, with the extrusion cut into final form by wires pulled through the extrusion. The streaking of the brick by the extrusion die causes lower porosity in the stiff-mud process. This lower porosity is advantageous for firebrick, which must withstand the penetration and attack of slag. The difference between these two methods is in the water content. The soft-mud process uses between 20% and 30% water, whereas the stiff-mud process utilizes between 12% and 15% water.

The dry-press process uses between 4% and 12% water. In this process, the clay mixture is formed in steel molds. Extruded products, such as bricks, can be recognized by the extrusion lines found on the sides of the product.

Once extruded or shaped in molds, the material is ready for drying. The clay product must be dried before firing to prevent excessive shrinkage and cracking. Drying may be natural or artificial and may take up to 2 days at proper temperatures and controlled humidity. The control of temperature is a function of the type of clay and the kiln used, the type of glaze used, and the conditions under which the glaze must be heated. If the product is heated too fast, voids or imperfections can be made in the product, rendering it unsuitable.

After the product is dried, a coating of ceramic glaze is sprayed on the dried clay. Glazing materials soften at temperatures lower than the drying temperatures of the clay. The glaze softens, fuses to the clay surface, and produces a glassy and permanent surface that is resilient and colorful.

Once the material is dried and glazed, it is burned. This process is accomplished in four stages: dehydration, oxidation, vitrification, and cooling, in that order. *Dehydration* takes place between 400° and 1200°F (206° and 654°C). Based on the materials used, the proper temperature is selected, and the excess water is driven off. The temperature is then increased to between 1000° and 1800°F (542° and 990°C). This is the *oxidation stage,* where the impurities are burned off the clay product. The temperature is then raised again to cause *vitrification,* where the material is hardened so that it will retain its shape after cooling. This process is slow and requires between 1 and 6 days to complete.

Once the first three stages have been completed, the *cooling* of the clay product begins. This cooling is closely monitored to avoid cracking, flaws, or color changes, which develop if the material is allowed to cool too fast. The cooling time varies from 1 to 7 days.

7.5.d Types of Bricks

Bricks are typically graded as common or face brick. Face brick is darker, denser, and stronger and is held to closer dimensional tolerances than is common brick. By increasing the firing temperature, the brick becomes darker, the porosity in the brick decreases, and the overall strength of the brick increases. Common brick absorbs approximately 15% water by weight compared to face brick, which absorbs only about 8% by weight. Face bricks are used on the outer surfaces, or faces, of buildings. Common bricks are used in the basic building of structures.

Other types of bricks and related products include glazed brick, paving brick, *terra-cotta*, and tile. Glazed bricks are coated with a hard, smooth, transparent coating to improve their wear and appearance. Paving brick is made from shale and is pressed into final shape before firing. Terra-cotta is a mixture of shale and clay. Terra-cotta is a strong material often used in drainage applications. It has a fine texture and a low coefficient of thermal expansion. It can be shaped by hand or carved, glazed, and fired to make decorative shapes. Tile is a clay material that doesn't exceed 75% solids by volume. Clay tile is extruded using the stiff-mud process. It is used as roofing, flooring, wall coverings, facings, and fireproofing material.

There are three grades of common brick: SW, MW, and NW. *Grade SW* bricks have a high resistance to wet environments and frost action, such as in underground applications. *Grade MW bricks* are used in relatively dry applications above ground where there may be a possibility of frost. *Grade NW bricks* are used in applications where there is no possibility of freezing. The particular grade used depends on the environmental conditions that the brick will be expected to endure.

Firing temperatures for bricks may range from 1800° to 2200°F (990° to 1214°C). Firebrick are often fired above 2400°F (1326°C). This results in greater density and lower porosity. A common brick weighs about 4 lb (about 2 kg). The denser firebrick weighs about 6 to 10 lb (about 3 to 5 kg). Standard size for a common brick is $8 \times 3^3/_4 \times 2^1/_4$ in. ($20.32 \times 9.53 \times 5.72$ cm). A common firebrick, also called a 9-in. straight, measures $9 \times 4.5 \times 2.5$ in. ($22.86 \times 11.43 \times 6.35$ cm). The reason for the difference is that the two types of bricks must be built into the same wall, the common brick on the outside and the firebrick on the hot face. Firebrick doesn't require mortar joints in the usual sense. Common and face brick, due to their dimensional variances, require layers of mortar. Building bricks are laid with 1/2-in.-thick mortar joints, and firebricks have virtually zero-thickness joints. The two surfaces can be bonded together, and the courses will match. A *course* of bricks is a horizontal row of bricks, including the height of the brick in the run plus one layer of mortar. Figure 7–15 shows a common brick and illustrates the process of laying up courses of brick.

Figure 7–15
Laying up bricks.

7.5.e Applications of Bricks

Bricks are generally laid to expose the $2^1/_4 \times 8$ in. face. This size allows for 3/4 in. of mortar between courses, making each course 3 in. tall. Hence, there are four courses per vertical foot. When the brick is laid with the $2^1/_4 \times 8$ in. face exposed, it is called a *running course*. Bricks laid with the $3^3/_4 \times 8$ in. face exposed form a *stretcher course*. Bricks laid with the $2^1/_4 \times 3^3/_4$ in. face exposed form a *header course*. Figure 7–16 shows the three types of courses discussed.

There are different methods of laying bricks to form various patterns in brick structures. The particular method used is called a *bond*. For instance, staggering stretcher courses produces a *running bond*. Layers that include five stretcher courses and a header course are known as a *common bond*. When stretcher and header courses are alternated, the result is called an *English bond*. There are variations in bonds that produce many different patterns.

Figure 7–16
Types of brick courses.

Header course

Running course

Stretcher course

7.6 TYPES AND PROPERTIES OF REFRACTORY MATERIALS

7.6.a Introduction

Recall that bricks are made from sand and clay or from sand and shale. Their physical properties vary according to the types and proportion of materials used, the production methods applied, and the time and temperature used in their production. Refractory materials are used in high-temperature applications, such as linings for furnaces, heat shields, and protective coverings. These materials include fireclays, high alumina, silica, and insulating firebrick.

7.6.b Refractory Materials and Cermets

The term *refractory* means capable of enduring heat. Refractory materials must be able to withstand high stresses at elevated temperatures. Refractories, for example, are generally used for furnace linings and to contain molten metals. The term *refractory* as used in this section refers to those materials suitable for the construction and linings of furnaces that operate at high temperatures.

Typical refractories are made of coarse oxide particles, called *grog*, bonded by finer refractory materials. Pure oxide refractories are both expensive and difficult to form. The finer refractory particles melt during firing to provide bonding. Porosity

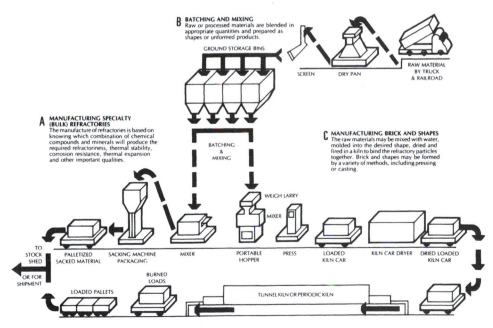

Typical refractory manufacturing flow sheet.

Courtesy of the Refractories Institute. From *Refractories*, Copyright 1987 by the Refractories Institute, Appendix 6, p. 43.

for refractories generally ranges from 20% to 25%, which provides increased thermal insulation. These materials include the aluminas (Al_2O_3), chromic oxides (Cr_2O_3), iron oxides (Fe_2O_3), magnesias (MgO), and silicas (SiO_2).

Oxide refractory materials are generally grouped into three basic types: acidic, basic, and neutral. Common acidic refractory materials include fireclays, high-alumina brick, and silica. Acidic firebrick may contain up to 80% alumina. Fireclay brick contains *kaolinite* ($Al_2O_3 \cdot SiO_2 \cdot 2H_2O$), in rough, prefered proportions of 40% alumina, 46% silica, and 14% water. After firing, the proportions are approximately 54% silica and 46% alumina. Fireclay bricks will not withstand heavy loads. They are intended as furnace linings to contain the heat. High-alumina brick contains more than 48% alumina. Increasing the alumina content by adding more kaolinite clay increases the refractoriness of the fireclay. High-alumina firebrick contains between 10% and 45% silica (SiO_2) and 50% to 80% alumina (Al_2O_3). Substantial amounts of mullite form in high-alumina clays, which provide a high resistance to temperature in addition to high hardness and mechanical properties. Mullite ($3 Al_2O_3 \cdot 2 SiO_2$) is a refractory material found naturally in the mineral buckite or from the fusion of silica sand and bauxite by prolonged heating at high temperature.

Silica is a good refractory material and is often used to contain molten metals. Silica brick retains its strength at high temperatures, allowing it to withstand moderate loads. Silica firebrick contains mostly silica, 95% to 97%. It may contain small amounts of finer oxide material, such as boron oxide, which melts and provides bonding.

Basic refractories (those with base properties compared to acidic) are mostly magnesium oxide. These include periclase (pure MgO), magnesite (83% to 93% MgO and 2% to 7% Fe_2O_3), dolomite (MgO and CaO), and olivine (Mg_2SiO_4). Basic refractories are more expensive than acidic refractories. They are generally confined to steelmaking and other high-temperature applications where compatibility with the material is of primary concern.

Neutral refractories contain 30% to 50% chromic oxide and up to 40% magnesium oxide. These refractories include chromite and chromite-magnesite materials, and they are used to separate the acidic and basic refractories so they do not attack each other. Chromite refractories generally contain between 3% and 13% SiO_2, 12% to 30% Al_2O_3, 10% to 20% MgO, 12% to 25% Fe_2O_3, and 30% to 50% Cr_2O_3. Chromite-magnesite refractories typically contain between 2% and 8% SiO_2, 20% and 24% Al_2O_3, 30% and 39% MgO, 9% and 12% percent Fe_2O_3, and 30% and 50% Cr_2O_3.

Other materials that are generally considered refractory materials include zirconia (ZrO_2), zircon ($ZrO_2 \cdot SiO_2$), and various other materials, such as nitrides, carbides, borides, and graphite. Several nitrides and cemented carbides are used to coat tooling inserts for high-speed, high-material-removal-rate conditions. Nitrides and borides have high melting temperatures and are less vulnerable to oxidation. Oxides and nitrides also find applications in such high-temperature applications as jet engines. Silicon carbide and aluminum oxide are both used in grinding wheels and abrasive sandpaper. Graphite is unique in the sense that as the temperature increases, so does its strength.

Carbides generally exhibit high hardness, high melting points, high compressive strength, and brittleness. Silicon carbide has long been in use as an abrasive. Tungsten carbide and tantalum carbide are used in cutting tool tips and inserts. Uranium oxide (UO_2) and boron carbide (B_4C) are used in the nuclear industry. Boron carbide (B_4C) is used as neutron shielding. Uranium oxide is widely used as nuclear reactor fuel.

Cemented carbide cutting tools are also an application of ceramic technology. They generally consist of tungsten, titanium, or tantalum carbide, embedded in a binder metal, typically cobalt. Carbide tips are of two general types: tungsten carbide and complex carbide. *Tungsten carbide* tends to weld itself to steels when cutting. Therefore, its use is generally confined to nonferrous metals, cast irons, and nonmetallic materials. *Complex carbides* are a combination of titanium and tantalum carbide. Certain combinations of oxides, carbides, nitrides, and similar materials are in a separate class of materials called *cermets*.

Cermets are composed of ceramic particles and powdered metal that have been fused together—for example, aluminum oxide mixed with chromium or a chromium alloy. In addition to chromium-alumina, cermets include such combinations as chromium carbides, iron-zirconia carbides, titanium carbides, tungsten carbides, and uranium dioxides. These combinations provide unique properties that are applied in such areas as space shuttle heat shields, silica-alumina fiber insulating blankets, superconducting magnets, and high-speed cutting tools. Cermet combinations used for cutting tools are generally those that result from combinations of chromium, titanium or tungsten carbide, and a nickel binder. Cermets are widely used as cutting tools due to their high shock and abrasion resistance, high hardness, and very good oxidation resistance.

7.7 ABRASIVES

All of the abrasives used in industry, with the exception of diamond, are synthetic. There are still some natural abrasives, such as garnet, flint, and emery, used in sanding and polishing materials. An *abrasive* is a hard material used to wear away a softer material. The hardest of these materials are the ceramics.

Diamond is the hardest material known and the only significant natural abrasive used. Typical applications for diamond abrasives are in brick and concrete saws, drawing dies, diamond drills, and machining cutters for hard materials.

After diamond, the next-hardest abrasive is silicon carbide (SiC). It is also called by its trade name, Carborundum. This versatile material is also used as a refractory. It is produced by mixing sand, coke, and sawdust; an electric arc is passed through the mixture, which reaches temperatures of about 4500°F (2502°C). The sawdust is used to produce porosity, which allows the gases produced to escape.

Aluminum oxide is also a synthetic abrasive; it can be distinguished by its lighter color. It is produced from bauxite, or aluminum ore. It is not as hard as silicon carbide, but it is more impact-resistant. It is, therefore, preferred for grinding the harder materials, because it wears away faster than silicon carbide, exposing new cutting surfaces more frequently. Silicon carbide abrasives are used for cast irons and the softer nonferrous metals.

These two abrasives are generally used for grinding wheels, where the grit is bonded together with a ceramic material such as sodium silicate or an organic material such as rubber or plastic cement.

7.8 SUMMARY

Ceramic materials are complex compounds and solutions that contain both metallic and nonmetallic elements. They are typically hard and brittle and exhibit high strength, high melting points, and low thermal and electrical conductivity. They have a wide variety of applications, including superconductive applications. Ceramic materials include glass, stone, clay, brick, and refractory materials. Each of these materials is unique in composition and application. Ceramics may be pressed, rammed, or extruded into final shape. A large part of the total cost of a ceramic product is the manufacturing cost. Most ceramics are made from low-cost raw materials.

Glasses are noncrystalline products of the fusion of inorganic materials, mostly oxides. They are amorphous, supercooled liquids that exhibit good mechanical properties and a wide range of thermal, electrical, and optical properties. They offer a wide range of applications in various fields, including communications, construction, and ornamentation.

Stone is one of the oldest building materials known. It offers a variety of properties based on the three main components of stone: silicon dioxide, silicates of alumina, and calcium carbonate. There are three basic types of rock—igneous, sedimentary, and metamorphic—each having its own properties and characteristics.

Clay is composed of fine particles that result from the disintegration of igneous rocks. When these fine particles are mixed with water, they become a plastic mass that can be molded or formed into shapes. After the clay product is formed, it is dried, fired, and glazed. Once fired, these shapes become hard and brittle. Glazing produces a smooth, impermeable surface.

Four general types of sedimentary clays are fireclays, glacial clays, residual clays, and sedimentary clays. These clays are used to manufacture products through the slip-casting, soft-mud, stiff-mud, and dry-press processes.

The term *refractory* means capable of resisting heat. Refractory materials must be able to withstand high stresses at elevated temperatures. Refractories are used for furnace linings and to contain molten metals. Refractory metals are those metals that have melting points above 3000°F (1650°C). Refractory materials include nitrides, carbides, oxides, borides, and graphite.

Most of the modern abrasive materials are synthetic, with the exception of diamond. Aluminum oxide and silicon carbide are two examples of abrasives. Aluminum oxide is preferred for harder materials, because it wears away faster, thus exposing new cutting surfaces more frequently. Silicon carbide is used for cast irons and softer nonferrous metals.

QUESTIONS AND PROBLEMS

1. Which is the cheapest glass to produce?
2. Which glass is commonly used in sliding glass patio doors?
3. Which glass offers the highest thermal resistance?
4. Which glass is the best choice for transmitting ultraviolet light?
5. Which glass has the highest refractive index?
6. Describe the difference between glass and ceramics.
7. If the index of refraction for a particular glass is 1.43 and the angle of incidence of the light source is 10°, what is the angle of refraction?
8. If a beam of light enters a piece of glass at 12° from the perpendicular and is bent to 9° from the perpendicular by the glass, what is the index of refraction for that glass?
9. If a beam of light enters a piece of glass at 15° from the perpendicular and the glass is known to have an index of refraction of 1.35, at what angle to the perpendicular will the beam of light exit?
10. Define the term *ceramic*.
11. What are the three main processes used to form clay products?
12. Define *sintering*.
13. What is *spalling*?
14. Describe the greenhouse effect.
15. List the seven general categories of glass.
16. What are the three basic types of safety glass?
17. Describe what happens when glass breaks.
18. List the seven basic building rocks described in the chapter.
19. List the nine clays used in general engineering applications.
20. List the four basic methods associated with clay manufacturing.
21. What are the three basic types of refractory materials?
22. List five common refractory materials.
23. Define the term *cermet*.
24. List three applications where ceramics have replaced other materials.
25. List five common applications for ceramic and refractory materials.
26. List and describe the application of the two common synthetic abrasives.
27. Is glass a liquid or a solid? Support your answer.
28. Can glass be alloyed? Why or why not?
29. List five applications of glass and provide the type most likely selected for each application.
30. What other materials are used as substitutes for glass?
31. How and why is glass tempered?
32. List five structural/industrial applications for ceramics. Why were these ceramics chosen for each of these applications?
33. What unique properties do ceramics have?
34. Select a product made from ceramics. Make a flowchart showing the process used to produce the finished product from raw materials.
35. What other materials can be substituted for ceramics?

CHAPTER 8
Cement, Concrete, and Asphalt

Objectives • Differentiate between the terms *cement* and *concrete*.
• Discuss the major components, uses, and properties of the different types of portland cement.
• Calculate the yield, cement factor, and quantities of the various materials used in concrete.
• List the types and uses of asphalt and asphalt concrete products.
• Explain the uses of reinforcing materials and prestressing to increase the strength of concrete.
• List and describe the uses of concrete additives.
• Describe the tests performed prior to mixing, after mixing, and in the laboratory to determine the quality of the concrete and its materials.

8.1 INTRODUCTION

The terms *cement* and *concrete* are often confused or incorrectly used interchangeably. Cements, as in organic rubber cements or inorganic portland cements, are adhesive materials that coat and bind the aggregate (bulk or filler material). Concrete has been defined as the product of bonding any aggregate together with a cementing agent that hardens into a solid mass. There are many cements available, but rock and stone are generally used for the aggregate. Concretes of various types have been used since the second century B.C., when the Romans used a volcanic ash found near the town of Pozzuoli, Italy. When mixed with lime, this *pozzuolana* made a cementing agent. Pozzuolana was mixed with stones, glass, and clay bricks to produce concrete for structures. Today's cement is primarily of the portland type. Asphalt is a petroleum product used as a bonding agent in asphaltic concrete, which is used in the paving of roads. Aggregates are also bonded together with special epoxies and super adhesives to produce various exotic concrete combinations. This chapter introduces the types and properties of various cement and concrete products and their uses.

Hydraulic cements, such as portland cements, set and harden by taking up water in a complex chemical reaction called *hydration*. Given an adequate source of water, portland cement may continue to harden for months. However, the design strength for concrete has been standardized to its strength after 28 days. As the four molecules of cement react with water, they all give off heat. This is known as the *heat of hydration*. When mixed with water, the four molecules react with the water to form a gelatin mass, which is actually composed of microscopic fibers. These fibers attach to the aggregate to form a matrix that binds the mass together. The mass sets in a matter of hours but will continue to harden, or *cure*, for months. The finer the cement powder is ground, the more surface area is exposed to water and the faster the cement will set.

8.2 MANUFACTURE OF PORTLAND CEMENT

When the use of pozzuolana declined after the fall of the Roman Empire, cement went into a dormant period until the middle of the nineteenth century. Earlier (1756), John Smeaton had performed some experiments with various limestones. Smeaton found that when limestone was heated in the presence of natural gas, an intense white light (limelight) was produced. In addition, he found that some limestone, when crushed and mixed with water, produced a solid mass upon drying. Not all limestone produced this effect, and only the quarries around Portland, England, seemed to work. Therefore, cement from this area became known as *portland cement*. The popular use of portland cement occurred overseas during the middle of the nineteenth century and was brought to the United States in 1871. The introduction of portland cement was important to the building trades because it made possible types of construction of which contractors were not previously capable.

Manufacture of portland cement.

Courtesy of Portland Cement Association.

The cement used in common construction practices is portland cement. This cement is a hydraulic silicate cement produced in several types. Corrosion-resistant concretes often use an alumina-based cement called *lumnite cement*. Lumnite cement is more expensive and is reserved for applications where its properties are essential. Portland cement is made from limestone and clay, with ferric oxide added as a flux. These materials are blended together in a powdered form and heated in a kiln to approximately 2900°F (1606°C). This procedure drives all the water from the mixture. The approximate percentages of the mixture by weight are

CaO	42.0%
CO_2	35.0%
SiO_2	16.0%
Al_2O_3	2.5%
MgO	2.5%
Fe_2O_3	2.0%

This mixture is drawn from the kiln in the form of a black slag called *clinker*. The clinker must be cooled rapidly to prevent it from reabsorbing moisture. To aid in the setup rate of the cement, 2% to 4% gypsum is added to the clinker. Without the gypsum, the concrete would set up too quickly. The clinker is then crushed down to about 200 grit (it will pass through a screen with 200 openings per linear inch) and bagged or stored in cement silos for future use.

Portland cements are made up of four compounds:

- Tricalcium silicate ($3CaO \cdot SiO_2$)
- Dicalcium silicate ($2CaO \cdot SiO_2$)
- Tricalcium aluminate ($3CaO \cdot Al_2O_3$)
- Tetracalcium aluminoferrite ($4CaO \cdot Al_2O_3 \cdot Fe_2O_3$)

The strength of the cement depends on the content of the two silicates, which make up 70% of the cement product. The Portland Cement Association (PCA) has labeled these four molecules as C_3S, C_2S, C_3A, and C_4AF as a shorthand method of identification.

There are five major types of portland cement. These types vary according to the percentages of the four molecules and their additives. The five types are labeled as follows:

Type I	Normal
Type II	Modified
Type III	High early strength
Type IV	Low heat
Type V	Sulfate resistant

The compositions of the five types are given in Figure 8–1.

Type I: Normal.   Used in general construction work.

Type II: Modified.   Generates heat more slowly than Type I. This is important in larger pours, where the large volume and smaller surface area reduce the cooling ability of the pour. It also has better sulfate resistance than Type I.

Type III: High early strength.   Used when high strength is required soon after pouring, such as when forms must be removed as soon as possible or in cold

Type	C₃S (%)	C₂S (%)	C₃A (%)	C₄AF (%)	Other (%)
I	45	27	12	8	8
II	46	29	6	12	7
III	53	19	10	10	8
IV	30	46	5	13	6
V	38	43	4	8	7

Figure 8–1
Typical compositions of portland cements.

weather to reduce the need to protect the concrete from freezing. Type III cement has slightly less strength than other cements.

Type IV: Low heat.   Generates less heat, more slowly. This type is used in very massive pours, such as dams and retaining walls. It requires more time to set up.

Type V: Sulfate resistant.   Sulfates found in soils and water will attack the cement and produce cracks.

The four constituents in portland cement take up water at different rates and in different amounts during the setting period of the cement. By adjusting the relative percentages of the four components, the properties of the cement can be altered significantly. The setting rate for the silicates is slow, whereas tricalcium aluminate sets very rapidly. The gypsum is added to prevent the tricalcium aluminate from setting too quickly. The tricalcium aluminate also generates heat.

8.3 CONCRETE

Cement is used as a binder for the aggregate in concrete. If there is not enough binder, the concrete will not be very strong. To produce the highest-strength concrete possible, there must be enough cement to cover each aggregate particle completely. Among the materials used for aggregate are gravel, stone, slag, and clays. In a standard concrete mix, about 75% of the volume is taken up by aggregates, typically stone and sand. This adds to the strength of the concrete and reduces the cost, because cement is the most expensive component of concrete. The two aggregate classes, *fine* and *coarse*, complement each other. The larger pieces contribute strength and bulk, whereas the finer particles fill the voids between the larger pieces.

The primary factor in strength for concrete is the *water-to-cement ratio*, measured in gallons per sack of cement. It takes a minimum of about 4 gal of water per sack of cement to mix the concrete completely. The 4-gal/sack ratio is very strong, and not much aggregate is needed. A 6-gal/sack ratio needs more aggregate but produces more concrete per mix. This mixture is weaker than the 4-gal/sack mixture. The

Concrete ingredients.

Courtesy of Portland Cement Association.

hardened concrete normally weighs about 150 lb/ft^3 and has a minimum compressive strength of 2000 to 6000 lb/in^2. The compressive strength is the maximum load in compression (being squeezed) that the concrete will withstand without failure.

Once the water/cement ratio is determined, the aggregate weights that will provide the proper *slump* are determined. Slump is determined through a slump test. A standard cone is filled with concrete and then removed, leaving a standing cone of concrete, which will tend to fall and spread out. The difference between the top of the concrete while in the cone and the top of the concrete once it has fallen, or slumped, is the slump value for that batch of concrete. The slump test for concrete is quick and easy and can be performed at the pour site. There are times when engineers will specify higher or lower slumps, depending on conditions. The higher the slump value, the thinner the mixture.

The specifications of a concrete batch include the following:

- Water/cement ratio
- Slump in inches
- Desired compressive strength
- Desired additives
- Aggregate size
- Cured specifications
- Delivery conditions

Two terms are often used in specifying concrete, *yield* and *cement factor*. The yield of a concrete batch is the number of cubic feet of concrete produced from a particular mix design. The cement factor is the number of bags of cement (94 lb/bag, typically) it takes per cubic yard of concrete. For example, what are the yield and cement factor for a concrete batch having 150 lb of aggregate, 30 lb of cement, and 2 gal of water? If the water weighs 128 oz/gal and the density of the concrete is 150 lb/ft^3, then

total weight $= 150$ lb $+ 30$ lb $+ (2 \cdot 128$ oz$)/(16$ oz/lb$) = 196$ lb
volume of concrete $= 196$ lb$/(150$ lb/ft$^3) = 1.31$ ft^3
volume per bag $= (1.31$ ft$^3 \cdot 94$ lb/bag$)/30$ lb $= 4.09$ ft^3/bag (yield)
cement factor $= (27$ ft^3/yd$^3)/(4.09$ ft^3/bag$) = 6.59$ bags/yd^3

The cement, fine and coarse aggregates, and water are all mixed at the concrete plant and loaded into a 6- to 9-yd^3 truck. These trucks have a rotating drum, which keeps turning on the way to the job site. This mixing keeps the coarse aggregate from settling to the bottom and the cement from rising to the surface. Once at the job site, the concrete is poured into forms with a minimum of handling. This may require pumping rather than pouring the concrete or using large crane buckets. For smaller jobs, the concrete is hauled in wheelbarrows.

Once the concrete is at the job site, many state and local agencies require tests to be made on the concrete batch. At a minimum, a slump test is performed on each batch, and three to six cylinders, each measuring 6×12 in., are taken. The cylinders are then taken to the laboratory, allowed to cure, and used for compression testing to determine if the batch meets the required strength.

The aggregates used for concrete must not all be of the same size. Aggregate sizes should vary so that the pieces fit together better. This fills in more area and provides greater strength. Aggregate takes up about 75% of the volume of the concrete. The remaining volume should be cement. Some air pockets and voids are always present in concrete. If aggregates of all the same size were used, larger gaps would exist to be filled with concrete. Because the concrete provides a binder for the aggregate filler, the result would be a low-yield, high-shrinkage, relatively weak concrete batch.

To fill in more volume and thus be more effective, aggregates of various sizes, from a large size (roughly 1 in. in diameter) down to sand, are mixed into the concrete. The various sizes fit together to take up a large volume of the concrete, and the cement then surrounds the aggregate and provides the adhesion. Properly graded aggregate allows a smooth troweled surface on the concrete, takes less concrete because more volume is taken up by aggregate, and provides greater strength because the concrete relies on the aggregate filler for strength.

To determine the size of the aggregate, a standard *sieve analysis* is performed. This test includes weighing an amount of dry aggregate and placing the test aggregate amount on the screen of a standard Tyler sieve series. This series has seven screens: a no. 4, no. 8, no. 16, no. 30, no. 50, no. 100, and a pan for the remainder. These screens are stacked in a cylinder form, from coarse to fine. The no. 4 screen has four openings per linear inch, and the no. 100 has 100 openings per linear inch. In order to perform a sieve analysis, a screen system is vibrated for 3 min, allowing adequate time for the smallest material to filter to the bottom of the system. After the sample has been sifted through, each screen is weighed, and the percentage of the total weight for each is calculated.

In addition to size tests, three other tests are made on aggregates. A *silt test* is made by filling a clear container with about 2 in. of aggregate. The container is filled about three-quarters full with water and shaken vigorously for 1 min. The mixture is then allowed to sit on a level surface for about an hour. If a layer of silt more than 1/8 in. settles on the aggregate, the aggregate is rejected or washed prior to using it in the concrete.

A 500-mL clear container is filled about two-thirds full with a 3% lye solution for the *calorimetric* test. This test is conducted to check for organic matter in the aggregate. Lye is sodium hydroxide (NaOH). The mixture is shaken for approximately 1 min and is allowed to sit for 24 hours. The lye above the aggregate at the end of 24 hours should be clear or a light yellow. The darker the color, the more organic matter the aggregate contains. Organic materials, usually sugars, prevent the concrete from setting up. The aggregate should be washed or rejected if there is excessive organic matter in the aggregate.

The third test checks the amount of water in the aggregate. Water in the aggregate is not accounted for in the mixture calculations. For example, if a person added 6 gal of water per bag of cement and the aggregate contained another gallon of water, the resulting concrete batch would end up having 7 gal of water per bag. The water in the rock is trapped and weakens the concrete. The ideal aggregate would neither soak up water from the mixture nor supply it with excessive water. A rudimentary test can be conducted by grabbing a handful of aggregate and packing it in the hand. A dry, crumbly mixture or one that leaves the hand wet is undesirable. The aggregate is saturated and suitable for use if it just crumbles after being packed in the hand but leaves the hand relatively dry. A more accurate test involves using a moisture probe. This is an analog meter that measures the electrical resistance between two points. The more water in the aggregate, the lower the resistance and the higher the reading. The moisture probe is also used for testing the moisture content of woods. There are other methods of testing the moisture content of aggregates, but these are two of the more common. Once the percent moisture content of the aggregate is known, it is added in the mixture calculations for the concrete.

The water in concrete does not dry out. The water must remain in the concrete for the concrete to cure completely. If the water evaporates too quickly, the curing process stops, and the concrete only partially cures. The curing process is difficult to restart once it has stopped. Great care must be taken to keep the concrete damp to ensure proper curing. This is done in several ways. Highway and road crews typically cover the fresh concrete with straw or plastic. Smaller jobs are usually kept wet with a spray hose.

When all of the necessary ingredients are available, clean, and tested, calculations are made to determine the proper amounts of aggregates, cement, and water to produce the desired quantity of concrete. These are estimated values based on conditions and experience. Factors such as the amount of water that evaporates from the concrete, the volume of aggregates, and the volume of voids in the concrete influence the mixture calculations. As a rule of thumb, the arbitrary ratios of components are 1 part cement, 2 parts fine aggregate, and 3 parts coarse aggregate. Water is added to provide the desired slump. This is a general recipe for the average, general-purpose batch. Reinforced concrete (discussed later) uses a 1:2:4 ratio, whereas larger pours may require up to a 1:3:5 ratio.

Although rule-of-thumb batches are good for small household jobs, more involved calculations are required for larger, more involved jobs. For example, if a particular job calls for a cement-sand-gravel mixture with proportions 1:2:4 and a water-to-cement ratio of 0.577 (6.5 gal of water per bag cement), the ingredients required would be as follows.

The bulk densities of the ingredients are

Cement	94 lb/ft^3
Sand	105 lb/ft^3
Gravel	96 lb/ft^3

The specific gravities of the ingredients are

Cement	3.10
Sand	2.65
Gravel	2.65

The unit volume of concrete is the sum of the absolute volumes of the constituents. These values are calculated by dividing the bulk weight by the solid weight. The bulk weight of each is calculated by multiplying its volume by its bulk density. For example,

$1.0 \text{ ft}^3 \cdot 94 \text{ lb/ft}^3 = 94.0 \text{ lb cement}$
$2.0 \text{ ft}^3 \cdot 105 \text{ lb/ft}^3 = 210.0 \text{ lb sand}$
$4.0 \text{ ft}^3 \cdot 96 \text{ lb/ft}^3 = 384.0 \text{ lb gravel}$
$6.5 \text{ gal} \cdot 8.345 \text{ lb/gal} = 54.2 \text{ lb water}$

Note: A bag weighing 94 lb yields 1 ft³ of cement. The solid volumes of the ingredients can then be found by dividing the bulk weights by the solid weights:

Bulk Weight	Solid Weight	Solid Volume
94 lb +	$(3.10 \cdot 62.4 \text{ lb/ft}^3)$	$= 0.486 \text{ ft}^3$ cement
210 lb +	$(2.65 \cdot 62.4 \text{ lb/ft}^3)$	$= 1.270 \text{ ft}^3$ sand
384 lb +	$(2.65 \cdot 62.4 \text{ lb/ft}^3)$	$= 2.320 \text{ ft}^3$ gravel
54.2 lb +	62.4 lb/ft^3	$= 0.870 \text{ ft}^3$ water

Summing the solid volumes, we get a total of 4.946 ft³. The ingredients required per cubic foot are

$0.486/4.946 = 0.098 \text{ ft}^3$ cement
$1.270/4.946 = 0.257 \text{ ft}^3$ sand
$2.320/4.946 = 0.469 \text{ ft}^3$ gravel
$0.870/4.946 = 0.176 \text{ ft}^3$ water

The weights of ingredients per cubic yard are

solid volume ·	solid weight · 27 ft³/yd³	= weight/yd³
0.098 ·	$(3.10 \cdot 62.4) \cdot 27$	$= 511.8$ lb cement
0.257 ·	$(2.65 \cdot 62.4) \cdot 27$	$= 1147.4$ lb sand
0.469 ·	$(2.65 \cdot 62.4) \cdot 27$	$= 2094.0$ lb gravel
0.176 ·	$62.4 \cdot 27$	$= 296.5$ lb water

The total weight in pounds for the mixture would be 4049.7 lb.

If there were 4% voids in the concrete, the overall weight would increase to

$$4049.7 \cdot 1.04 = 4211.7 \text{ lb}$$

These are some of the calculations performed to determine the amounts of materials to be used in concrete mixtures. Concrete is usually ordered in cubic yards, and aggregates are ordered in tons. These calculations assume ideal conditions and are given as a demonstration of common calculations performed in estimating materials for concrete mixtures.

8.4 REINFORCED AND PRESTRESSED CONCRETE

Concrete has a very low tensile strength; in other words, it fails quickly when pulled apart under a tensile load. It is very strong in compression, the tendency to be squeezed together. Steel has a high tensile strength. It makes sense to add steel to concrete where the concrete must withstand forces other than compres-

Reinforced concrete and reinforced overpass.
Courtesy of Portland Cement Association.

sion and place the concrete where the compressive forces are. This idea is the principle behind *reinforced concrete*. If you have ever witnessed the pouring of road sections, you can see the reinforcing bars (rebar) of steel laid out in a grid within the concrete. Steel wire or fencing is used in smaller jobs. Placement of the reinforcement is critical. The purpose of the reinforcement is to absorb the stresses in the concrete. When a floor section, such as a garage floor, is poured, a skeleton of rebar is constructed. Then the concrete is poured around it. When a slab is poured, the rebar is laid out, and the concrete is poured over it. The rebar is lifted to the approximate center of the slab, where it provides the maximum strength.

To overcome the weaknesses of concrete in tension, it makes sense to avoid allowing the concrete to go into tension at all. This principle is the basis for *prestressed concrete*. In prestressing, the concrete is kept in compression. One method of prestressing concrete requires placing steel rods or cable under tension in position as the concrete is poured over them. After the concrete has set up, the tension on the rods is released. This places the concrete in compression. The tensile stress in the prestressed rods in the concrete exerts a compressive force on the concrete. For the concrete to fail in tension, the tensile load on the concrete must exceed the compressive force exerted by the tension in the rods. Prestressed concrete beams have several times the strength of simple reinforced beams.

8.5 CONCRETE ADDITIVES

Concrete contains a certain percentage of air. The amount of *entrained air* depends on the conditions under which the concrete was mixed, the type of concrete used, and other related factors. The amount of entrained air can be determined by placing a known amount of concrete in a sealed, pressurized container and observing the amount by which the concrete can be compressed. The air in the concrete is compressed by the pressure in the container. The amount of entrained air is another common test conducted on concrete batches in the laboratory.

Concrete will absorb water. This fact causes some difficulties for the contractor. In winter, the water that the concrete absorbs freezes. When the concrete thaws, small patches of the concrete flake off. Several cycles of freezing and thawing cause roads, driveways, and other concrete constructions to be destroyed. One way to combat this is by air entrainment. Types I, II, and III portland cements are air-entraining. They contain about 4% to 7% very small air bubbles. They resist damage from the freeze-thaw cycle. When concrete is mixed, additives can be included in the mixture that cause air bubbles to be produced in the concrete. These air bubbles provide room for the water to expand during freezing. These air-entrained concretes and additives are used in areas where the freeze-thaw cycle would cause damage. The disadvantage to air entrainment is that it weakens the concrete. Concrete may be weakened up to 20% by entrained air.

Accelerators and retarders are also concrete additives that advance or delay the initial set of a concrete. An accelerator can be used in cold weather for early removal of forms or to protect the concrete. Calcium chloride in amounts up to 2% is often used. Retarders are often used in hot weather to allow sufficient time to work the concrete. The retarder delays hydration of the cement, which retains the water for workability. Starch or sugar in amounts up to 0.05% is used to retard hydration for up to 4 hours. Hardeners and colorizing agents are also used in concrete. Colorizing agents are used to produce various patterns in the concrete. Hardeners are used in concrete floors to provide better abrasion and wear resistance. They may be added to the mixture or worked into the surface.

Fly ash is the most widely used admixture in concrete. Fly ash and silica fume are called *pozzolans*. *Pozzolans* are siliceous or aluminosiliceous materials that have no cementitious value themselves, but when finely ground and in the presence of water, they react with the calcium hydroxide released by hydrating portland cement and exhibit cementitious properties.

Cementitious materials are substances that have hydraulic cementing properties (set and harden in the presence of water). Fly ash is a by-product released from the combustion of coal in electric power plants. Upon ignition, most of the volatile matter and carbon in the coal are burned off. The mineral impurities in the coal remain and are fused together. These spherical particles are carried from the combustion chamber by the hot exhaust gasses. The fly ash can then be collected by electrostatic separators or bag filters. Generally, the fly ash can be used in this form, without further processing. It is basically silicate glass containing silica, alumina, iron, and calcium, with other trace elements.

Silica fume, also referred to as microsilica or condensed silica fume, is another pozzolanic admixture. This gray, powdery product is the result of the reduction of high-purity quartz with coal when producing silicon or ferrosilicon alloys in an electric arc furnace. Silica fume rises as a vapor when the temperature rises above 3630°F (1979°C). The vapor cools, condenses, and is captured in huge cloth bags. The condensed fume is further processed to remove impurities and to control particle size. Condensed silica fume is primarily silicon dioxide in spherical shape.

Ground, granulated blast-furnace slag made from iron blast-furnace slag is a non-metallic material that consists primarily of silicates and aluminosilicates of calcium. The molten slag is rapidly cooled from 2730°F (1484°C) with water to form a glassy, granulated material similar to sand. These mixtures are three very common admixtures for cement and concrete.

Fly ash and slag are used to reduce the water requirements of concrete mixes. Typically, they can reduce the water requirement from 1% to 10%, providing the same slump. Silica fume increases the necessary water for the same slump. Fly ash and slag generally increase the workability of a concrete mix. Fly ash and slag are used to reduce the heat of hydration of the cement. They also generally retard the setting time of the concrete. Fly ash, ground, granulated blast-furnace slag, and silica fume are all used in addition to, or as a replacement for, some of the cement in concrete. Fly ash may be used in quantities up to 15% to 20% of the cement by weight. Slag is generally used in quantities of up to 40% of the cementing material in the concrete mixture. Silica fume is typically used in amounts between 5% and 10% of the cement by weight. These three materials generally reduce the amount of cement required, increase the strength of the concrete, increase the pumpability of the concrete, and generally improve the concrete's resistance to sulfate attack. These advantages are in addition to reducing the overall cost of materials.

8.6 ASPHALT AND ASPHALTIC CONCRETE

Asphalts are organic materials composed of the residue or leftovers after the lighter constituents of petroleum have been distilled off. There are also some natural rock asphalts. The terms *bitumen* and *bituminous* are used to describe these materials. Coal tars are often included in this category, although they are extracted from coal. Coal tars are often called *pitch*. These materials are generally dissolved in turpentine, petroleum solvents, and carbon disulfide. All of these materials are generally used for waterproofing membranes such as roofing shingles, felt papers, and bituminous paints. However, the primary application here is in combination with aggregates, where they are used for surfacing roads, driveways, and parking lots, to name a few applications. The asphalt provides the necessary adhesive to bind the materials together.

Hot asphalts are applied after heating. They soften, not melt, at a temperature of approximately 300°F (149°C). Straight-run asphalts make up the majority of the total asphalt production in the United States, being used in runways and road construction. The asphalt is mixed with hot stone, poured hot, and then rolled to produce a smooth, hard surface. Air-blown asphalts are oxidized to obtain desirable properties. The primary use for air-blown asphalts is in built-up roofs. Blown asphalt is generally harder, less tacky, and more rubbery. Emulsion asphalts are emulsions of small droplets of asphalt dispersed in a water base. They are used primarily for waterproofing walls and painting surfaces. They are also mixed with latex and used for patching roads. Cutback asphalts are dissolved in solvents such as petroleum oils or naphtha. Cutback asphalts are no longer used, due to environmental and health concerns.

Bitumens have many uses today. In general, they are widely used to provide waterproof coatings for walls and other structures. They are used to pave roads, which accounts for approximately 85% of the asphalt produced in the United States. Asphalt in the pavement acts as a binder for the aggregates, much as cement does in concrete. Before it is mixed with the gravel, sand, or crushed stone, the asphalt is heated or emulsified with water. The asphalt "cures" by solidifying or by evaporation of the water. The adhesiveness, durability, and waterproofing, along with reasonable cost, of asphalt make it useful as a building material. Asphalt products are used in flooring, roofing, shingles, and insulation. Roofing products, such as shingles and roll roofing, are manufactured from asphalt-soaked felt papers coated with additional asphalt and a crushed mineral aggregate.

8.7 SUMMARY

Cements are materials used to bind other materials together to form a solid mass. Concrete is the product of cementing materials together. Portland cement is made from limestone and clay, with ferric oxide added as flux. Portland cement is available in five common types, each of which has special properties and applications. Hydraulic cements, such as portland cement, dry and harden by hydration. Aggregates are added along with cement and water to form concrete. The primary factor for strength in concrete is the water-to-cement ratio. The yield of a concrete batch is the number of cubic feet of concrete obtained per sack of cement. The cement factor is the number of bags of cement (94 lb/bag, typically) it takes per cubic yard of concrete. Various tests are performed on concrete materials prior to mixture, during pouring, and later in the laboratory, including the slump test, silt test, calorimetric test, and test for entrained air.

Concrete is weak in tension. To overcome this, concrete is typically reinforced or prestressed. Reinforcing bars are added to boost the tensile strength of concrete or the concrete is prestressed to avoid having it ever come under tension.

Various additives are used in concrete to enhance its properties. Air-entraining additives are used to reduce the effects of the freeze-thaw cycle. Colorizing agents are used for decorative effects. Accelerators and retarders are used to advance or delay the initial set of a concrete. These products allow more or less time to work the concrete, depending largely on environmental conditions. Hardeners are sometimes used to enhance the abrasion and wear resistance of concrete floors, for example.

Asphalts are organic materials composed of the residue or leftovers after the lighter constituents of petroleum have been distilled off.

QUESTIONS AND PROBLEMS

1. Explain the difference between cement and concrete.
2. What function does ferric oxide serve in the manufacture of cement?
3. List and provide a short description of the five types of portland cement.
4. What is a clinker?
5. What is hydration of cement?
6. What is the primary factor in determining the strength of concrete?
7. What are the yield and cement factor of a test batch of concrete in which 180 lb of aggregate, 30 lb of cement, and 256 oz of water were used if the batch has a density of 144 lb/ft^3?
8. Calculate the quantities required for a 5 yd^3 batch of concrete if the desired percentages are 24% water, 18% cement, and 58% aggregate.
9. What is the water-to-cement ratio for the batch in question 8 in gallons per bag?
10. If a concrete has a density of 144 lb/ft^3, what would a cubic yard weigh?
11. A rectangular garage floor measuring 24↔28 ft is to be poured to a depth of 6 in. Specifications for the concrete require a cement factor of 5 bags per cubic yard, a water/cement ratio of 6 gal per bag, and a density of 150 lb/ft^3. How many yards of concrete are required?
12. List the four molecules present in portland cement.
13. Describe the procedure for a slump test.
14. Describe the silt test for aggregate.
15. Describe the calorimetric test for aggregate.
16. Explain the uses of accelerators and retarders in concrete.
17. Why is entrained air necessary in concrete?
18. Describe the process for reinforced concrete.
19. Explain the process of prestressing concrete.
20. Could concrete be poststressed; if so, how?
21. List five applications of concrete that you encounter in everyday life.
22. What other materials may be substituted for concrete in these applications?
23. You are pouring a concrete pad for a 24'↔32' garage at a 4 in. depth. Determine the amount of concrete needed and fully specify the concrete batch to be delivered.
24. List an application for each type of portland cement and the factors that go into your selection.
25. What factors do you think go into the selection of aggregates for concrete? How do you think these selection criteria influence the properties of the concrete product?
26. Why is concrete so prevalent as an engineering material?
27. What advantages and disadvantages do asphalt and its products present? How do these influence its selection as an engineering material?
28. Asphalt is commonly used as a roofing and paving material. What substitutes can you recommend for these applications?

CHAPTER 9
Composites

Objectives • Describe the components and different types of composites currently available.
• Explain the different types of composite construction and the reasons behind them.
• Describe the various manufacturing methods used to produce composites.
• List the different reinforcing materials used in composites.
• List the various matrix materials used in composites.

9.1 INTRODUCTION

There are situations in which no single material has the required properties and characteristics to suit an application. In these cases, two or more materials are combined to form a *composite*. Composites are combinations of two or more materials that exhibit improved properties over their individual components. Therefore, in composites the whole is greater than the sum of the parts (synergy). Wood, celery, bamboo, and corn are all examples of nature's composites, in which two materials combine to reinforce and bind together.

All of these composites generally have one thing in common: a matrix or binder combined with a reinforcing material. If the fibers are directionally oriented and continuous, the material is termed an *advanced composite*. Reinforced concrete is a good example of a composite material. When concrete is reinforced with steel rebar, the concrete becomes the matrix, which surrounds the reinforcing fiber, the rebar. Another example is reinforced fiberglass products such as fishing rods. Glass fibers are set in a thermosetting resin matrix. This produces a strong, lightweight, flexible fishing rod. Other fibers are produced from aramid (Kevlar and Nomex), boron, carbon, graphite, and ultrahigh-molecular-weight polyethylene (Spectra). The matrix for these materials is typically a thermosetting epoxy resin. These materials provide some exceptional increases in mechanical properties, sometimes three to six times greater than steel.

9.2 TYPES OF COMPOSITES

The natural composites have already been introduced. Other common varieties of composites include these combinations: fiber-resin, fiber-ceramic, carbon-metal, metal-concrete, metal-resin, and wood-plastic. Most of the contemporary advanced

Graphite fishing rod.

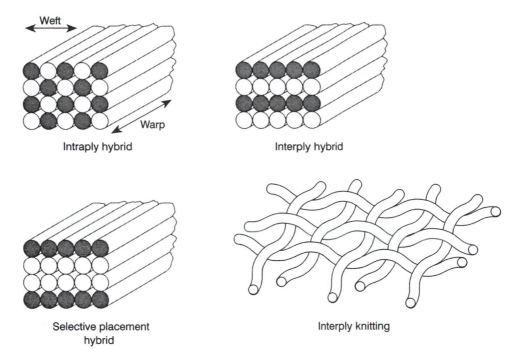

Figure 9–1
Fiber placement in composites.

composites use glass, Kevlar (an aramid), or one of the various types of graphite fibers. Graphite fibers are really not graphite, but carbon, because they do not have the hexagonal-close-packed structure of graphite. All of these fibers exhibit high strength and light weight. Initially, the fibers used in composites were made by heating asphalt and pulling carbon threads through a die. Today, carbon fibers are produced by heating a rayon or other organic fiber up to 1800°F (990°C) in an inert atmosphere. Other fibers produced for composite use include black fibers for graphite, yellow for Kevlar, and white or clear for glass.

Composite design involves careful selection of the reinforcing fiber. Each fiber material has advantages and disadvantages. Glass is a strong fiber, but it is heavier than other fibers. Graphite and Kevlar are more expensive, yet lighter weight, than glass. One solution is to use more than one type of fiber in the matrix. This *hybrid* is a combination of fibers, giving the best of each to the composite. Hybrids are available in various forms: intraply, interply, selective placement, and interply knitting. These four forms are shown in Figure 9–1.

Interply forms have two or more different fibers, or cloths or mats made from the fibers, stacked in alternating layers. *Intraply forms* have the different fibers intermixed within the same layer, or *ply*. Fibers may also be selectively placed within the matrix. This allows each fiber to be placed where it will contribute the best, according to its properties. For example, expensive carbon fibers can be placed where their

Figure 9–2
Common polymers used in composites.

Thermoplastics	Thermosetting
ABS	Phenolic
Nylon	Epoxy
Polycarbonate	Melamine
Polyethylene	Polyester
Polypropylene	Silicone
Polystyrene	Urea
Teflon	Urethane
Vinyl	

strength is needed most, and cheaper glass fibers can be used elsewhere. This produces a better, cheaper, stronger product. Another form used to weave reinforcing fabric is the *interply knitting form*. Strands of Kevlar may be woven to strands of graphite by fine filaments of Kevlar. Two to five plies are sewn together in this fashion. This reduces the lay-up time while providing added strength.

The direction in which a fiber form is woven can influence its properties. For example, if stronger fibers are used in one direction and weaker fibers are used perpendicular to that, the fabric will exhibit the greatest strength in the direction of the stronger fibers. The direction the stress or load is applied is called the *warp*; the direction perpendicular to the stress is called the *weft*. Between the warp and the weft are the bias directions.

The resins used for composites can be almost any polymer. The specific properties required of the composites and the application determine which polymer to use. Some of the more common polymers used are given in Figure 9–2.

9.3 COMPOSITE CONSTRUCTION

Reinforcements can be in the form of laminates, mats, particulates, roving (continuous fibers), woven fibers, chopped fibers, flakes, or honeycomb. These are common forms; the list is not all-inclusive. The purpose of adding these reinforcements to the polymer matrix is to enhance the mechanical properties of the material. Continuous reinforcements, such as fibers, fabrics, and honeycomb structures, tend to distribute the applied load throughout the entire structure and provide longer-term stability.

Composite use dates back to roughly 1915, when a phenolic-paper laminate was produced for electrical insulation. The most important discoveries in the field of reinforced plastics (RP) or fiber-reinforced plastics (FRP) were epoxies, polyester resins, and reinforcement by glass fibers. These discoveries revolutionized the field of composites and opened up applications such as storage tanks, boat hulls, furniture, and other similar applications. Reinforced plastic composites have been increasingly used in the automobile and aircraft industries.

Composites are generally grouped into three categories: particulate, fiber, and laminar. Examples of these three groups include concrete, a mixture of cement and aggregate, which is a particulate composite; fiberglass, a mixture of glass fibers imbedded in a resin matrix, which is a fiber composite; and plywood, alternating layers of laminate veneers, which is a laminate composite.

One type of composite structure is the *dispersion-strengthened composite*. This type of composite contains small particulates or dispersoids, which increase the strength of the composite by blocking the movement of dislocations. The dispersoid is typically a stable oxide of the original material. A common example is sintered aluminum powder (SAP). SAP has an aluminum matrix, which contains up to 14% aluminum oxide (Al_2O_3). This composite is produced with the *powder metallurgy process*, where the powders are mixed, compacted at high pressures, and sintered together. Sintering involves heating a material until the particles of the material fuse together. Only the edges of the particles are generally bonded together; the whole particle does not melt. In terms of sintered ceramics, the product is a strong, rigid, brittle product that exhibits good compressive strength, high melting points, and good heat resistance. Examples of dispersion-strengthened composites include Ag-CdO, used as an electrical contact material; Pb-PbO, used in battery plates, and Be-BeO, used in nuclear reactor and aerospace components.

Particulate composites contain larger amounts of coarse particles that produce certain properties rather than specifically add strength. The particulates used include metals, ceramics, and polymers. Cemented carbides, called *cermets,* contain ceramic particles in a metallic matrix. These include the tungsten carbide (WC) inserts commonly used in tooling for modern machining operations. Tungsten carbide is extremely hard, stiff, and brittle and can withstand the high temperatures during rapid cutting. Tungsten carbide particles are often combined with cobalt to improve the properties for cutting. As the tool wears, the tungsten carbide particles fracture or pull out of the matrix, leaving a sharp edge for continued cutting.

Fiber-reinforced composites use strong fibers imbedded in a softer matrix to produce products with high strength-to-weight ratios. The matrix material transmits the load to the fibers, which absorb the stress. Fiber-reinforced composites have been used for centuries—straw has been used in the making of mud bricks since the Pharaohs. Steel rebar has been used for many years in reinforced concrete structures. Fiberglass cloth and resin have been used to repair many automobile fenders. These applications are among the more familiar.

The strength of these composites comes from the bonding between the reinforcement fibers and the matrix. The length-to-diameter, or *aspect*, ratio of the fibers used as reinforcement influences the properties of the composite. The higher the aspect ratio, the stronger the composite. Therefore, long, continuous fibers are better than short ones for composite construction. However, continuous fibers are more difficult to produce and place in the matrix. Shorter fibers are easier to place in the matrix but offer poor reinforcement. Some trade-off is made when shorter, discontinuous fibers are used with aspect ratios greater than a specified minimum value. The greater the number of fibers, the stronger the composite. This holds true up to about 80% of the volume of the composite, where the matrix can no longer completely surround the fibers. Examples of fiber-reinforced composites include borsic-aluminum, used in aerospace applications; Kevlar-epoxy and Kevlar-polyester, used in aerospace, boat hulls, sporting goods, and bulletproof vests; graphite-polymer, used in aerospace, automotive, and sporting goods applications, and glass-polymer, used in lightweight automotive, marine, and sporting applications.

Laminar composites are generally designed to provide high strength and low cost at a lighter weight. A familiar laminar composite is plywood, where the veneers are joined by adhesives, typically phenolic or amine resins. The individual odd number of plies are stacked so that the grain in each layer runs perpendicular to that of the layers above and below it. This technique offers plywood that is strong and yet inexpensive. Some safety glass is a laminate structure, where an adhesive such as polyvinyl butyral is used between two outer layers of glass to keep the glass from flying when broken. Formica is another common laminate used for countertops. Clad metals such as U.S. coins are metal-metal laminate composites. Laminates require two or more layers be bonded together.

Sandwich structures have thin layers of facing materials over a low-density honeycomb core, such as a polymer foam or expanded metal structure. A familiar sandwich-structured composite is corrugated cardboard. The corrugated paper core is covered by two faces of thin paper. In structures of this type, the facing material serves to fix the inner core in place. The core provides the strength. This honeycomb structure finds wide use in industries such as the aircraft industry, where higher strength and lower weight are important factors. These structures are lightweight, stiff, and strong and can be filled to provide sound and vibration damping. The honeycomb structure is shown in Figure 9–3.

9.4 FABRICATION

Fabrication of composite materials can be accomplished through hand lay-up, injection molding, lamination, filament winding, or pultrusion. These methods were presented in the chapter on plastics (Chapter 5). The oldest method of composite man-

Figure 9–3
Honeycomb structure.

Figure 9–4
Resin transfer molding.

ufacture is *hand lay-up*. In it, the resin is poured into a mold or applied to the surface, and the fabric is then rolled into the resin or applied to the surface. The hand lay-up technique is used in boat building, auto repair, and home projects, because it requires minimal equipment and the result cures at room temperature.

Prepreg fibers are woven fibers impregnated with resin prior to manufacture. The prepreg fabric is laid in place, heated while it is being laid, and cooled by a blast of

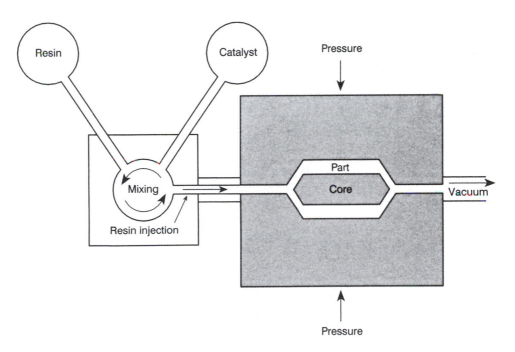

Ultimately reinforced thermoset
reaction injection process (URTRI)

Figure 9–5
URTRI process.

cold air to set it in place initially. The formed product is then cured at elevated temperatures and pressures.

Resin transfer molding (RTM) is a two-piece matching cavity-mold system used for molding composites. This process is shown in Figure 9–4. The reinforcement material is placed in the mold cavity. A chopped mat is often draped along the inside of the mold. The mold halves are clamped together and the resin is introduced into the cavity. RTM offers fast curing times and is less labor-intensive than other methods.

The *ultimately reinforced thermoset reaction injection* (URTRI) process allows the use of cores in the molding process. A core is cast from a high-temperature epoxy foam. The core is wrapped with layers of graphite fabric and is positioned in the mold cavity. Epoxy resin is pumped into the heated mold and cured for about 5 min. The mold is then opened, and the part is extracted. This process is illustrated in Figure 9–5.

The primary drawback of composites is the time it takes to manufacture them. Parts that are stamped out of metal every few seconds take several minutes to complete if made of composites. The increased time often increases the cost of manufacturing. However, with the ever-increasing technology, composite production has increased, and applications have broadened. For example, more automotive parts are being made from composites. It is thought that in the near future there will be a totally composite car on the market. Figure 9–6 illustrates the strength of composites versus steel.

9.5 SUMMARY

There are situations in which no single material has the required properties and characteristics to suit the application. In these cases, two or more materials are com-

Figure 9–6
Comparison of tensile strength of composites versus metals.

Material	Tensile Strength lb/in² [MPa]	
Metals		
Aluminum alloys	13,000–24,000	90–165
Brasses	40,000–120,000	276–827
Bronzes	40,000–130,000	276–896
Cast iron	18,000–60,000	124–414
Cold-rolled steel	84,000	580
Magnesium alloys	20,000–45,000	138–310
Monel	100,000	690
Stainess steel	85,000–95,000	586–655
Titanium	95,000	655
Zirconium	24,000–40,000	165–276
Fibers for Composites		
Glass	500,000	3,448
Graphite	300,000–410,000	2,069–2,827
Polyaramid (Kevlar)	410,000	2,827
Steel	60,000–320,000	414–2,206

bined to form a composite. Composites exhibit improved properties over their individual components. Composites are available in many varieties, including fiber-resin, fiber-ceramic, carbon-metal, metal-concrete, metal-resin, and wood-plastic. These composites are generally formed by suspending reinforcing fibers in a binding matrix. The reinforcing fibers are typically made from glass, Kevlar, or a graphite material. These fibers are woven into fabrics by intraply, interply, selective placement, or interply knitting. Composites are available as dispersion-strengthened composites, particulate composites, fiber-reinforced composites, or laminate composites. The aspect ratio and volume of the reinforcing material decide the strength of the composite. The matrix may be one of a number of materials, including thermoplastics (T-plastic), thermosets (T-set), ceramics, and metals. The matrix is required to surround and bind to the reinforcement material, giving the composite its strength.

Composites are manufactured in a variety of shapes using a variety of methods, including hand lay-up, prepreg, resin transfer molding (RTM), ultimately reinforced thermoset reaction injection (URTRI), filament winding, and pultrusion. Composites are replacing other materials, such as wood and metal, in many applications and will continue to do so as manufacturing costs and times are reduced.

QUESTIONS AND PROBLEMS

1. Define the term *composite* as it is used in this chapter.
2. What are advanced composites?
3. What factors influence the final properties of composites?
4. List and describe the four methods of weaving fabric for composites.
5. Describe the lamination process.
6. Explain the sandwich structure for composites.
7. List and describe the processes described in this chapter for manufacturing composites.
8. What is a hybrid?
9. Define *warp* and *weft*.
10. Define *aspect ratio* and discuss how it affects composite strength.

11. List five common applications of composite materials.
12. Of these applications, how many are "new" applications? How many have used composites as a substitute for another material? What material was replaced in these applications?
13. What factors influence the decision to select a composite for an application?
14. Why is wood and lumber so widely used in the construction industry?
15. Are there substitutes for wood in construction? List these and provide advantages and disadvantages for each.

CHAPTER 10
Adhesives and Coatings

Objectives
- Define the terms related to the field of adhesives.
- Recognize types of adhesives, including glues, cements, pastes, thermoplastic adhesives, and thermosetting adhesives.
- Describe the various properties related to adhesives and uses of adhesives.
- Describe the materials used for sealants and caulking compounds.
- Define the terms related to organic and inorganic coatings.
- Recognize common coatings, including paints, enamels, lacquers, and varnishes.
- Describe the processes commonly used to apply coatings.

10.1 INTRODUCTION

Adhesives play a major part in contemporary society. Bonding with animal and vegetable adhesives has been used for thousands of years. The discovery of many thermoplastic and thermosetting resins has changed the field of adhesives. Adhesives that have higher strengths than some metals have been developed. In fact, they have become so advanced and commonplace that adhesives have replaced welding and riveting of various light-gage components. They have a wide range of applications for bonding similar and dissimilar materials. Adhesives have been used in many applications to replace solder, such as in wrapping metal cans. The new super glues bond quickly and exhibit excellent strength. This chapter introduces various glues, cements, and adhesives. Included are their uses and their particular properties, strengths, advantages, and disadvantages.

Also presented in this chapter are the paints and coatings used to beautify, preserve, and protect other materials. Included in this discussion are the various types available, their applications, their properties, and their benefits.

10.2 ADHESIVES

10.2.a Introduction

Everyone has used an adhesive at some time, but have you ever stopped to think what happens when you bond two pieces of material together? One fundamental requirement for adhesion is intimate contact between the surfaces to be bonded. In addition, careful surface preparation is important. The two surfaces should be clean and free from dirt, grease, oil, and other contaminants. The *adhesive* is then generally placed on one or both of the surfaces to be bonded, the *adherend* materials. The adherend surfaces are then brought into contact, and the adhesive is allowed to *set* and then *cure*. The term *set* is used to describe the process of a liquid adhesive converting to a hardened condition. If the conversion involves cross-linking by polymerization, the process is referred to as *curing*.

There are two theories on how adhesives bond materials. One theory supports the idea of mechanical bonding of the adhesive to the adherend materials. All surfaces have peaks and valleys on their surfaces, no matter how polished they are. In mechanical bonding, the adhesive fills these irregularities on the surface and mechanically "locks" the pieces together. The theory is that roughening the adherend surfaces tends to increase bond strength. There are those who disagree with this theory and believe that the bonding mechanism is composed of molecular forces called *van der Waals forces* between molecules in the adhesive and the atoms or molecules in the adherend materials. Adhesion is probably a combination of these two actions and results from a mechanical locking of the adherend materials through the adhesive and the action of molecular forces within and between materials.

The function of an adhesive is to wet both surfaces completely, eliminating all air and voids that may exist between the surfaces in contact. If the forces between the

Figure 10-1
Wetting the surface.

adhesive and the adherend materials are greater than the forces between the adhesive molecules, the adhesive will flatten out or spread over the adherend, as shown in Figure 10–1. This is known as *wetting* the surface. *Wettability* refers to the ability of an adhesive to wet the surface of the adherend materials. This state occurs when forces between adhesive molecules and base molecules are greater than between the adhesive molecules. Ideally, all of these forces should be the same (within the adhesive, between the adhesive and the adherend materials, and within the base material). Two terms, *adhesion* and *cohesion*, are used to describe the forces at work in joining together two materials. Adhesion is the joining of two dissimilar materials and involves the adhesive molecules and the adherend molecules. Cohesion is the joining together of similar materials, such as the adhesive material itself.

The surfaces to be joined should be clean and free of contaminants, because these contaminants prevent the complete wetting of the surfaces. Dirt or any other contaminant results in a weak point in the adhesive layer at the point of contamination once the surfaces are joined.

Adhesives have some disadvantages compared to welding, soldering, and mechanical fasteners. One disadvantage is the need for surface preparation. As stated previously, the surfaces to be joined must be clean and relatively free of contaminants; otherwise, the joint is weak. Another disadvantage is the curing time. In addition, the curing process may require additional heat and/or pressure. Finally, adhesives are generally confined to working temperatures of less than 300°F (150°C).

Adhesives offer advantages over other types of fastening. They tend to distribute the stress over the entire joint rather than concentrating the stress at the point of fastening, such as in riveting or bolting. Adhesives can be used to join dissimilar materials, such as rubber to glass, paper to glass, paper to rubber, and metal to rubber. Adhesive joints offer excellent fatigue resistance and are often used for electrical insulation and moisture barriers. In contrast to other joining methods, adhesives may be used to join small, thin, and soft materials that cannot be easily joined using other methods.

10.2.b Types of Adhesives

Chemical composition is the most commonly applied classification system for adhesives. Figure 10–2 illustrates the classification system for general adhesives. This system is the generally accepted classification method used by manufacturers.

Glue is a term generally applied to materials that occur naturally, such as those from animal, vegetable (starch), casein, and soybean origins. *Adhesive* is a generic term applied to synthetic bonding agents. In the generic sense, an adhesive is any

Figure 10–2
Classification of adhesives.

Natural Resins

Proteins
Natural rubber base
Starches and dextrine

Synthetic Resins

Synthetic rubber base
Thermoplastic resins
Thermosetting resins

Inorganic Materials

Silicates

material that causes one body to stick to another. *Cements* are those used for a specific purpose, such as rubber cement, plastic cement, glass cement, stone cement, and leather cement. *Pastes* are generally water-soluble products used for paper and paper products. *Pressure-sensitive adhesives* do not wet the surface to which they adhere. Tape is one example of a pressure-sensitive adhesive. The adhesive used remains permanently tacky and does not set. On the back side of the tape is a release coat. This coating allows the tape to be unrolled without stripping the adhesive from the tape itself.

10.2.b.1 Natural Adhesives   Animal glues, also called *protein-based adhesives*, are made from by-products of the animal packing industry, especially hides and bones. These are among the oldest adhesives known. They are usually organic high polymers of colloidal form, generally derived from collagen. Collagen is a protein found in the skin, muscular tissue, and bones of animals. They are available in flake or granulated form or dissolved in water at temperatures around 150°F (65°C). These glues are applied as either solutions or emulsions, and evaporation of the vehicle (usually water) results in hardening. (An emulsion is a suspension of fine liquid particles in another liquid in which the first liquid is not soluble in the other.) These glues do not resist water and are fairly weak when exposed to mold, fungi, or insects.

Mucilages are made from *vegetable glues*. Vegetable glues or gums are often used for envelopes, labels, and stamps. Vegetable glues are made from cassava (tapioca) flour dissolved with caustic soda. Corn or potato starch is also used in vegetable glues. The least expensive of all glues are those that deteriorate when exposed to the environment. Starch pastes deteriorate with time and exposure to the atmosphere. They are generally mixtures of starch or dextrine, water, and a strengthener such as another glue, resin, or gum. Vegetable glues are often used in lower-grade plywood and furniture. They exhibit excellent strength after drying but are weakened by moisture, mold, and fungi.

Casein glue is made from dried milk curds, lime, and other chemical ingredients. It is prepared by acidifying milk to separate the desirable protein material as curd and is washed and dried. One ton of skim milk will produce approximately 65 lb of casein. This glue is used for plywood and furniture (wood to wood) but is susceptible to mold and fungi attack. It is more expensive than vegetable glues but exhibits similar properties. Casein glue is often used to seal the seam of cigarette papers.

Soybean glue is produced from soybean meal, after the oil has been removed. Soybean meal, as a dry powder, is mixed with lime, caustic soda, silicate of soda, and other ingredients. This produces a low-cost, moderately water-resistant glue used in Douglas fir plywood and similar applications.

Albumin glue is made from dried animal blood, with additives. It is based on the protein material albumin, which is found in blood. It is highly moisture-resistant but deteriorates with age and suffers from attack by mold and fungi. It has excellent strength and durability when cured in plywood joints at temperatures around 170°F (77°C).

Sodium silicate glues are inorganic glues used as quick-setting glues in paper and packaging industries. Many of the glues presented here find use in these industries. Paper has low tear strength, low moisture, and heat resistance, which imposes no problem for these glues. Sodium silicate glues are also used in abrasive grinding wheels and the flux coating of welding rods.

Pastes are mixtures of starch or dextrine with water and a strengthener, such as glue, resin, or gum. Starch is a polymeric substance that occurs naturally in materials such as wheat and potatoes. Starch is a polymer of glucose, one of the carbohydrates. Starch-based pastes are widely used in the paper industry and in such applications as wallpaper, decorations, cardboard boxes, and tapes.

Emulsion adhesives consist of dispersed globules of adhesive in a water base. Bonding takes place through evaporation of the water carrier. Latex adhesives are often chosen for their lower shrinkage rate. In addition, they can often be removed with rubbing. Other applications that are suitable for pastes and emulsion adhesives include corrugated cardboard packaging, paper bags, envelopes and stamps, bookbinding, labeling, paper tapes, and laminations.

10.2.b.2 Cements Elastomeric cements, or *rubber cements*, are mixtures of uncured rubber and a chemical solvent, such as naphtha or benzol. This mixture is a viscous liquid that can be used to cement rubber materials permanently or cement paper, cloth, or similar materials temporarily. The latex type of rubber cement is often used to join fabric and fibrous materials. It is frequently used in shoe construction. Natural rubber cements have high bond and cohesive strengths and good initial tackiness. They also have good initial strength and are waterproof. However, they disintegrate with age.

Pyroxylin cements are made from scrap celluloid (nitrocellulose film used in movies) or other pyroxylin materials dissolved in a chemical solvent, such as ether alcohol. These are the common "household" cements. They bond almost any surface and harden by evaporation of the solvent. They are also used in the shoe industry.

Magnesia cements are a mixture of calcinated magnesia and magnesium chloride. When cured, they harden to a stonelike mass and are used to bind organic materials and repair masonry structures.

10.2.b.3 Thermoplastic Adhesives Thermoplastic adhesives are generally made from cellulose, acrylics, and polyvinyls. These products are dissolved in a volatile solvent, applied, and then cured by evaporation of the solvent. This tends to cause shrinkage (about 30%), and stresses arise from the voids left in the joint. These stresses produce a weakening effect in the joint. White household glues are commonly solvent-release polyvinyl acetate resins.

Polychloroprene (neoprene) adhesives are the most common of the thermoplastic adhesives. They are available in liquid form and are generally used with a filler, which adds strength, and a curing agent. These materials are mixed just prior to use. An activator may be added to allow the adhesive to cure at room temperature.

Polyvinyl alcohol adhesives are resealable. They are often used to bond leather, paper, and cloth. Although they are moisture-sensitive, water-soluble forms of polyvinyl adhesives are available.

Polyvinyl acetate adhesives are often used when bonding porous materials to nonporous materials, such as combinations of metals, glass, wood, and other products. They are used in resealable tapes, shipping cartons, and cellophane.

Acrylic adhesives are used in solution to bond acrylic products, such as Lucite or Plexiglas. Ethyl acrylic is especially useful as a pressure-sensitive adhesive due to its flexibility and tackiness.

Hot melt adhesives are a recent adhesive application based on an established principle. Asphalts have been used in hot melt applications for building up roofs for many years. Hot melt adhesives are mixtures of polymers that are heated and applied to one of the surfaces to be bonded. The other surface(s) to be bonded are then pressed into the adhesive while it is still hot. Once applied, the hot melt adhesive wets both surfaces and begins to cool. When cool, the adhesive is set. The softening temperature of the polymers used determines the application range of the hot melt adhesive.

10.2.b.4 Thermosetting Adhesives   Thermosetting adhesives consist of products made of epoxy, melamine, phenolic, and urea resins. When subjected to heat and pressure, partially cured polymers cross-link to form a hard bond. This is the basis for most of the thermosetting adhesives.

Epoxies are among the most expensive of the adhesives discussed. They are chemically curing adhesives that are reliable, high in strength, and abrasion- and water-resistant and that have a low viscosity, low shrinkage, and excellent wetting characteristics, even on difficult surfaces. Due to their low shrinkage rate, they may be used as fillers and adhesives. In contrast to other adhesives, where the bond strength decreases with increasing bond-layer size, epoxy bond strength is independent of the thickness of the bond layer. They are among the best bonding agents, particularly for metal-to-metal joints, and are used to bond glass, ceramics, metals, and wood. They are often used as a two-part adhesive.

Phenolic resins are combined with formaldehyde and partially polymerized to form thermosetting adhesives. With applied heat and pressure, bonding and complete polymerization is accomplished. Phenolic resins are available in powder, liquid, and film forms. The liquid form sometimes requires a catalyst, although most phenolic-based adhesives are made to cure at room temperature. They are used in the production of sandpapers and emery cloth. Phenolformaldehyde and resorcinol-formaldehyde are two types of thermosetting adhesives based on phenolic resins.

Urea-formaldehyde adhesives are not as strong as phenolic adhesives, but they offer a shorter curing period and require lower curing temperatures. They are water-soluble and offer excellent durability under protected conditions. They disintegrate rapidly when exposed to severe moisture conditions, but the addition of starch makes them waterproof.

Melamine-formaldehyde adhesives cure under applied heat and pressure only. They offer better durability than the ureas but less than the phenolics. They are comparatively expensive but are often combined with urea resins for use in plywood construction.

Polymeric alloys are combinations of two or more of the thermoplastic, thermosetting, or elastomeric substances. These alloys produce the highest-quality joints, often at very high costs. Modified epoxies are placed in this category. They are applied as solvents, as materials to be cured under heat and pressure, or as catalytically cured materials at room temperature and pressure.

Cyanoacrylate adhesives, often called super glues, quickly bond surfaces by absorbing the moisture from the adherend materials. Isocyanates, such as toluene

Pressure-sensitive		
Rubber	Polyvinyl ether	Silicone
Hot melt		
Polyamide	Polyester	Polyethylene
Polyproprylene	Polyvinyl acetate	
Two-part Component Adhesives		
Epoxies	Phenolics	Polyester
Polysulfide	Polyurethane	Silicone
Moisture Reactive		
Silicone	Urethane	Cyanoacrylate
Heat-activated		
Polyamide	Epoxy	Polyvinyl acetals
Phenolic	Rubber	Some urethanes
Water- or Solvent-based		
Butyl rubber	Butadiene-styrene	Nitrile rubber
Neoprene	Acrylic	Cellulose ester
Phenolics	Polyamide	Vinyls
Some urethanes		

Figure 10–3
Types and applications of adhesives.

diisocyanate, diphenylmethane diisocyanate, and triphenyl methane triisocyanate, are very reactive and will actively react with such compounds as alcohols and hydroxyls. They react readily with water to form carbon dioxide. This is one reason why they bond skin instantly. A drop or two of these adhesives will quickly and permanently bond materials such as rubber, metals, and other nonporous materials. They are generally not used to join porous materials such as wood and fabric. They exhibit excellent strength and durability but are generally expensive and pose some handling difficulties. Cyanoacrylates are formed from alkylcyanoacetates and formaldehyde in combination, which yields the cyanoacrylate monomer. Thickeners, plasticizers, and inhibitors are added to yield suitable properties for general-purpose adhesive applications. Cyanoacrylate ester is a common type of adhesive. These adhesives exhibit properties and bond strengths that compare with epoxy resins.

Figure 10–3 separates the various types of adhesives into categories based on their unique properties and applications.

10.2.c Properties of Adhesives

Adhesives and their properties cannot be fully understood without a working knowledge of the terminology associated with them. The following are general terms associated with adhesives and their properties and will aid in your understanding of the properties.

Adherend: Base materials that are joined by adhesives.

Adhesive: A general term used for any substance that bonds materials by joining their surfaces. Included in this category are glues, cements, and related products.

Bond strength: The measured resistance of an adhesive to separation of the joined surfaces and a general statement of an adhesive's strength.

Coverage: This term is more often associated with paints and other coatings but is also applied to adhesives. It is designated as the area of adherend surface that a standard unit weight or volume of adhesive will cover in application. Coverage is generally based on the pound or gallon of adhesive.

Cure and set: Set is the conversion of an adhesive from the tacky state into the hardened state. Cure is the set that occurs as a result of polymerization or cross-linking. Catalysts are sometimes used to accelerate curing time.

Density: Related to viscosity and measured with a hydrometer. Density is the weight per volume of a material.

Peel strength: The measured resistance of an adhesive joint to stripping. Many adhesives are strong in shear (exceeding 5000 lb/in.^{2}); however, they are often limited in peel strength. Therefore, most adhesive joints must be protected against peeling. Figure 10–4 illustrates the stressing of joints by shear and peeling.

Pot life: Applies to two-part adhesives only. In a two-part system, the components are mixed prior to use, and the adhesive is applied. Curing or cross-linking of the material begins immediately, and the useful period of the mixture depends

Figure 10–4
Shear and peeling stresses.

on the useful pot life of the mixture. It is generally based on the rate of change in viscosity or bond strength or both.

Pressure-sensitive adhesive: Adhesive that adheres to a surface by applied pressure, such as adhesive tapes.

Storage life: Refers to the amount of time an adhesive can be stored under certain conditions and still be useful. Separation, skinning, and other detrimental effects may occur. These effects may depend on temperature, humidity, and related storage considerations.

Tack: The property of an adhesive that causes the surface coated with an adhesive to form a bond on contact with the other surface to be joined. It is the stickiness of the adhesive.

Viscosity: The resistance to liquid flow of a material. It is measured with a viscometer and is an important factor in assessing the suitability of an adhesive for an application.

When using an adhesive for a lap joint, there is an optimum film thickness above or below which the joint will weaken. Optimum film thicknesses typically vary between 0.002 and 0.005 in., depending on the adhesive used. Generally, it is better to use the least amount of adhesive possible without causing gaps or voids in the joint.

Bond strengths are directly proportional to the width of the lap joint but are independent of the length or overlap. Joint stresses are generally concentrated in the first inch of overlap and, as the length of the lap increases, the unit strength decreases.

Pressure-sensitive adhesive tapes, such as electrical tape, masking tape, and friction tape, rely on their tackiness for application. The adhesives used in these tapes must be permanently tacky so that they can be peeled off without leaving a residue or destroying the adherend material. Pressure-sensitive tapes do not wet the surface and, therefore, cannot be used in high-stress applications. The adhesives used in pressure-sensitive tapes are generally rubber cements with additives to maintain their tackiness. On the top side of the tape, a release coat is applied so that the tape does not stick to itself when unrolled. The peel strength of a pressure-sensitive adhesive is very important in tape applications. The tape must hold firmly but be allowed to peel off the surface without destroying it. A good example of a pressure-sensitive adhesive developed for this purpose is the Post-it note.

10.2.d Sealants and Caulking Compounds

Sealants are a special type of adhesive. Sealants and caulking compounds are used to close gaps and weatherproof joints in buildings and structures. They must remain pliable for years under varying environmental conditions, remain elastic over time, adapt to varying hot and cold weather conditions, bond to a wide variety of materials, and resist attack by various elements, including weather, water, mold, fungi, and other natural elements.

Polymer	Advantages	Disadvantages	Uses
Acrylic	U-V stability	Heat resistance	Glass, plastic
Cellulosics	Solubility	Cost, moisture	Leather, paper, wood
Cyanoacrylate	Fast-setting; bonds hard-to-bond plastics	Cost	"Super" glues
Epoxies	Versatility; strength; solvent resistance	Peel strength, cost	High-strength bonding
Olefins	Flexibility; ease of handling; cost	Creep, heat resistance	Laminations, book binding, packaging
Phenolics	Cost; heat and weather resistance	Brittleness	Plywood; abrasive wheels
Polyamides	Flexibility; oil and water resistance	Cost	Metal can seams; hot melt
Polyesters	Flexibility	Cost, limited adhesion	Foils, plastics, shoes
Polysulfides	Weatherability; temperature range	Strength	Elastomeric adhesives, sealants
Polyurethane	Versatility	Care in handling; heat resistance	Wide range of materials
Rubber	High tackiness	Strength	Pressure-sensitive tapes, general-purpose adhesives
Silicone	Temperature range	Cost, strength	Sealants, pressure-sensitive tapes
Urea	Cost, curing temperature	Moisture sensitivity	Plywood, furniture
Vinyls	Versatility	Moisture sensitivity	Household glues, furniture, shoes

Figure 10–5
Adhesive and sealant polymers.

Contemporary sealants include polysulfides, silicones, polyurethanes, butyl, polybutene, polyisobutene, acrylic rubbers, and polychloroprene. These products are designed to last from 10 to 25 years in service. Virtually all of the elastomers can be used as caulking compounds. The polysulfide (Thiokol), acrylic, silicone, butyl, and chloroprene compounds exhibit the best weathering characteristics. Polysulfides, urethanes, and silicones give the best low-temperature performance.

When using a sealant, the maximum expected movement of the joint or gap is considered. Many superior grades of sealants withstand 50% elongation over extended periods of time. Beyond 50%, wider sealant fillets or wider initial gaps must be used. Elongations of 100% or more will develop a corresponding stress in the joint, which tends to pull the sealant from the substrate.

Figure 10–5 lists various polymers that are used as adhesives and sealants and the specific advantages, disadvantages, and uses of each.

10.3 COATINGS

10.3.a Introduction

Most materials can be coated with one of the various types of organic coatings: paints, lacquers, varnishes, and enamels. These organic coatings use oils and synthetic resins as liquid binders or vehicles that typically harden by cross-linking. This action produces a hard, solid, thermosetting film.

Paints, dyes, and stains were used primarily for aesthetic purposes until the Industrial Revolution. The use of metals to replace wood necessitated the use of paint to preserve and protect the metal machinery and structures from corrosion. With the introduction and growth of polymer technology about 50 years ago, many quality paints and adhesives were developed. The basic ingredient of most organic coatings is a polymer or mixture of polymers, the principal functions of which are as follows:

- Protect materials from corrosion, exposure, and weathering
- Improve visibility through luminescence or reflectivity
- Provide electrical, thermal, or acoustic insulation
- Improve appearance through decorative effects
- Enhance marketability and advertise the product
- Improve safety (warning)
- Inform users (traffic lanes, road signs, and so on)
- Camouflage or hide certain features (hunter's clothing)
- Identify persons or products (color coding)

Organic coatings may contain the following: (1) a binder or vehicle; (2) coloring agents or pigments; (3) solvents and additives to control viscosity; and (4) additives to act as inhibitors, stabilizers, or thickeners and to alter physical or chemical properties. This section deals with the formulation of these coatings, the functions of their ingredients, and the uses of these coatings.

10.3.b Types of Coatings

Different types of finishes or coatings are required for different applications. There are different types of paints, varnishes, enamels, and lacquers for each application. The primary difference among these are the vehicles used. *Paints* are a mixture of a binder and pigment that typically hardens by cross-linking. *Enamels* are a blend of a paint and a varnish. They harden through cross-linking, where the agent is oxygen available in the environment. *Lacquers* contain a binder dissolved in a solvent that dries by evaporation of the solvent to produce a thermoplastic finish. *Varnishes* form a thermosetting film but contain no pigment.

Organic coatings contain a vehicle and frequently contain pigments and driers as well. The finely ground particles enclosed in a coating are technically known as the *pigment.* These particles may or may not be colored. The liquid portion of the coating is called the *vehicle.* The vehicle contains a binder material, which forms a film and holds the pigment. The vehicle also contains a thinner, which makes application easier. The binder is typically a liquid polymer that coats the surface and holds the pigments and additives during application. The binder causes the coating to adhere to itself as well as the surface to which it is applied. Binders include natural oils, polymers, and various types of resins.

Pigments are used to alter the physical and sometimes chemical properties of the film. Pigments are used to control color, rust properties, gloss, and smoothness of finish. Red and blue lead pigments and zinc and lead chromates are used as rust inhibitors. Pigments are finely divided particles in the binder that do not dissolve in the solvent. The ratio of pigment to vehicle and the pigment particle sizes determine whether a paint is flat or glossy. Pigments are also used to hide imperfections or absorb light. The hiding power of a paint depends on its ability to absorb light. They do react with chemicals in the environment and decompose with time and exposure to the elements. Paint fading can be observed in older cars that have not been properly cared for.

Vegetable oils such as tung oil and linseed oil are common vehicles that are composed of unsaturated fatty acids. These acids combine with the oxygen in the air to form polymers through addition polymerization. Deterioration of the paint occurs through continued oxidation and eventual fragmentation of the polymeric molecules.

Resins are another type of vehicle. With the exception of shellac, which is of animal origin, all the useful natural resins are of vegetable origin. Natural resins are unsaturated acidic substances capable of polymerization, oxidation, and reaction with basic organic substances. Synthetic resins are also used, sometimes alone and sometimes in addition to natural resins. Excluding shellacs and lacquers, the curing of the resins takes place through oxidation and/or polymerization. Shellacs and lacquers cure through the loss or evaporation of the solvent.

There are three types of solvents used in binders: solvents, diluents, and thinners. *Active solvents* are used to dissolve the binder and lower the viscosity of the coating. *Diluents* are used to weaken the active solvent and increase the overall amount of coating. *Thinners* are used to increase the amount of coating material without reducing the effectiveness of the solvent. Suitable thinners commonly used for organic coatings are alcohols, petroleum solvents, coal-tar derivatives, and turpentine.

In addition to solvents, diluents, and thinners, there are a number of additives that can be used to enhance or reduce certain properties of the coating. These additives are used in small quantities to produce specific characteristics in the coating. They will affect the chemical as well as the physical properties of the material. The general categories of additives are antiskin agents; biocides and fungicides; catalysts; coalescing agents; defoamers; driers; extenders; flow modifiers; freeze-and-thaw stabilizers; pigments; plasticizers; stabilizers, and thickeners.

Antiskin agents are used to prevent the coating from forming a skin on top of the material during storage. This skin forms on top of the coating when it is left in storage for a long period of time. Antiskin agents rise to the top of the material during storage and prevent skins from forming.

Biocides and fungicides inhibit the growth of microorganisms in coatings. Microorganisms thrive in the presence of water and oxygen at temperatures between 72° and 100°F (22° and 38°C). These microorganisms tend to break down the polymer chains during storage. Biocides inhibit growth under moist conditions, whereas fungicides inhibit growth during dry conditions.

Catalysts are used to aid cross-linking of polymers. They reduce the energy required to cross-link. Coalescing agents are used to enhance the ability of coatings to coalesce (combine or merge) after the vehicle (water) has evaporated from water-soluble coating materials.

Defoamers are used to reduce the tendency of a coating to foam by increasing the surface tension of the coating material. Defoamers are used because, without them, the coating tends to form bubbles and imperfections, which leave craters in the finish.

Driers are usually soapy materials made by replacing the hydrogen in some organic materials with lead, manganese, cobalt, iron, zinc, or calcium. These act as catalysts for polymerization and/or oxidation. They are not useful in organic coatings, which dry through evaporation.

Extenders control the gloss and adjust the viscosity of the coating. Some common extenders are barium sulfate, diatomaceous silica, gypsum, kaolin, and calcium carbonate. Calcium carbonate is used to enhance glossiness, whereas diatomaceous silica is used to reduce the glossiness of a coating.

Flow modifiers are used to control the viscosity of the coating material. They are used to aid in brush, roller, and similar applications, where the material should be easy to spread but viscous enough to stay where it is applied without running.

Freeze-and-thaw stabilizers are applied in water-based materials, where freezing will render the coating useless. These stabilizers reduce the freezing temperature of the water in the coating to prevent freezing.

Pigments are added to produce the desired color. They may also be used in combination with powders and flakes to produce various effects, such as metal flake and metallic colors. Figure 10–6 is a brief list of pigment materials commonly used in coatings.

Figure 10–6
Typical pigments and extenders used in industrial coatings.

White	**Red**
Titanium oxide	Cadmium
Zinc oxide	Iron oxides
White lead-carbonate	Toluidine red
White lead-sulfate	Phthalocyanine red
Yellow and Orange	**Green**
Lead chromate yellow	Chrome green
Toluidine yellow	Chrome oxide
Chrome orange	Phthalocyanine green
Zinc chromate yellow	**Metallic and Metal Flake**
Blue	Aluminum powder
Iron	Lead powder and flake
Ultramarine	Nickel flake
Phthalocyanine blue	Stainless steel
Extenders	
Magnesium silicate	
Calcium silicate	
Calcium carbonate	
Barium sulfate	
Aluminum silicate	

Plasticizers increase the flexibility or plasticity of a coating material, thus allowing for expansion and contraction in service. Stabilizers, when added to coating materials, help prevent damage by heat, chemical action, and ultraviolet rays. They are often added to reduce the effects of particular environmental or service conditions, such as those listed. Thickeners are used in emulsion-type coatings to increase the viscosity of the coating and thus to provide an even flow during application.

As mentioned previously, paints are a mixture of a vehicle and pigment. The binder in the vehicle provides the hardened surface, whereas the pigment provides the color. Water-based paints contain a water emulsion as the vehicle. Drying time for a paint is based on the evaporation rate of the solvent or the rate of oxidation of the vehicle. Sprayed paint takes less time to dry than paint that is brushed or rolled.

Paint is a term often used in place of enamel. Originally, the term referred to pigmented coatings in which the binder was a drying oil, such as linseed oil. Paints were originally based on natural unsaturated oils, such as linseed oil and tung oil. These oils cross-link with oxygen in the air to harden. Today's paints contain little or no natural oils.

Water-based paints use water as a thinner. These paints are generally odorless, are less flammable, are quick-drying, are easy to clean up, and are for interior use. The binder for these paints is generally an oil and resin combination. These paints are subject to freezing when exposed to low temperatures and spoilage when exposed to higher temperatures.

Water-based paints come in two classes: oil/water emulsions and latex emulsions. The *oil/water emulsion* is a suspension of oil-resin binders in a water phase. Latex paints are water-based paints. Similar to aerosol paints, where a liquid is dispersed in a gas, a *latex emulsion* is a suspension of fine particles of spherical organic resin in water. When latex paints are applied, the water evaporates, leaving the resin particles to form a film coating. Latex paints form a film through coagulation of the particles. Latex applications provide low odor and low fire hazard, use water as a thinner, and are easy to clean up with water. The resins used in latex paints are typically acrylics, butadiene-styrene, and vinyls. Latex paints are used extensively for interior and exterior house paints. The two most widely used polymers are polyvinyl acetate and acrylics. Polyvinyl acetate latexes are cheaper than acrylics and can "breathe," thus reducing the tendency of the paint to blister.

A typical composition for a commonly used house paint is as follows:

Typical Composition of White Exterior House Paint

Pigment by weight	60%
Vehicle by weight	40%

Pigment composition by weight

Titanium dioxide	15%
35% leaded zinc oxide	50%
Magnesium silicate	35%

Vehicle composition by weight

Linseed oil	67%
Mineral spirits	33%

Latex is a dispersion of a polymer in water. Organisols and plastisols are two special types of dispersions. Organisols are polymer particles dispersed in an organic solvent, and plastisols are polymer particles dispersed in a plasticizer.

Varnish is similar to paint without the pigments; also, a synthetic or natural resin is added to the binder. Natural resins include batu, copal, hauri, and manila gums and pitches. Synthetic resins include those mentioned earlier: alkyds, epoxies, melamines, phenols, silicones, and ureas. These resins can be used individually or in combination to produce the desired physical and chemical effects. When synthetic resins are used, the coating is termed an *alkyd paint*. Alkyds are primarily polyesters. The drying oil, such as castor, corn, cotton, fish, or soybean oil, is partially oxidized and polymerized at high temperatures. The resin is then dissolved in the oil, and the solvent is added. The drier is sometimes added during different stages of this process. Varnish is an "oil-based" paint in that the coating is formed by oxidation of the oil and by evaporation of the solvent. Oxygen from the air reacts with the oil vehicle to induce cross-linking of the polymer. The gloss of a varnish depends on the oil-to-resin ratio used. Oil-based paints are used primarily for exterior wood and interior decorative finishes.

Shellac, often confused with varnish, is a natural resin produced from the secretion of the lac bug. This natural resin is thinned with alcohol and used as a coating, which can later be removed by alcohol.

Enamels are mixtures of varnish and colorizing agents, having vehicles similar to varnishes but with the addition of a pigment. Synthetic resin-based enamels are replacing the oil-based varieties. Synthetic enamels harden by polymerization to produce a very smooth surface finish. Enamels can be allowed to air dry or can be baked on, depending on the application.

There are several types of enamels currently available, including acrylic, alkyd, epoxy, phenolic, silicone, urethane, and vinyl. Acrylic enamels offer good color retention, even at high temperatures. Alkyd enamels also offer good color and gloss retention and are resistant to moisture and chemical attack in a variety of applications. Epoxy enamels generally require baking and offer flexibility and resistance to chemical attack. Phenolic enamels are brittle coatings, but they offer resistance to oil, water, and chemicals. Straight silicone enamels offer excellent heat resistance. Silicone can also be used in combination with other resins to increase the heat resistance of the enamel. Urethane enamels are abrasion-resistant. Vinyl enamels are moisture-resistant. A typical composition for an enamel is as follows:

Typical Composition of Flat Black Enamel by Weight

Pigment	21%
Vehicle	79%

Pigment composition

Carbon black	24%
Silica and silicates	76%

Vehicle composition

Resin	10%
Vegetable oils	12%
Mineral spirits	78%

Lacquers are generally solutions of cellulosic resins in volatile organic solvents. They are high-molecular-weight polymers that remain on the surface after the solvent has evaporated. Plasticizers and other resins can be added to improve film quality and various properties. Acrylic resins added to lacquers increase the water and chemical resistance. Butadiene-styrene also increases the water and chemical resistance and increases the adhesive properties. Cellulose acetate additives increase the heat resistance of the lacquer. Vinyl resins increase the abrasion and oil resistance. Cellulose nitrate is the polymer used in many high-quality automotive lacquers.

There are a number of resins used today in paints. These include the following:

Acrylonitrile-butadiene
Alkyds: amino, phenolic, styrenated
Amino
Bituminous
Butadiene-styrene

Cellulose nitrate
Cellulose butyrate
Chlorinated polyester
Chlorinated rubber
Epoxies: amines, phenolics, polyimides
Ethyl cellulose
Neoprene rubber
Phenolics
Polyimides
Polyesters
Silicone
Urethane
Vinyls: acetate, butyral, chloride

Alkyd resins are utilized in many synthetic finishes. Many materials can be used in conjunction with the alkyd molecule, such as alcohols, various oils, and many fatty acids. Alkyd resins also improve the appearance of other coatings and are often blended with butadiene-styrene, acrylics, phenolics, and polyvinyl acetate. This mixture enhances the adhesion, flexibility, and water resistance of the finish. Many aerosol paints are alkyds mixed with drying oils. They are good general-purpose paints. Depending on the percentage of oil (usually tung or linseed oil), alkyd-resin coatings with a higher percentage of oil (roughly 60%) are used in marine and architectural applications, whereas alkyd-resin coatings with a lower percentage of oil (roughly 40%) are used as appliance and furniture coatings. The amount and duration of ultraviolet exposure determines the loss of appearance and protection of alkyd-resin coatings.

Similar to the alkyd- and epoxy-resin coatings, acrylic-resin coatings also are applied in automotive and appliance finishes because they retain their gloss over time. They have good hiding qualities and offer satisfactory protection at a nominal price.

Bituminous paints or varnishes are produced by dissolving coal tar in mineral spirits. They are often used as waterproofing materials. They have low porosity, which makes them an excellent choice for waterproofing, but they shrink and crack when exposed to sunlight, even for short periods of time.

Epoxy resins adhere well to most surfaces and are resistant to most solvents and cleaning fluids. In addition, they do not require a primer, are resistant to abrasion, offer good flexibility, and have working temperatures up to 300°F. They are widely used for machine and industrial equipment coatings.

Cellulose nitrate is used primarily in quick-drying lacquers. Ethyl cellulose has the best flexibility with good resistance to acid and alkali attack. Cellulose acetate butyrate offers the best resistance to hydrocarbons.

Phenolic coatings are durable with good chemical and temperature resistance. They are available as two-part catalyst or oxidation-drying varnishes. They are used as tank linings and in similar chemical-resistant applications.

Overall, urethane enamels provide superior surface coatings. They provide excellent bonding, abrasion resistance, and resistance to moisture and chemical attack. However, all urethane coatings degrade when exposed to the ultraviolet radiation in sunlight. Polyurethane coatings can cure by catalyst in a two-part application, by chemical reaction with moisture in the air, and by oxidation with urethane oils and alkyds. The two-part system is often used in industrial coatings and high-quality finishes. Chemical reaction and oxidation are commonly used for clear wood finishes. Polyurethane coatings dry to a hard, long-lasting, resilient finish.

Vinyls—in the form of polyvinyl chloride, a copolymer, PVC and vinyl acetate, or vinyl butyrate—are generally used in coatings to provide chemical resistance. Vinyl paints are generally lacquers used on outdoor structures. They are also available in latex house paints. A common use for vinyls is in organisol and plastisol coatings of fixtures, tools, and other similar products.

10.3.c Inorganic Coatings

Most of the coatings previously discussed are applied to the surface of a product and adhere to the surface. Inorganic coatings or films react with the surface to which they are applied and become a part of that surface. Five basic types of inorganic films are anodizing, chromizing, siliconizing, and oxide and phosphate coatings.

The *anodizing process* is used to convert the surface of a product to an oxide. The product, composed of a metal, is immersed in a bath of electrolyte such as boric acid, chromic acid, oxalic acid, or sulfuric acid. Sulfuric acid is the most common. Oxalic and boric acids are used to produce special effects on the metal surface. Oxalic acid produces a yellowish tint, whereas boric acid is used largely in the production of capacitor dielectric foils (anodized film has a very high dielectric strength). Sulfuric acid is typically used at a concentration between 15% and 25%.

When anodizing a metal product, the part is lowered into a lead-lined tank containing the anodizing solution. The lead lining in the tank acts as the cathode (−) for the reaction. The solution becomes the anode (+). An electric current is applied at voltages of 12 to 24 V. The surface of the part is typically oxidized to thicknesses up to approximately 0.005 in. (0.13 mm). Thicknesses up to 0.010 in. (0.25 mm) can be reached at lower temperatures. There is a practical limit to the film thickness. As the thickness increases (from the inside out), the electrolyte tends to dissolve the film already produced. This reaction increases as the film thickness increases, until they reach an equilibrium. This equilibrium rate is the practical film thickness for that temperature. Other variables that affect the properties of the anodic film produced are the alloy being anodized, the type of electrolyte, the voltage used, the temperature of the electrolyte bath, the concentration of the electrolyte, and anodizing time. The thickness of the anodic film is controlled chiefly by the anodizing time. Typical anodizing time is 30 min. The film color can also be altered by chemical dyes to produce certain color effects. The anodizing process is chiefly used on aluminum, magnesium, steel, and zinc products.

Chromizing is another process used to deposit an inorganic film coating. It involves heating a ferrous metal part in a closed container in which chromium and hydrogen are present. The metal part absorbs approximately 30% of the chromium to depths of 0.0005 to 0.005 in. (0.013 to 0.13 mm). Chromizing produces a corrosion-, abrasion-, and mildly acid-resistant surface. *Siliconizing* is basically the same process as chromizing, but it is done in the presence of silicone carbide and chlorine. It typically penetrates to a depth of 0.100 in. (2.5 mm).

Oxide coatings are used as protection against corrosion. They are produced by exposing surfaces to oxidizing gases or solutions under elevated temperatures. The gases or solutions oxidize the surface of the metal part. Common oxide coatings are black oxide, dichromate coatings, and gun-metal finishes. *Phosphate coatings* are produced by dipping or spraying the metal part with phosphoric acid and manganese oxide. The manganese produces a green color in the coating. Phosphate coatings are generally used on ferrous metal, aluminum, tin, and zinc parts. Commercially, this process is known as *Parkerizing*.

10.3.d Electroplating

In comparison to organic finishes, which adhere to the surface, and inorganic film coatings, which become part of the surface by reaction, electroplating is an electrochemical process used to plate materials. Plating is done by immersing a current-carrying material (a metallic object, generally) in a solution that contains the plating material. In this arrangement, the object becomes the cathode (–), and the solution or another electrode becomes the anode (+). Once current is applied, the plating material is deposited on the cathode. Materials that are plated are generally made of cadmium, chromium, copper, lead, nickel, tin, zinc, and the precious metals, such as gold and silver. Figure 10–7 illustrates the electroplating process.

Electroplating is generally used to increase the appearance of objects, provide protection against corrosion, increase the abrasion- or wear-resistance of a part, and increase the dimensions of worn parts. It can be performed on any of the common metals.

Nickel plating is used on steel as a protective coating against corrosion. It can also be used to plate aluminum, copper, and steel for decoration as well as corrosion protection. Also included in this application are lead, cadmium, and zinc. Cadmium is often used on small pieces, but it is not used on food-processing items because it is toxic.

Figure 10–7
Electroplating.

Chromium is available for plating as commercial, or hard, chromium. Commercial chromium is used as a coating over pieces that have already been coated with copper or nickel. Nickel tarnishes over time, but a coating of chromium provides the best protection and decoration. Hard chromium is used on steel and aluminum products to provide a hard, corrosion-resistant surface.

Copper and brass are also used as undercoatings for parts that will later be coated with chromium or gold. Brass is often plated and polished for decorative effect as well as for the protection it offers. It may be covered with an organic finish to protect against tarnish.

Tin is used to plate such items as food cans and cooking utensils because it is nontoxic and will not tarnish or corrode. It is sometimes combined with copper, nickel, zinc, or a plating material.

The precious metals are chiefly used for decorative purposes. They can also be used for protection against chemical attack and oxidation in electronics applications, as an example. They can be applied in very thin layers without decreasing their advantages in terms of protection and appearance.

10.3.e Dipping, Plating, and Spraying Coatings

Coatings such as aluminum, lead, silicates, tin, zinc, and some polymers are applied to surfaces using dipping, plating, spraying, and fluidized bed processes.

Cold dipping is a process by which the plating material displaces the base metal on the surface of a part when it is dipped into a bath of plating material. This result is possible when the electrochemical potential of the base metal is higher than the material used to plate the part. For example, nickel is often used to cold-plate steel and aluminum parts.

Hot dipping is also a process where a metal is dipped into a bath of plating material. The part that is dipped must have a higher melting point than the plating material. For example, steel or brass parts can be hot-dipped in baths of aluminum, lead, or tin. Galvanized steel is a zinc-coated steel produced by the hot-dip process or by electroplating. The zinc coating protects the base steel from corrosion. Under constant wet conditions, the zinc coating turns white as it oxidizes. This is known as *white rusting*.

Hard surfacing is a process used to weld metal to the surface of a part. Parts are generally made of high-alloy carbon steel or carbide, cobalt, boron, and/or nickel alloys. Surfacing is used to build up worn parts or to increase the abrasion and corrosion properties of the part. Once the buildup of the surface reaches more than 3% of the base metal, the process is called *cladding*. Clad steels often contain a mild steel core, which provides strength and ductility, clad with another metal that provides superior abrasion or corrosion resistance. For example, stainless steel cookware is often a clad steel, where a mild steel core is clad with stainless steel. Copper is another metal often used to clad steel surfaces, such as electrical wires and conductors.

Fluidized bed-coating processes involve immersing heated parts into a tank filled with the coating material, typically a thermoplastic polymer such as PVC. The polymer coating is in a granular form and can be colorized. Pressurized air is blown in the container of coating material to keep the coating stirred during application. The part is heated prior to application and is then lowered into the container full of coating, and the air is applied. The air keeps the coating stirred up to help ensure an even coating. Any material that can be heated can be coated with the fluidized bed process.

Metal spraying is performed by blowing an atomized metal against a surface to be coated. As the sprayed particles hit the surface, they flatten out into flakes. These flakes lock into the irregularities in the surface of the part. The surfaces to be coated are often sandblasted or gritblasted prior to spraying to roughen the surface and provide better bonding. The depth of the sprayed surface does not usually exceed 0.005 in. (0.13 mm). Metal-sprayed parts can be fired after spraying to diffuse the flakes and produce a better finish. Any material that can be melted or atomized can be sprayed.

Plastisol and organisol dipping involve immersing the part to be coated in a bath of coating compound. This process is used to coat such items as tool handles and kitchen utensils. The part may be heated and then dipped into the thermoplastic (PVC) coating. The longer the part remains in the bath, the greater the buildup of the coating.

Vacuum plating is a process used to deposit thin vaporized metal films onto base surfaces under a high vacuum. Under a high vacuum, these film deposits have high vapor pressures and low boiling points. Capacitors are produced using vacuum plating. The zinc or aluminum dielectric is vacuum-deposited onto paper surfaces. Vacuum deposition is also used for coating plastic parts with metal.

Vitreous enameling involves fusing a glass coating onto a base metal surface. Refractories, fluxes, and opacifiers are mixed and melted in a furnace. The mixture is then cooled rapidly, producing glass particles. Pigments, chemicals, and certain clay materials are then added to the glass for application. Once the coating is applied to the part, it is fired to fuse it with the base material and produce a glassy surface. This process applies a thin coating to the base. Additional coating depth is accomplished by spraying. This glassy coating is resistant to heat, abrasion, and chemical attack. Products that are vitreous-enameled include refrigerators, ovens, bathtubs, and many kitchen utensils.

10.4 SUMMARY

Adhesives must be formed of molecules that have a greater attraction for molecules of adherend materials than they do for other adhesive molecules. The surfaces to be joined must be clean and wetted thoroughly by the adhesive. Many polymeric substances exhibit excellent adhesive properties; among these substances are epoxy resins.

Paints, lacquers, enamels, and varnishes are used principally to protect the materials to be coated from the environment and to enhance their appearance. The coatings contain a vehicle that is a film-forming agent and a diluent. The binder, or film-forming agent, includes unsaturated compounds, soluble resins, latex, and water dispersions of oil-resin combinations. It may also contain dyes or pigments. Pigments can be used to enhance hiding power and to improve the physical and chemical properties of the coating. Catalysts can be used to promote polymerization or oxidation and are referred to as driers.

Paint is a rather broad term. It specifically refers to a drying oil, thinner, pigment, and drier. Varnish consists of a boiled drying oil, resin, thinner, and drier. Enamels are paints containing varnish. Water-based paints are oil-resin emulsions or latex for which water is the thinner. Lacquers are solutions of cellulosics and other resins with plasticizers in volatile organic solvents. Shellacs consist of a solution of a natural animal resin in alcohol.

Organic coatings prevent air, moisture, and other materials from reaching the base material to which they are applied. They provide good protection against various forms of attack by adhering to the surface of the base material. Inorganic coatings include many polymeric enamels and resins. These coatings also protect the base surface from attack by becoming part of the surface through chemical reaction or polymerization. Aluminum, cadmium, chromium, copper, lead, nickel, tin, and zinc are among the inorganic coatings used to protect base surfaces.

Coatings are applied through various processes. These processes include direct applications by brush or spray, dipping, electroplating, use of a fluidized bed, and other similar processes. The primary purpose of any coating is to beautify and/or protect the base surface.

QUESTIONS AND PROBLEMS

1. Define the following terms:
   a. Adhesive
   b. Adherend
   c. Cure
   d. Set
   e. Tack
   f. Peel strength
2. Define the term *wettability*.
3. Describe what happens when an adhesive "wets" an adherend surface.
4. Describe a pressure-sensitive adhesive.
5. Describe the effects that thickness, length, and width have on bond strength in lap joints.
6. Define the following terms:
   a. Glue
   b. Cement
   c. Paste
7. List and describe the six types of glues identified as natural adhesives.
8. List and describe the three types of cements described in this chapter.
9. List and describe the five types of thermoplastic adhesives described in this chapter.

10. List and describe the six types of thermosetting adhesives described in this chapter.
11. List the eight basic materials used in sealants and caulking compounds.
12. List the four basic ingredients in an organic coating.
13. List five basic functions of a coating material.
14. List five of the seven types of enamels described in this chapter.
15. Define the following terms:
   a. Paint
   b. Enamel
   c. Lacquer
   d. Varnish
16. List three types of solvents discussed in this chapter.
17. Define the following terms:
   a. Solvents
   b. Diluents
   c. Thinners
18. Describe latex paint. How does latex "dry"?
19. Describe the inorganic coatings described in this chapter.
20. What is the difference between organic and inorganic coatings?

21. Define the following terms:
    a. Anodizing
    b. Chromizing
    c. Oxide coatings
    d. Siliconizing
22. Describe and illustrate the electroplating process.
23. Describe the following processes:
    a. Hot dipping
    b. Cold dipping
    c. Cladding
    d. Metal spraying
    e. Fluidized bed coating
    f. Galvanizing
    g. Plastisol coating
24. Look around you and describe five applications each for adhesives and coatings.
25. Write a specification for a plywood adhesive. Be as specific as possible.
26. Describe the difference between adhesion and cohesion.
27. Describe the purposes of paints and applied coatings.
28. Referring to your list of purposes, write a specification for the coating you described.
29. List five applications for inorganic coatings, such as platings, enamelings, oxides, and other such coatings.

CHAPTER 11

Fuels and Lubricants

Objectives • Recognize common types of fuels and lubricants.
 • List the common properties of fuels and lubricants.
 • Describe various manufacturing processes related to the discovery and refinement of fuels and lubricants.
 • Recognize the uses of common fuels and lubricants.

11.1 INTRODUCTION

This chapter provides a basic explanation of the nature, uses, and limitations of commonly used fossil fuels and petroleum-based lubricants.

A fuel is any substance or substances that, when combined with another substance, release useful energy. This includes all substances that are involved in *exothermic reactions*. Exothermic reactions are reactions that produce energy in the form of light, heat, and/or electricity. Reactions that absorb heat from the environment during reactions are termed *endothermic reactions*. Substances react differently, depending on their atomic weight, concentration present, and the presence or absence of a *catalyst*. A catalyst is any substance that aids a reaction while not being consumed in the reaction.

Reactions that produce useful energy from fuels are generally *oxidation* reactions. In the oxidation reactions in which we are interested, the fuel is oxidized. Some of the more common fuels used are wood, coal, coke, naphtha, gasoline, kerosene, fuel oils, and natural gas. In addition, by-products or refinements of these and other products are used as fuels. These by-products include paper, coal gas, ethanol (grain alcohol), and liquefied petroleum gas.

Petroleum derives its name from the Greek word *petra*, meaning *rock*, and the Latin word *oleum*, meaning *oil*. Petroleum is produced by high temperatures and pressures, which act on decomposed organic material (plants and animals) over very long periods of time. These petroleum pockets are found in rock layers that form the upper strata or crust of the earth.

It is not uncommon for petroleum to seep to the surface. The Egyptians noticed this petroleum seepage and used the black, sticky substance for embalming practices. Other people have used petroleum differently. For instance, in 1814 Samuel Kier, a Pittsburgh apothecary, collected the seepage and sold it as a remedy for most afflictions. In 1852, he began distilling it for use as a lamp fuel.

The first petroleum company was formed in 1854 in Titusville, Pennsylvania, by George Bissel. Bissel dug into the ground, hoping that the seepage would be adequate to fulfill his need. This method of collecting seepage didn't work then and doesn't work today either.

In 1859, the first drill rig was used to drill for oil. This well produced 20 barrels of petroleum a day. As a result, petroleum distillates were used as substitutes for the animal and vegetable oils used at that time to lubricate machinery.

The oils used for lubrication remain after all the lighter products have been removed through the distillation process. Petroleum samples from different sources around the world differ in physical appearance, chemical properties, and performance, and not all petroleum deposits are suitable for use as fuels or lubricants. Consideration should be given to the amount of investment in both time and money needed to produce a useful fuel or lubricant from the raw materials as opposed to the amount of useful energy that can be gained from its utilization. In addition, the consequences on the environment, future societies, and international dependence from economic and military standpoints must be recognized and reconciled.

11.2 TYPES AND PROPERTIES OF FUELS

In prehistoric times, people were limited to tasks that could be performed by hand, using only muscle power and the simple tools that were available. Transportation was limited to walking or to riding the few available domesticated animals. Even though they worked very hard, the fruit of people's labors was just enough to feed, cloth, and shelter them.

The first fuel used by humans must have been the wood used to build the first fire. As humans learned to control fire, life became more comfortable and more productive. Originally, fire would have been used for heating and cooking. It would be many years before people learned to harness the energy released by fire. Once the combustion process was discovered, people began using it to improve their lives and their lifestyles. They used wood and coal to run steam engines, gasoline and diesel fuels to power automobiles, and natural gas to heat their homes.

In general, fuels can be separated into three categories: solid fuels, liquid fuels, and gaseous fuels. Three fuels commonly used today that fit these categories are coal, oil, and gas.

11.2.a Combustion and Burning

Antoine Lavoisier, a French chemist, discovered in 1776 that the oxygen in the air enables substances to burn. When a substance burns, the oxygen in the air combines with the fuel. This combination process is known as *oxidation.* When the hydrocarbons in the fuel mix with the oxygen in the air, the by-products of the reaction are carbon dioxide (CO_2) and water (H_2O). The oxidation process produces heat energy, although sometimes very slowly, as in the case of the oxidation of iron or rust. Rust is simply the oxidation of the iron in ferrous products during which the heat energy developed dissipates slowly.

For fuels to burn, they must first be at or be raised to a temperature that breaks the molecular bonds holding the complex molecules together, forming simpler molecules such as CO_2 and H_2O. As the bonds are broken, energy is released in the forms of heat and light. The more energy released, the better the fuel. The approximate molecular weights for several elements and compounds often found in combustion equations are shown in Table 11–1.

For the elements carbon (C) and hydrogen (H), the complete combustion equations are

C	$+O_2$	$=CO_2+175{,}000$ Btu
12 lb	32 lb	44 lb
$2H_2$	$+O_2$	$=2H_2O+246{,}000$ Btu
4 lb	32 lb	36 lb

For CH_4 (methane) the combustion equation is

$$CH_4 + (x)O_2 = (y)CO_2 + (z)H_2O$$

Table 11–1
Approximate molecular weights for various substances used as fuels.

Gas		Molecular Weight (Approx., in grams)	BTU/ft³ (Standard Temperature and Pressure)
H_2	(Hydrogen)	2	325
O_2	(Oxygen)	32	—
N_2	(Nitrogen)	28	—
CO	(Carbon monoxide)	28	323
CO_2	(Carbon dioxide)	44	—
H_2O	(Water)	18	—
CH_4	(Methane)	16	1,010
C_2H_4	(Ethylene)	28	1,640
C_2H_6	(Ethane)	30	1,760
C_3H_8	(Propane)	44	2,500
C_4H_{10}	(Butane)	58	3,275
C_8H_{18}	(Octane)	114	6,300
CH_4O	(Methanol)	32	—
C_2H_6O	(Ethanol)	46	—

where x, y, and z are constants representing the number of molecules required for complete combustion. To balance both sides of the previous equation requires that $x = 2$, $y = 1$, and $z = 2$. Therefore, for the complete combustion of one molecule of methane to take place, two molecules of oxygen must be present. The products of the complete combustion of methane are one molecule of carbon dioxide and two molecules of water.

This combustion equation shows that 1 ft³ of methane requires 2 ft³ of oxygen for complete combustion. The resulting reaction from the combination of methane and oxygen produces 1 ft³ of carbon dioxide and 2 ft³ of water. Assigning approximate weights to these components results in the following: 1 lb (454 g) of methane combined with (2·32/16, or 4, lb (1816 g) of oxygen produces 44/16, or 2.75, lb (1248 g) of carbon dioxide and 36/16, or 2.25, lb (1022 g) of water.

We know from the previous discussion that oxygen is required for combustion, but how much air is required to provide that much oxygen? The air around us is composed of approximately 78% nitrogen and 20% oxygen by volume. The remaining percentage is made up of several trace gases. The volume of air required for complete combustion of methane is the quantity of oxygen required (as found in the previous equation) divided by the percentage of oxygen in the air. For methane (CH_4), this number is (4 lb)/0.20, or 20 lb (9080 g). The volume of air required is the volume of oxygen required divided by the percentage of oxygen in the air, or (2 ft³)/0.20 = 10 ft³. These values are minimum figures—the minimum amount of oxygen required for complete combustion to occur.

One quantity associated with fuels and combustion is heat. Heat cannot be measured directly; it must be measured indirectly by its effects. The device most commonly used to measure temperature is the thermometer, graduated in degrees Fahrenheit or degrees Celsius. However, the unit most commonly used to describe heat energy is the British thermal unit (Btu). One Btu is the heat energy required to raise the temperature of 1 lb of water 1°F.

The heat energy value or heating value for solid fuels such as coal is measured in Btu per pound (gram). The heat energy value released by liquid fuels such as gasoline is measured in Btu per gallon (liter). Finally, the heat energy value obtained from gaseous fuels such as natural gas is measured in Btu per cubic foot (cubic meter). As a comparison, 1 ton of coal (13,000 Btu/lb) will produce the same amount of heat energy as 26,000 ft^3 of natural gas (1000 Btu/ft^3), 208 gal of gasoline (125,000 Btu/gal), or 3250 kWh of electricity.

Any fuel containing hydrogen yields water as a combustion product. When the temperature of combustion drops below 120° to 140°F (49° to 60°C), water is allowed to condense out of the combustion mixture. This causes the vaporized heat of any condensed water to be liberated. The heat value of the combustion mixture is calculated based on two factors: (1) all the water vapor was condensed, and (2) no water vapor condensed. These two values are the high heat value (HHV) and low heat value (LHV), respectively. In the United States, the HHV is used in calculating the efficiency of internal combustion engines.

11.3 COAL

The earliest fuel used was wood. Due to availability and lack of refinement technologies, for a long time it was the only fuel used. However, with the growth of industrialization came a need for new and better fuels. Wood cannot be replaced fast enough to sustain a modern industrial society. As fuel needs increased, people turned to coal as an alternative to wood.

Coal had been known for years as a fuel, but it was not known how to burn it cleanly or efficiently. The first uses of coal were by the poor, because it was cheaper than wood, and by blacksmiths, who used it in the forges. Coal was also used in the conversion of iron ore to steel. Coal is the basic ingredient of coke, which is required in the production of steel. Coal is a complex substance produced from the decomposed remains of vegetation. Over a long period of time, the remains were changed by biochemical action, submersion, high pressures, and temperatures into various grades of coal. Due to variations in the degree of metamorphic change found in coal, no two coal samples are ever identical in all aspects. Coal is generally classified according to grade, variety, size, and intended use.

Coal is ranked by the amount of gaseous matter it gives off when heated. The coal that produces the most matter is known as *lignite;* the middle class of coal is called *bituminous,* or *soft, coal;* and the coal that produces the least by-product matter is called *anthracite,* or *hard, coal.* There are no clear distinctions between these categories. These terms are relative terms used to grade, or rank, coal. Figure 11–1 illustrates how coal is formed.

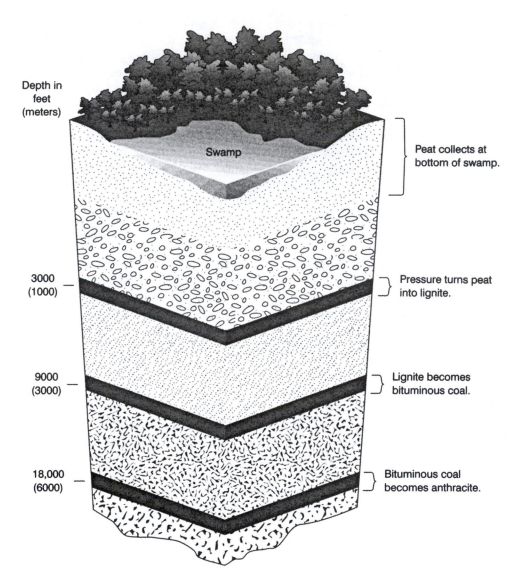

Depth in
feet
(meters)

Swamp

Peat collects at
bottom of swamp.

3000
(1000)

Pressure turns peat
into lignite.

9000
(3000)

Lignite becomes
bituminous coal.

18,000
(6000)

Bituminous coal
becomes anthracite.

Figure 11–1
How coal is formed.

Peat was formed as large trees and dense vegetation grew and died, falling or being swept into swamps or marshes by rivers and streams. After a period of time, they were covered by sand and mud, which cut them off from the atmosphere. If they had been left to decompose in the atmosphere, they would have formed carbon dioxide and water. While under water, bacteria (both anaerobic and aerobic) started to attack the submerged matter. This decomposition or partial decay caused the formation of peat.

The quality of the resultant coal depended on the toxicity developed during the decomposition and aeration of the surrounding water during the decomposition process. Highly toxic, less aerated water tended to preserve the decaying plant material and thus produced brighter, softer coals, whereas less toxic, aerated water helped to break down the plant material, thus forming duller, harder types of coal.

Peat consists primarily of cellulose and is not considered to be a form of coal. Special circumstances or conditions are required for peat to become coal. Peat can be used for fertilizer, plant bedding, or insulation and was used as a primitive building material. Peat can also be burned as a fuel. It yields about 1.5 times as much heat as wood.

Shifts in the earth's crust formed depressions, or basins. Dead vegetation collected in these depressions, and rivers flowed through the depression, dropping sediment on top of the decaying material. Over time, these sediments became sedimentary rock formations such as limestone, sandstone, and shale. Due to the increased heat and pressure produced by the weight of the sedimentary rock, the peat was chemically transformed into lignite.

Lignite is considered to be an immature coal and contains a greater carbon content than peat. Lignite has a high moisture content, often as high as 30%. Upon exposure to air, most of this moisture evaporates, causing shrinkage. Lignite, when used as a fuel, has only half of the heat energy value of bituminous coal. Therefore, lignite is mined in very small amounts.

Bituminous coal is lignite that has been transformed by elevated temperatures and pressures over time. The resulting chemical changes increase the carbon content of bituminous or soft coal to approximately 80%. It is a valuable source of fuel in the United States. It appears in greater quantities at less depths than anthracite coal. Bituminous coal is categorized according to its percentage of volatile matter: low volatile (14% to 23% volatile matter), medium volatile (23% to 33%), and high volatile (over 33%).

Bituminous coals burn rapidly and often produce large quantities of smoke unless proper precautions are taken. Low- and some medium-volatile bituminous coals are commonly used to heat homes. High-volatile coals are used extensively in industry. Some of these coals tend to coke or fuse into lumps when heated and are, therefore, called *coking coals*. These coking coals are used to produce coke for steelmaking. Bituminous coals are located all over the world.

Anthracite coal is formed through additional heat and pressure. Due to changes and shifts in the earth's crust, land covering the decomposed vegetation folded upon itself, or buckled, to form hills and mountains. This brought the necessary temperatures and pressures to bear upon the bituminous deposits to form anthracite coal. Anthracite contains only about 5% volatile matter while having 90% or more carbon content.

Anthracite is a hard, dense coal that is brittle and lustrous. It tends to burn with a short, blue flame, which produces very little smoke. It is better suited for home heating use, but because it is deeper in the earth, it is more expensive than bituminous coals. Anthracite coal is found at greater depths, as illustrated in Figure 11–1. Therefore, it is more difficult to mine and more expensive to produce. Only a small percentage, about 10%, of the coal produced in the United States is anthracite.

Table 11–2
Attributes and requirements of
the various coals.

Coal	Percentage of Carbon
Anthracite	90–95%
Bituminous	80–90%
Lignites	60–80%

Approximate Ages for Various Coals in Millions of Years	
Anthracite	200–250
Bituminous	100–200
Lignites	20–50
Peat	1–5

Table 11–2 illustrates the approximate carbon content and the age requirements for the production of the various coals.

The temperatures required for coal production do not exceed 572°F (300°C) in normal coals. Pressures required range from a few pounds to many hundreds of pounds per square inch (kilograms per meter).

As coals increase in rank, they do the following:

• Decrease in moisture content
• Increase in carbon content percentage
• Decrease in their amount of volatile matter
• Increase in heat energy value

11.3.a Producing Coal

Coal typically lies deep in the earth in the form of veins or seams. These are found in different orientations. Some are horizontal, some are tilted at steep angles, and some have been exposed by rivers and glaciers, which have removed the soil that covered them. Those veins that lie deep in the earth are mined by surface or strip mining. Veins that lie deep in the earth are mined through underground mining techniques.

Surface or strip mining involves removing or stripping the earth's layers that cover the coal deposits. The earth is removed by earth movers, bulldozers, and power shovels. Once the earth is removed and the coal is exposed, the coal is scooped into rail cars, large trucks, or conveyor belts.

Strip mining is economical, requiring fewer miners, no lighting or ventilation systems, and no shoring. It is less hazardous because there is less danger of explosion and little danger of asphyxiation from harmful gases that collect underground. It allows easier access for machinery and equipment, but these machines and equipment can be used only during relatively fair weather. Strip mining can be a viable option when the land is replaced using soil conservation and proper ecological techniques.

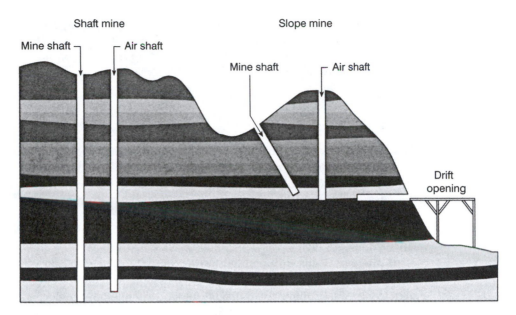

Figure 11–2
Three types of underground mines.

Most coal deposits lie deep in the earth's surface, too far below the surface for strip mining to be practical or economical. The method used to recover the coal is depends on the location and angle that the deposits make with the surface. If the coal is located in the side of a mountain, a horizontal tunnel is dug to reach the coal. This horizontal tunnel is called a *drift*. If the coal deposits are located below the earth's surface, a slope or a shaft is dug. *Slopes* are tunnels dug at an angle and used as an entrance ramp into the coal deposits. *Shafts* are vertical openings to the surface. Figure 11–2 shows the orientation of the three types of mining.

All mines have at least two additional openings, which provide both ventilation and an emergency exit in case of an accident.

Mine safety is important because the coal miner is exposed to a number of hazards, including poisonous gases, explosions, flooding, cave-ins, and the risks of every occupation, such as tripping, falls, strains, and similar occupational hazards. To help reduce the risk taken by miners, mining engineers have suggested and implemented many safety precautions. The Bureau of Mines has also implemented safety standards and procedures to help ensure that miners are safe on the job.

Mining safety procedures include the following: Fans and blowers on the surface force air down into the mine shafts to help dispel dangerous gases that may collect and explode or poison miners. Water tends to collect in the bottom of the mine and cause additional hazards. When the mine tunnel is cut, it is cut on an incline to allow for drainage. The tunnels then drain into a central well, where the water can be pumped out by large pumps.

Various electronic measures are used to "sniff" for dangerous gases, either poisonous, explosive, or both. Mine tunnels are constantly inspected for potential cave-ins. Proper shoring and support beams help reduce the potential for cave-ins. For further information concerning mine safety, contact the U.S. Bureau of Mine Safety or your local state or district Bureau representative.

11.3.b Refining Coal

Coal, as it comes from the mine, contains impurities and foreign matter. It may contain quantities of clay, shale, slate, and rocks that have fallen into the coal during mining. Most of the impurities result from sediments that were deposited along with the vegetation when the material was forming the coal. These impurities must be removed, because the foreign matter reduces the heating value of the coal, increases the ash content during burning, and melts when heated to form clinkers (partially vitrified masses such as ash, which tend to clog the furnace).

In the past, coals with high impurities were not mined, because it was expensive to separate the impurities. Developing technology has made it more economical to mine coals of less quality. After separation or cleaning, the coals are graded according to size, depending on their intended use or purpose.

The coal is run through a variety of screens, which vary in size from approximately 3 in. (75 mm) mesh to roughly 1/8 in. (3 mm) mesh. Machines may also be used to separate the denser impurities from the less dense coal. The coal is less dense and, therefore, lighter than the impurities so it can be floated above and separated from the denser impurities. Table 11–3 lists the steps taken in preparing and separating coals.

11.3.c Quality of Coals

Consumers buy coal by the ton. They are interested in the heat energy value for the coal they purchase. Good-quality anthracite yields approximately 13,000 Btu per

Table 11–3
Grading procedures of coals.

1. Coal is separated at site into rough categories of size and value.
2. Coal is passed through screens that separate the different physical sizes.
3. Larger lumps (> 6 in.) are separated for further processing.
4. Large sizes are crushed to provide the smaller sizes required by industry.
5. Coal is cleaned to remove as many of the impurities as possible.
6. Coal is allowed to dry or is mechanically dried.
7. Different coals are blended to produce the type of fuel best suited for its intended purpose.

pound, whereas good-quality bituminous has slightly higher yields. In addition, consumers are interested in the ash content, sulfur content, moisture content, and percentage of volatile matter in the coal. The presence of these elements has an impact on the heat energy value of the coal and also creates problems such as the need for special precautions.

As stated previously, impurities such as sand, clay, and other foreign matter won't burn properly, so it costs more to remove the ashes produced. Accordingly, coal with low ash content demands a higher price and higher demand.

When heated, the ashes of some coals tend to stick together, or form clinkers, which tend to clog the furnace. Coals that have higher ash-melting temperatures are therefore preferred, because they will tend not to form clinkers.

Sulfur poisons the environment as well as forming clinkers and attacking the metals used in coal furnaces. Sulfur-free coal is sought for these reasons and because sulfur adds undesirable properties to metals.

Moisture content is important in grading the quality of coal. Not only is the consumer buying the weight of the water, but when coal burns, the water in the coal is changed to steam. The conversion of water from its liquid state to its vaporous state requires a great deal of heat energy. The energy expended in the conversion of water to steam is lost to the consumer. This cost increases as the moisture content of the coal increases.

Coals have these advantages:

- No special storage requirements, such as tanks, constant-temperature units, pipelines, and so on, are necessary. All that is required is a hard, dry, flat surface.
- In home heating, quality coals have less sulfur content than many fuel oils, which makes them less of a pollutant.
- Coal can be converted on-site to a usable form.
- Coal reserves exist in many countries, making them less dependent on foreign petroleum.

Of all nonrenewable fuel reserves known today, coal will probably last the longest. The U.S. Bureau of Mines estimates that half of the world's supply of coal is located in the United States. However, air-pollution standards have reduced the economical advantages of using coal. This pollution represents not only a hazard to the community but also a large amount of wasted energy and wasted coal. To help reduce this danger, coal can be converted into gasoline, diesel fuel, lubricating oils, and coal gas. Therefore, coal may become an even more important fuel in the future because of its flexibility and availability.

11.4 PETROLEUM

Petroleum and petroleum products are infused into our nation, our industry, and our lives. Petroleum products are used to power our cars, trains, planes, and ships. They are used to heat our homes, produce electricity to illuminate our offices, and pave our roads. Petroleum is more than just a fuel. It can be used as a lubricant or as a base for such products as antifreeze, paints, cold creams, hand lotions, and shampoo, as well as in the production of synthetic rubber.

People have been using petroleum for more than 6000 years, but the petroleum industry as we think of it today started in 1859 in the town of Titusville, Pennsylvania. Colonel E. L. Drake supervised the first drilling operation, which struck oil at a depth of 69 ft. Since then the industry has undergone a phenomenal growth.

11.4.a Source of Oil

Most of our oil reserves are hidden deep in the earth. Thus, very sophisticated means of detection must be used to find these oil pools. It is generally believed that petroleum comes from the decomposed remains of animals and plants that were covered millions of years ago. At that time, the earth was covered by a large percentage of water. Marine life and plant life lived and died in those waters. The remains of those plants and animals settled on the bottom and were covered with sediments dumped there by different tributaries. Cut off from air, the remains decomposed and were transformed into petroleum and gases through biochemical processes involving the aerobic and anaerobic bacteria and the enzymes present.

Changes occurred in the earth's surface. The earth buckled, cracked, heaved, and bulged, creating mountains, valleys, and features that were not previously present. The sediments and fossil remains were formed into sandstone, shale, and limestone, which are known as sedimentary rocks.

When fine particles of mud and clays are compacted and hardened under great pressures, the salt water, oil, and gas within them are squeezed out and forced into tiny crevices in the adjacent layers of rock and sand. If possible, they work between layers until their migration is halted by dense stone. Once they are surrounded by dense stone, an oil pool is said to have been formed. Oil not trapped in pools is lost. It continues to migrate until it escapes or seeps onto the surface. An oil pool consists of water, gas, and oil, which have been trapped within the pockets or tiny crevices located in porous rock such as sandstone. The trick is to find out where these pools are.

11.4.b Locating Oil Pools

Many different methods are used to locate oil. Some of them are less than scientific. These latter methods include hit-or-miss drilling, divining or holding a bent twig and drilling where it bends or points toward the surface, throwing a piece of straw into the air and drilling where it lands, and similar methods that have little chance of success. These methods are just gambles at best based on experience, frustration with other methods, or tradition.

Modern scientific methods rely on geological information collected over many years. The rocks and formations in an area are studied. Delicate instruments are used to take soundings and to study the various forces acting in the rock structures. Probably the most widely used and successful device for locating oil pools is the *seismograph*.

The seismograph is used to map the layers of rock beneath the earth's surface. It does this by measuring the vibrations produced and reflected by the localized use of

Table 11–4
Methods of locating petroleum.

1. Visual methods include observing seepage from the ground and collecting fossils found in the strata.
2. Geological methods include mapping the age of rocks, their nature, and the number and types of formations present.
3. Geophysical methods include
    a. Gravimetric: Measuring the variations in the earth's crust with very sensitive instruments.
    b. Seismic: Measuring the amplitude of reflected shock waves passed through the earth's crust.
    c. Magnetic: Measuring local variations in the earth's magnetic field, which shows the distribution of different types of rock in the earth's crust.
    d. Exploratory drilling: Taking core samples and examining them for fossil formations and evidence of porous or nonporous rock formations.

an explosive such as dynamite. The procedure is to drill a hole in the area under study and then explode a charge of explosive, causing a small earthquake in the area. The vibrations produced in the earth radiate outward, similar to the ripples produced when a rock is thrown into a body of standing water. Where the vibrations hit rock formations, they are reflected back. Weak reflections show soft formations, whereas strong reflections show hard rock formations.

A series of seismometers is placed around the area to be studied, and their readings are recorded for further study. After the information is gathered and studied, the geologist can then recommend to the driller where to begin drilling. Even though the area may be promising, it doesn't necessarily mean that oil will be found there. After seismometer readings have been taken and the geologist has made a recommendation, it is still a gamble. The odds are better, but the game remains the same.

In general, there are five methods commonly used to locate oil deposits. These include visual, geological, and geophysical methods, as given in Table 11–4.

11.4.c Drilling for Oil

The two basic types of drilling are the rotary and the cable tool, or percussion, methods. The *cable tool method* is the oldest of the two methods. In cable tool drilling, a hole is punched in the earth's surface by repeatedly lifting and dropping a heavy, chisel-shaped tool bit that is suspended from a long cable. This apparatus is contained in a tall tower or derrick made of steel, wood, or a combination of the two. An engine is used to lift and drop the tool. The force of dropping the tool bit shatters the rock and drives the tool bit downward into the earth's surface. When a sufficient amount of debris has accumulated, a bailer is used to clean out the hole. As the hole gets deeper, sections of steel pipe or well casing are lowered into the hole to prevent the hole from caving in and to prevent water from entering the hole.

The cable tool drilling method is rarely used today. It is best suited for shallow wells where little casing is needed. Its advantage is that the equipment used is relatively inexpensive, but its main disadvantage is lack of speed. It is a very slow process.

The more common practice is *rotary drilling*, where a drill bit is used. This drill bit is similar to the common metal or wood bits used to drill through metal or wood. The equipment used for rotary drilling is more expensive than the cable tool apparatus, but it is faster and is, therefore, much more common.

The rotary drill consists of long sections of pipe with a sharp bit attached to the lower end. The pipe is rotated; as the bit rotates, it grinds the rock into a powder. Additional sections of pipe are added as the hole gets deeper. Bits are replaced as they get dull.

Figure 11–3
Rotary drilling.

Rigging block

Swivel

Mud hose

Drill works

Mud pump

Drill pipe

Casing

Bit

Coolant is pumped through the pipe, providing cooling and lubrication for the bit during drilling. This coolant is usually mud. As the coolant leaves the drilling area and rises to the surface, it carries drilling samples with it. These samples are collected and sent to geologists, who can examine and make recommendations based on their findings.

When the drilling reaches oil, a control head is tapped onto the casing. Tubing is run through the casing to allow the oil to travel to the surface. Valves at the top control the flow of the oil to storage tanks, transportation facilities, or refinement areas. The rotary drilling method is illustrated in Figure 11–3.

11.4.d Oil Production

Usually the pressure developed within the surrounding rock formation is enough to force the oil to the surface. In one situation, pressure may be exerted by the gas trapped above the oil in the pocket formed by the rock; also, the gas may be dissolved in the oil due to the pressure exerted within the pocket. When the pocket is pierced by the drill bit, the pressure attempts to escape. It is much like shaking a bottle of soda and then taking the cap off. As the gas escapes, the liquid is taken with it. A third situation involves water trapped beneath the oil, which pushes the oil out of the pocket. Because oil is lighter than water, it floats on the top and is pushed out through the tubing.

In some situations, the pores of the rock formation are too small to allow the oil to flow to the surface. In these situations, the rock formations may be broken up by explosives (when in sandstone) or dissolved by acids (when in limestone). When the rock layers are opened up, the oil flows to the surface.

As the oil is pushed or escapes from the pocket, the pressure decreases. Through proper control methods, a large percentage of the oil can be recovered. If the gas is allowed to escape uncontrolled, less than 20% of the oil is extracted. When the pressure drops to a level where it no longer pumps oil out of the well, the oil must be pumped out through the tubing. This is more expensive, but it is necessary to recover as much of the deposit as possible.

Crude oil, which comes directly from the well, must be transported to a refining facility. Oil fields are located in remote areas far away from refineries. Crude oil is transported by tankers, pipelines, or tank trucks to the refineries.

Crude oils are viscous liquids that vary in color from clear to almost black. Whatever the color, all oils contain compounds, called *hydrocarbons*, that consist mainly of hydrogen and carbon. Hydrocarbons differ in the number, proportion, and arrangement of the hydrogen and carbon atoms. Thus, each hydrocarbon has different properties. Hydrocarbons differ not only in their internal structure, but also in the way they form linkages, or chains.

Petroleum oils are generally classified into three main groups: the paraffin, naphthene, or asphalt series. *Paraffin-series crude oils* are often used to produce lubricating oils and also produce low-octane gasolines and fuel oils. Paraffin-series crude oils are used to produce paraffin wax as well. Other hydrocarbons are characteristic of the naphthene series. *Naphthene-based crude oils* are intermediate in that they can be used to produce adequate gasoline but are not a direct source of high-grade lubricating oils. *Asphalt-series crude oils* contain little or no paraffin wax but usually leave a residue of asphalt. Distillates of the asphalt series include higher-octane gasolines and general-purpose lubricating oils. When crude oil contains more than one series of hydrocarbons, it is known as a *mixed-base crude oil.*

Crude oils are separated at the refinery into gasoline, kerosene, lubricating oils, fuel oils, asphalt, and similar products. These products are called *fractions*, because they are parts of the whole. Crude oil is separated by a process based on the varying boiling points of different factions because of their diverse hydrocarbon characteristics.

Because of the different boiling points within the hydrocarbon groups, the distillation process can be used to separate the fractions. Petroleum distillation uses the principles of vaporization and condensation to separate the fractions. The same distillation process can be used to desalinate water. For example, a mixture of salt and water can

Refinery Flow Chart This is a highly simplified flow chart showing the conversion of crude oil to products useful to consumers.

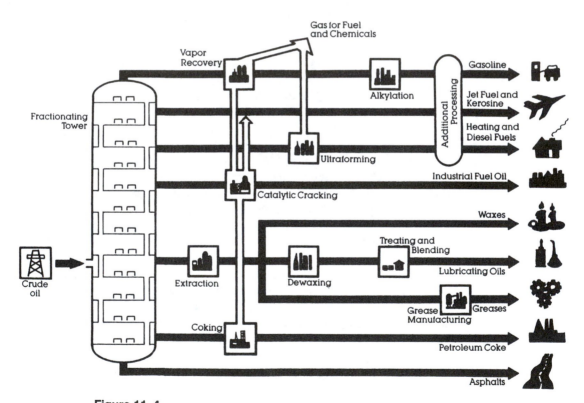

Figure 11–4
Distillation.
Courtesy of American Petroleum Institute.

be heated until the water is boiled away, leaving a salt residue, which has a higher boiling point, in the bottom of the container. To capture the water, a piece of coiled tubing could be attached to the top of the salt-water solution container while it is being heated. When the solution is heated, the steam produced will escape through the coiled tubing, or *condenser*. The condenser cools the steam and condenses the steam back into water. The condensed water is known as the *distillate*. Distilled water can be produced in this fashion. A container is used to capture the distillate, and the impurities are left in the solution container. All impurities that have a boiling point lower than water can be removed using this process. Figure 11–4 shows the process just described.

Fractional distillation is used to separate the hydrocarbon groups, or fractions, in the different petroleum series. The fractions with lower boiling points vaporize first, then the fractions with the next-highest boiling point, and so on, down the ranks of boiling point.

One method of fractional distillation is shown in Figure 11–5. This is called the *shell oil still*. Smaller and older refineries may still use shell oil stills to refine crude oil. Crude oil enters the first horizontal cylinder, where a modest heat vaporizes the first fraction. The crude remaining enters the second cylinder, where the next fraction is vaporized and removed. The crude is then allowed to enter the third cylinder, and the process continues until all fractions have been removed.

Another faster, more accurate method of separating the different fractions is based on the principle that substances boil and condense at the same temperature

Figure 11–5
Shell oil still.

given the same conditions. Thus, instead of separating the fractions by vaporizing them one by one at gradually increasing temperatures, fractions are condensed individually at successively lower temperatures. This process uses a single structure rather than individual stills.

Crude oil is first heated in a still. The resulting vapors are then piped into a fractionating column or bubble tower. This is a vertical steel chamber approximately 25 ft in diameter and 125 ft high. Within the tower, there are several floors or individual chambers, where a steel plate, partially filled with condensate of the particular fraction, separates chambers. Each plate is located at approximately the condensing temperature of the fraction it is used to collect. The condensing temperature ranges from the highest at the bottom to the lowest at the top.

A continuous stream of vaporized crude oil enters near the bottom. Heavy fractions immediately condense and fall to the bottom of the tower. Those vapors that remain continue to rise through the first plate. Here they pass through vapor risers and are deflected downward by bubble caps on the risers. The hydrocarbons with boiling points equal to or higher than the liquid found in the plate remain there. The remaining vapors have lower boiling points and continue to rise. This process continues toward the top of the tower.

At the top of the tower, the temperature is kept high enough to prevent the condensation of the gasoline there. Gasoline vapors travel out through the top of the tower, where they pass through a condenser, in which the gasoline is separated from noncondensable gases. The rising vapors heat the plates, and the condensates cool them sufficiently to keep the temperature relatively constant. Fractions are piped off the different plates, where they will continue to be processed further into gases, gasoline, naphtha, kerosene, fuel oils, heating oils, diesel fuel, lubricating oils, road oil, and asphalt.

Fractional distillation produces about 18% gasoline. This is not enough to satisfy the demand for gasoline. If fractional distillation production were increased, other fractions would be produced in larger quantities than needed to satisfy demand. Alternatives to fractional distillation were found to provide more gasoline without overproducing other fractions.

Cracking is used to break down heavier fractions, composed of larger, more complex molecules, into simpler ones. Some of these simpler molecules can be used to produce gasoline. In the cracking process, heavier fractions under pressure are heated in the presence of a catalyst. Recall that the purpose of a catalyst is to induce or increase the rate of a chemical reaction without being used up in the process.

The lighter fractions that escape through the top of the fractionating column but are not used to produce gasoline can be induced to polymerize into heavier, more complex molecules of gasoline. Using catalysts, these materials can be polymerized under heat and proper pressure conditions. This type of polymerized gasoline is used in the field of aviation.

Gasoline can also be produced synthetically. Petroleum products can be produced from coal. The proportion of hydrogen and carbon in coal is much lower than in petroleum. If, in the presence of a catalyst, hydrogen is combined with powdered coal, the coal is converted into a liquid fuel. This liquid fuel can be distilled into gasoline and other fuel and lubricating products.

11.4.e Grading Petroleum Products

The primary grading system used for motor fuels is the octane rating. The octane rating of a motor fuel is a measure of the knocking qualities of that fuel. The higher the octane rating of the fuel, the smoother and more efficiently the engine will perform.

Knocking or pinging in an engine is the spontaneous ignition of the fuel before the spark passes through it. This premature detonation limits the amount of power developed in the engine. Detonation results in overheating and a loss of power. One of the most important factors connected with detonation is the compression ratio of the engine. For any given combination of engine design and fuel, there are limits to the compression ratio. The engine design and compression ratio cannot easily be changed by the average person, so the octane rating or antiknock rating of the fuel supplied to the engine is changed.

The octane number of a fuel is the isooctane percentage by volume of the fuel mixture with the same detonation tendency as the fuel in question. Four methods are employed to determine the octane number for a fuel using a standard engine and compression ratio: the research, motor method, aviation lean-mixture, and aviation rich-mixture methods. The *motor method* is used for low-octane numbered fuels; the *research method,* for high-octane numbered fuels; and the *aviation methods,* for aviation-grade fuels.

To assign an octane rating to commercial pump gasolines, engineers perform two tests and average them. If you look on a gas pump, you will see the octane number that was assigned to the gas in the pump. Next to this number you will see $(R + M)/2$. This means that the research number gained was added to the motor number obtained and the sum was averaged.

Gasoline distilled from crude oil has an octane rating of between 50 and 75, depending on the grade of crude and the refining process used. Refineries use several methods to increase the octane rating of their gasolines. One method is to add the hydrocarbons gained from cracking, which produces higher-octane gasolines. The most common process is to add tetraethyl lead. The most obvious drawback to using lead is that it is poisonous. It also poisons the environment. Alternatives to lead addition have been found to increase the octane rating of gasoline.

The addition of small quantities of lead to gasolines increases their antiknock ratings. Tetraethyl lead used as a gasoline additive is a solution of $(C_2H_5)_4Pb$, ethylene dibromide, and a dye. The maximum quantity used as a gasoline additive is about 6 cm^3/gal. Some gasolines respond to the addition of lead better than others. Performance numbers for aviation-grade fuels often reach 160 or better. Gasohols, or mixtures of gasolines and alcohols, react negatively to the addition of lead. The addition of lead to gasolines has been reduced significantly and is all but outlawed in most regions. Lead-free gasolines obtain sufficient octane ratings without the use of tetraethyl lead additives.

11.4.f Diesel Fuels

Diesel engines are *compression-ignition* engines. Gasoline engines are typically spark-ignition engines. In gasoline engines, a spark is produced as the piston is compressing the air-fuel mixture that has previously been introduced into the cylinder. In diesel engines, fuel is injected near the top of the compression stroke, where the fuel spontaneously ignites and the cylinder pressure is increased by fuel combustion to several hundred pounds per square inch.

Diesel fuels that have higher spontaneous ignition temperatures produce what is called *diesel knock*. Diesel knock is caused by violent combustion taking place, with a sudden increase in pressure, after most of the fuel has been introduced into the cylinder. As in detonation in gasoline engines, this combustion results in overheating and a loss of power. Paraffin-series hydrocarbons have the lowest spontaneous ignition temperatures and are, therefore, the more suitable choice for producing diesel fuels.

Diesel fuel has three grades: no. 1, no. 2, and no. 4. Number 1 is used in situations where frequent speed and/or load changes are expected. Number 2 is used for heavy industrial service. Number 4 is used in low- to medium-speed engines, typically stationary engines. Different additives are used to produce several characteristics in all fuels. In particular, additives are used in diesel fuels to prevent rust, inhibit corrosion, remove water, and weatherize the fuel during winter.

High-octane spark-ignition fuels make poor-quality compression-ignition fuels. In fact, the lower the octane number for compression-ignition fuels, the better. The figure of merit for compression-ignition fuels is the *cetane number*. The cetane number of a diesel fuel is the percentage by volume of cetane ($C_{16}H_{34}$) that has the same overall performance of a standard compression ignition engine using the same cetane numbered fuel. Cetane numbers for common diesel fuels are as follows:

High-speed diesel fuels	<50
Medium-speed fuels	<35
Low-speed fuels	>15

11.4.g Other Liquid Fuels

The next grade of petroleum product (descending in the fractionating column) is for kerosene and fuel oils. Fuel oils include those fuels used for lighting and heating, in heavy-equipment applications, in gas turbine fuels, and in ram jet fuels.

Kerosene has gained popularity as a heating fuel in kerosene heaters. These heaters are wick-type heaters used for space heating. Fuel oils have been used for years for heating homes. Fuel oils are rated from no. 1 to no. 6, no. 1 being the easiest flowing and most readily burning. Number 6 flows with difficulty and is difficult to vaporize. Number 6 burns well and has a high heat value, but it requires preheating before it will flow to the burner. Number 2 is the most popular for home heating.

Ethyl alcohol (grain alcohol) or ethanol may also be used as a motor fuel, especially when blended with gasoline; it is sometimes called *gasohol*. Five percent to 25% of anhydrous alcohol (alcohol with all water removed) may be used in different blends. Advantages include an increased octane rating, cleaner burning, and cooler burning temperature. Disadvantages stem from the lower heat-energy value of the alcohol, which means that more fuel must be burned or the burn rate must be increased to produce the same power. It may also be harder to start in cold weather.

11.5 GASEOUS FUELS

Gaseous fuels can be used to heat and power homes and factories cleanly and efficiently. They have been and are used as a fuel for furnaces and illumination. However, gaseous fuels can be more expensive than solid or liquid fuels, depending on the application and availability of the products.

The four primary types of gaseous fuels are natural gas, coal gas, water gas, and liquefied petroleum gas. *Natural gas* is the principal gaseous fuel used in this country. Natural gas accompanies petroleum wherever petroleum is found. The gas is drilled in much the same way oil is drilled, only the gas on top of the oil is removed. This gas is primarily methane, which has a heat energy value of approximately 1000 Btu/ft^3.

Coal gas is manufactured through the distillation process and involves the volatile-matter by-products of coke production. Coal heated in the coking process transforms into coke and volatile matter. The volatile matter is a mixture of several substances. When the volatile matter is passed through water, the coal tars and light oils condense on the surface, ammonia dissolves in the water, and coal gas bubbles through the water. Coal gas has a heat-energy value of approximately 600 Btu/ft^3. The coke produced during the manufacture of coal gas is too soft for use in metallurgical situations and is, therefore, used to produce water gas.

Water gas has a lower heat-energy value, about 300 Btu/ft^3. Air is forced through glowing coal or coke until the fuel is white hot; then the air is shut off, and steam is sprayed on the coals. The white-hot carbon combines with the oxygen in the steam to form carbon monoxide and hydrogen. Carbon monoxide and hydrogen are both inflammable gases, which have a high heat-energy value. When the temperature drops below 1900°F (1038°C), no more gas is produced by this reaction, so the steam is turned off and the air is turned back on. Water gas is sometimes referred to as *blue gas,* because carbon monoxide and hydrogen burn with a blue flame. However, carbon monoxide is poisonous and dangerous to use.

Liquefied petroleum (LP) gas is a combination of propane (C_3H_8) and butane (C_4H_{10}). A mixture of propane and butane is pumped into a storage tank under enough pressure to turn them into liquids. When the tank is opened, the liquid escapes and returns to the gaseous state. The heat energy value of LP gas is typically between 2500 and 3000 Btu/ft^3.

11.6 TYPES AND PROPERTIES OF LUBRICANTS

In addition to its use as a fuel, petroleum can be used to produce lubricants, which can be used to reduce friction in machines and other applications. A film of lubricant between the moving parts reduces the friction and reduces the power required to move these parts.

Lubricant oils come from a variety of sources, including animal, vegetable, and mineral origins. Animal and vegetable oils were used long before the discovery of petroleum. The primary animal oils used came from whales. The more important vegetable oils were olive, castor, and cottonseed oils. Most of the lubricating oils used today comes from the crude oil taken from the earth. This crude oil is distilled into various products, including gasoline, fuel oils, and paraffin.

Greasy substances are often referred to as lubricants. The word *lubricant* comes from the Latin word *lubricus*, meaning *slippery*. *Lubrication* is primarily concerned with reducing friction between moving parts. Any substance used to accomplish a reduction in friction or to change the frictional properties of a system is called a lubricant. The more common lubricants include oils and greases, although many liquids and gases may be used for lubrication. The particular application dictates which material is best suited for the desired performance.

In order for a lubricant to be effective, it must have enough viscosity to hold the moving parts apart but not be so viscous as to slow the parts down. *Viscosity* refers to the ease with which a lubricant flows. Heavier lubricants flow more slowly and are said to be more viscous, or to have a higher viscosity. One example is the difference between water and honey. Honey is more viscous. Viscosity is probably the most important property of a lubricant and is influenced by temperature, pressure, and fluid motion.

One method used to measure viscosity requires a person to pour a known quantity of oil through a standard-size funnel and record the time required for the oil to pass through the funnel. The oil may be held at an elevated or reduced temperature, depending on the purpose of the test. The less time it takes to pass through the funnel, the less viscous the oil. The Society of Automotive Engineers (SAE) has assigned numbers to the different times required for lubricants to pass through standard funnels. The lower the SAE number, the lower the viscosity.

Lubricants typically become more viscous as the temperature is lowered. Conversely, lubricants become less viscous as the temperature increases. A lubricant selected for one temperature range may not perform satisfactorily at a temperature outside of that range. One option is to buy a multigrade oil. These oils have been blended to acquire better cold- and warm-temperature operating characteristics. For instance, a 10W-40-weight oil exhibits 10-weight lubricant characteristics at lower temperatures, and it exhibits 40-weight lubricant characteristics at higher temperatures.

The Saybolt standard universal viscometer is the standard instrument used for determining the viscosity of various lubricants between the temperatures of 70° and 210°F (21° and 100°C). The results of such tests are Saybolt Seconds Universal (SSU). The SAE viscosity numbers are based on Saybolt viscometer results. SAE numbers with a W suffix are based on a Saybolt viscosity at 0°F (–18°C), whereas those without the W suffix are based on values at 210°F (100°C). For example, a 10W-40-weight oil exhibits the viscosity of a 10-weight oil at 0°F and the viscosity of a 40-weight oil at 210°F (100°C).

The viscosity index (VI) is an empirical system used for expressing the rate of change in viscosity due to temperature. The index uses a scale from 0 to 100 to grade oils according to viscosity-temperature sensitivity. To determine the VI for an oil, the viscosity of the sample product is measured at 100° and 210°F (32° and 100°C) and compared to a standard table according to ASTM D 567-53 and D 2270-64. A direct reading is taken from the VI chart.

Flash-point test apparatus.

Other test procedures for determining the usefulness of petroleum products include specific gravity, density, flash point, heat energy or calorific value, cloud and pour points, odor, color, and ash content. Viscosity is probably the most important property of a lubricant. A lubricant's viscosity can be influenced by its temperature, pressure, or fluid motion. Sometimes soap or detergent is added to an oil to increase its viscosity. Some oils are listed as being detergent or nondetergent for this reason. Detergent-type oils tend to wash internal parts clean, which is sometimes undesirable. A lubricant's density correlates with its viscosity.

The *specific gravity* of a liquid is the ratio of the density of the liquid to the density of water at the same temperature. The reference value used is 60°F (15°C), with tables used for correction at temperatures other than 60°. The most rapid and convenient method of determining the specific gravity of a liquid is by using a *hydrometer*. A hydrometer is placed in the sample and is allowed to come to rest. The specific gravity for the material is shown on the hydrometer's scale, which coincides with the surface of the liquid. Specifications on testing the specific gravity of a lubricant are given in ASTM D 1298 and the API (American Petroleum Institute's gravity is given in ASTM D 287). Some lubricants, mainly oils, exhibit a separation of the wax dissolved in them at temperatures slightly above the point where they solidify under specific conditions. The temperature at which this occurs is known as the *cloud point*. When a lubricant does not separate or it is impossible to tell when it separates, no cloud point can be determined for that lubricant (see ASTM D 97).

The temperature at which a lubricant will just barely pour under prescribed test conditions is known as its *pour point* (see ASTM D 97). Pour-point additives can be added to lower the temperature that an oil will flow. This allows better protection during colder months or colder conditions.

The *flash point* is the temperature at which an explosive air-vapor mixture is formed. In order to determine the flash point of a specimen, the sample is heated in a brass cup over a gas flame that is mounted in an air atmosphere. The air-fuel mixture is heated to approximately 25°F (14°C) below the expected flash point. Then, a flame is dipped into the vapor produced by the air-fuel mixture (a stirrer is used to aid in mixing the air and fuel). The temperature of the fuel in the cup is raised 2°F, and the flame is again introduced. This continues until a distinct flash is seen when the flame meets the vapor in the cup.

A bomb calorimeter is used to ascertain the heat energy or calorific value of a fuel or lubricant. Typically, to determine the amount of residue, or *ash content*, of a fuel, a sample of the fuel is weighed and placed in a dish. The dish is then heated, and a flame is applied to the material. When the material's combustion is complete, the material remaining is reweighed, and the percentage of ash content is determined.

11.6.a Lubricants

Following is a list of the various engine lubricants:

Society of Automotive Engineers Viscosity Classification of Engine Lubricants

SAE Number	Maximum Viscosity at 0°F (–18°C)	Maximum Viscosity at 210°F (100°C)
5W	up to 4,000	3
10W	up to 12,000	—
20W	up to 48,000	—
30	—	up to 58
40	—	up to 70
50	—	up to 110
80	up to 100,000	—
90	—	up to 120
140	—	up to 200
250	—	above 200

Following are some examples of typical additives and viscosity results.

Examples of Selected Lubricant Additives*

Additive	Purpose
Oils	
Viscosity index improver	Lowers the rate of change in viscosity due to temperature change

*Viscosities in Saybolt Seconds Universal(SSU)

Detergent-dispersant	Keeps insoluble substances in suspension to maintain cleanliness
Oxidation inhibitor	Reduces oxidation of oils which leads to breakdown
Rust inhibitor	Helps reduce rusting of metallic parts
Extreme pressure	Prevents seizure of surfaces
Foam inhibitor	Decreases tendency to foam
Greases	
Filler	Increases bulk
Oxidation inhibitor	Reduces oxidation in application
Anti-wear agent	Reduces wear of metal parts
Extreme pressure	Reduces friction and prevents seizures
Stabilizer	Increases operating temperature range

11.7 SUMMARY

Fuels and lubricants play an important role in any transportation system. The combustion engine relies on fuels and lubricants to operate. Understanding the fuels and lubricants used in common transportation equipment helps develop an understanding of the importance of these products in our everyday lives. Today's engines require these materials and will continue to rely to some extent on these materials in the future. Until a proper economic replacement has been found, fuels and lubricants will continue to play an important role in any mechanized society.

QUESTIONS AND PROBLEMS

1. What is oxidation?
2. Describe the combustion process.
3. Describe the process required to start a wood fire.
4. What conditions must be met before complete combustion is achieved?
5. How is heat energy measured?
6. What is a Btu? Describe the use of Btus.
7. List the advantages and disadvantages of coal as a fuel.
8. List the advantages and disadvantages of liquid fuels.
9. List the advantages and disadvantages of gaseous fuels.
10. List the three major types of coal and the advantages and disadvantages of each.
11. Name three hazards particular to the coal mining industry and what measure you would take to reduce the risk.
12. How did the petroleum industry begin?
13. Where is petroleum typically found?
14. What is a seismometer? How is it used to find oil?
15. Describe the two methods used to drill for petroleum.
16. What forces can be used to push petroleum to the surface?
17. Of what is oil made?
18. Diagram and explain the process of fractional distillation.
19. How is ethanol used as a fuel? From where does it come?

20. What is viscosity? Diagram and describe an original design for measuring viscosity.
21. How is coal gas produced? Water gas? LP gas?
22. Using the ASTM standards, diagram the apparatus used to determine SSU. What does the figure given for SSU mean?
23. Describe three applications for each of the following: solid fuels, liquid fuels, and gaseous fuels.
24. You are interested in locating fuel reserves for your investors. Describe the process by which you would locate coal, oil, and natural gas reserves in the United States.
25. Describe three applications for each of the following: oils, greases, and solid lubricants.
26. In each of these applications, describe the factors that influenced the selection of that particular lubricant, in terms of service conditions, viscosity, environment, and such factors.
27. Why do some fuels have higher heat values than others? What makes a better fuel?

SECTION TWO
Principles of Mechanical and Nondestructive Testing

CHAPTER 12
Mechanical Behavior

Objectives
- List and define the basic mechanical properties of materials.
- Recognize and describe the terminology related to mechanical behavior and mechanical testing.
- Describe the effects of different conditions on test results.
- List and describe the use of equipment related to mechanical testing.
- Graph and describe the data and attributes of stress-strain curves.
- List the various categories of strength and properties that reflect differing strengths.
- Define energy capacity and hysteresis.
- Recognize and describe different types of mechanical failures.

12.1 INTRODUCTION

This chapter introduces the fundamental concepts related to the properties and testing of materials. These concepts include the mechanical properties of materials, testing conditions and their effects on test data, stress and strain under a variety of loading conditions, slip, elastic and plastic behavior, strength, energy capacity, and failure. Along with these fundamental concepts, the terminology and formulas needed to understand these concepts are given. These terms and formulas are used in the following chapters on testing to provide you with some background and experience in materials testing.

12.2 FUNDAMENTAL MECHANICAL PROPERTIES

Strength, in the broadest sense of the term, refers to the ability of a material or group of materials to resist the application of a load or loads without failure. Failure, in this instance, may be in the form of rupture caused by excessive stress or excessive deformation. The general properties of materials that relate to this problem are known as mechanical properties.

Mechanical properties are those that deal directly with the behavior of materials under applied forces. These properties are usually described by terms that relate to stress, strain, or both stress and strain. Basic mechanical properties were listed in Chapter 1.

Materials testing is an area of research that deals primarily with the methods and procedures used to measure and reliably, precisely, and accurately determine the mechanical properties of materials. These are primarily (1) a determination of the applied force or load and (2) the test specimen's change in length based on the applied force. These two measures are termed *stress* and *strain*, respectively.

Basic mechanical properties include strength, stiffness, hardness, elasticity, plasticity, and energy capacity.

The *strength* of a material is determined by the maximum stress at which failure occurs. Failure, in this situation, is considered to occur when a material is no longer elastic or it ruptures. There are many different types of strength, depending on the type of load or loads applied.

Stiffness deals with the resistance of a material to deformation under load while in an elastic state. Thus, stiffness is measured by the modulus of elasticity, which is defined later. Regarding stiffness, we can think of a mattress as being considered stiff if it doesn't give much when we lay down on it.

Hardness is a measure of a material's resistance to indentation or abrasion of its surface. Hardness may also be considered an indication of a material's strength. As an illustration, try scratching the surface of a piece of wood and a piece of steel. The wood scratches easily, whereas the steel is more difficult to scratch. We can say that the steel is harder than the wood.

Elasticity is the ability of a material to deform without taking a permanent set when the load is released. A rubber band is a good example of an elastic material under normal conditions. A rubber band can be stretched several times its normal length, short of its breaking point. Once we let go of it (remove the load), it returns to its original condition.

Plasticity refers to a material's ability to deform outside of the elastic range and yet not rupture. Blowing a bubble from a piece of bubble gum is a test of the plasticity of the bubble gum. When we apply pressure by blowing into the gum, a bubble forms. If we release the pressure at this point, the gum collapses and does not return to its original state. Therefore, the gum has gone beyond its elastic limit, yet it remains plastic, below the breaking strength of the material.

Energy capacity is a function of a material's strength and stiffness. It measures the ability of a material to absorb energy or have work done on it. The term *resilience* is used for energy capacity within the elastic range, and the term *toughness* is used to describe the energy required to rupture a material.

12.3 MECHANICAL TEST CONSIDERATIONS

No one test determines all of the properties contained in the previous section. Therefore, to simulate the conditions under which a material would be used in service, a number of tests are desired as well as necessary. The principal factors related to testing conditions concern three main areas: (1) the manner in which the load is applied; (2) the condition of the material specimen at the time of the test, and (3) the surrounding conditions (environment) during the actual testing.

Loading, or the method of applying the load, is commonly used to select and classify a mechanical test procedure. Three factors used to define the method of load application are (1) the kind of stress induced; (2) the rate at which the stress is developed, and (3) the number of cycles of load application.

There are five primary types of loading: tension, compression, torsion, shear, and flexure. In *tension* and *compression tests,* an axial load is applied to determine the effect on the specimen's cross section. In *torsion testing,* the load is applied through a twisting motion. In *shear testing,* the load is applied to shear, or cut, the material, similar to cutting a piece of wire with a pair of wire cutters. In *flexure testing,* both tensile and compressive forces are developed on the bottom and top of the material specimen, respectively. These tests are described in greater depth in the chapters that deal with them specifically. Figure 12–1 illustrates the basic types of loading.

Regarding the rate of load application, there are three main groups: static test, long-time tests, and dynamic tests. If the load is applied quickly, yet not fast enough to induce inertial effects on the test's results, the test is called a *static test.* Static tests usually range in length from a few minutes to several hours. Most mechanical tests fall into this category.

Special exceptions in this category involve sustaining the applied load over a longer period of time. They are known as *long-time* tests, of which a creep test is an example.

If the load is applied very quickly and in a manner such that the test results are affected by inertial forces, the test is called a *dynamic test.* There is one special dynamic test called an *impact test,* which is an important indicator of the toughness of a material.

Figure 12–1
Basic types of loading.

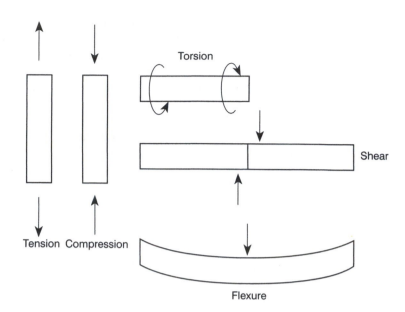

Tests involving the number of times the load is applied fall into two groups: single-load application and endurance, or fatigue, tests. In *single-load application tests,* the load is applied once, and the effects of this single application are studied. The stress and strain developed are the most critical data recorded. In *endurance,* or *fatigue, tests,* the effects of repeated loading are studied to determine the endurance, or fatigue strength, of a material. The stress and strain are also recorded for endurance tests. In addition, the number of cycles of load application is an important factor and must be recorded.

12.4 TESTING CONDITIONS

Conditions of the test refer to the condition of the material specimen at the time of the test and the environmental factors that relate to the test. These conditions include such items as form, dimensions, finish, temperature, barometric pressure, and humidity.

There are three classes of tests dealing with temperature: (1) those conducted at room temperature (which includes a majority of tests); (2) those conducted at elevated temperature (such as of jet engines, rocket boosters, heat shielding, insulation, and so on), and (3) those conducted at low temperature (of brittleness, ductility, insulation, and so on).

Some materials react differently based on the humidity or moisture conditions of the environment. Wood, for example, is hygroscopic, meaning that wood absorbs moisture when it is exposed to it and water evaporates from wood when it is in a dry environment. Tests where moisture content in the atmosphere may affect the results specify necessary environmental conditions at the time of the test. This specification for testing is required so that tests conducted under specified conditions are standardized, meaning that similar tests made under similar conditions should yield comparable results. In addition, tests may be conducted under a variety of environmental conditions, which must be specified and standardized for the tests to be reliable. These conditions include salt sprays for paint and corrosion testing; baths to simulate corrosive or underwater conditions, and different reagents, which are used to simulate a variety of environmental variables and conditions while the material is in service.

When specifying testing conditions, the physical condition of the specimen, including physical dimensions, chemical composition, and visual characteristics, as well as how to conduct the test should be detailed. In regard to conducting the test, holding, gripping, and any external support that the specimen requires during the test must be specified.

The aim of standardized test procedures is to keep all conditions constant except those under investigation. The variation in the conditions under investigation are then measured and recorded in standardized form so that anyone reading the data can immediately identify the results of the tests.

12.5 STRESS AND STRAIN

When testing materials, a load is applied, and the effects are measured. The *load* is typically measured in units of force, such as pounds-weight (lbf) or newtons (N), except in the case of torsion tests, where the load is a moment measured in foot-pounds (ft-lb) or newton-meters (N·m).

Figure 12–2
Basic types of stress.

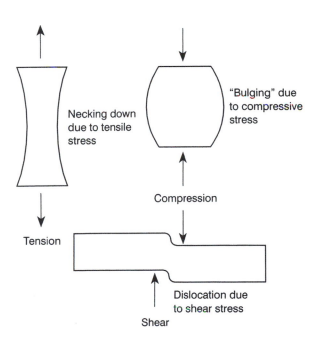

Necking down due to tensile stress

Tension

"Bulging" due to compressive stress

Compression

Dislocation due to shear stress

Shear

Stress is defined as the intensity of the internally distributed forces or components of forces that resist a change in the form of a body. Stress is commonly measured in units dealing with the force per unit area, such as pounds per square inch (lb/in.^{2}) or megapascals (MPa). There are three basic types of stress: *tension, compression,* and *shear*. The first two, tension and compression, are called *direct stresses*. Figure 12–2 shows the three basic types of stress.

Stress is calculated based on the original cross-sectional area of the specimen. In tension and compression testing, where the specimen is subjected to uniformly distributed forces, the stress is calculated by dividing the measured load by the minimum original cross-sectional area. The following formula is used to calculate the stress developed in a specimen:

$$\sigma = F/A$$

where: σ = stress developed in specimen (lb/in.^{2} or MPa)
          F = applied load (lb or N)
          A = original cross-sectional area of specimen (in.^{2} or mm^2)

The force must be converted to newtons before writing to megapascals. One megapascal (1 MPa = 1×10^6 Pa) is equal to 145 lb/in.^{2}, or 1 lb/in.^{2} is equal to 0.006895 MPa.

Example

A piece of wire 12 in. long is tied vertically. The wire has a diameter of 0.100 in. and is used to support a load of 100 lb. What is the stress developed within the wire?

$$\sigma = (100 \text{ lb})/[\pi \cdot (0.05)^2] = 12{,}739 \text{ lb/in.}^2, \text{ or } 87.86 \text{ MPa}$$

The term *deformation* is used to indicate the change in form of a body due to stress, thermal change, moisture absorption, or any other cause. Deformation is considered to be a linear change in dimensions and, therefore, is measured in units of length. Exceptions to this are the flexure and torsion tests. In *flexure testing,* the deflection is measured from some specified original starting position. In *torsion testing,* the angle of twist (in radians), or *detrusion,* is measured to determine the deformation.

Strain is defined as the change per unit length in the linear dimensions of a body accompanying a coincident change in stress. Strain, therefore, is the physical change in the dimensions of a specimen that results from applying a load to the test specimen. It is calculated by determining the ratio of the change in unit length to the original length and is, therefore, a dimensionless number that is typically quite small. The formula used to calculate the strain that accompanies a coincidental stress is

$$\varepsilon = (l_f - l_o)/l_o$$

where: ε = strain
          l_f = final gage length
          l_o = original gage length

Example

A 0.100 in. wire that is 12 in. long and supports a 100-lb weight stretches a total of 0.100 in. What is the strain for the wire?

$$\varepsilon = (12.100 \text{ in.} - 12.000 \text{ in.})/12.000 \text{ in.} = 0.00833$$

Strain is generally thought to be a unitless number. When units are given for strain, they are typically mm/mm or in./in., which are somewhat confusing. Strain is measured in the direction of the applied load and is measured parallel to the specimen under stress. An exception to this is shearing strain, which is measured parallel to the shearing force. Shearing strain is calculated based on the dimension perpendicular to the shearing force and is expressed in radians. Three types of strain are illustrated in Figure 12–3.

Permanent set is the change in form of a body or deformation remaining once the load is released. For instance, buckling—a permanent set—occurs in an aluminum soda can if a 200-lb person steps on it. Once the person gets off the can, the deformation is still present. The strain is the change in height of the can resulting from the person standing on it divided by the original height of the can.

The stresses and strains previously discussed are called *nominal stresses* and *nominal strains,* because they are based on the original dimensions of the specimen. "True" stresses and "true" strains are computed based on instantaneous dimensions when the specimen is subjected to given load conditions.

When test specimens are subjected to tensile or compressive stresses in a given direction, the strain developed in that direction is called the *axial strain*. However, strain also takes place in the direction perpendicular to the given direction of tensile or compressive stress; it is known as *lateral strain*. While the test specimen is within the elastic range, the ratio of lateral to axial strain, with loading taking place along a

Figure 12–3
Three types of strain.

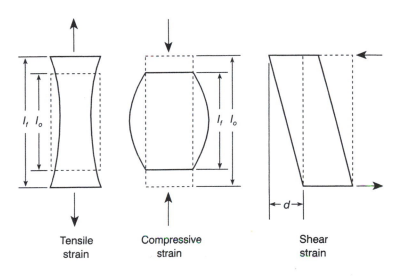

Tensile
strain

Compressive
strain

Shear
strain

single axis, is called *Poisson's ratio*. Poisson's ratio (P.R.) is found by dividing the lateral strain by the axial strain:

$$P.R. = (\text{lateral strain})/(\text{axial strain})$$

Example

What is Poisson's ratio for a material that exhibits a lateral strain of 0.002 and an axial strain of 0.005?

$$P.R. = 0.002/0.005 = 0.4$$

When axial extension takes place, a coincident lateral contraction also occurs, and vice versa. This can be seen by stretching a rubber band. While an axial extension occurs (the rubber band is stretched), a lateral contraction also occurs (the rubber band gets thinner). The range of Poisson's ratio for most common structural materials is typically 0.25 to 0.7.

12.6 STIFFNESS

The *stiffness* of a material is a measure of its ability to resist deformation under load. It is determined by the rate of stress to given strain. The greater the stress required to produce a given strain, the stiffer the material is said to be.

While under simple stress within the proportional limit, the ratio of stress to corresponding strain is known as the *modulus of elasticity*, which is designated by E. The modulus of elasticity refers to the stiffness of a material. It corresponds to the three types of stress: tensile, compressive, and shear. Therefore, there are three moduli of elasticity: the modulus in tension, the modulus in compression, and the modulus in shear. Under tensile stress, the stiffness of a material is often called *Young's modulus*. Young's modulus is the relationship between the stress and strain within the proportional limit:

$$E = \sigma/\varepsilon \text{ or stress/strain}$$

Example

Calculate Young's modulus for the wire used in the previous examples. The stress at the proportional limit is 300 lb/in.^{2}, and the strain is 0.005.

$$E = 300 \text{ lb/in.}^2 / 0.005 = 60,000 \text{ lb/in.}^2, \text{ or 414 MPa}$$

The higher Young's modulus is for a given material, the stiffer the material will be. The modulus of elasticity is the same as Young's modulus. It can also be found by calculating the slope of the stress-strain curve for a given material within the range of its linear proportionality between stress and strain. This situation is illus-

Figure 12–4
Calculating Young's modulus.

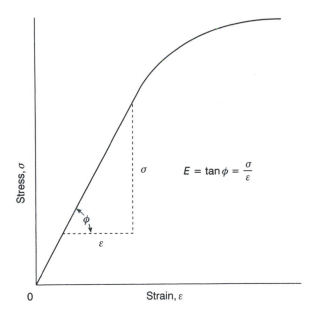

$$E = \tan \phi = \frac{\sigma}{\varepsilon}$$

trated in Figure 12–4. The term *modulus of elasticity* is somewhat confusing, how-ever, because it is a measure of the stiffness of a material, not its elasticity. Table 12–1 lists some common values of Young's modulus for a variety of materials. The third modulus associated with the elastic range is the *modulus of rigidity*, which is the stiffness of a material under shear stress.

Figure 12–5
Initial, tangent, and secant moduli.

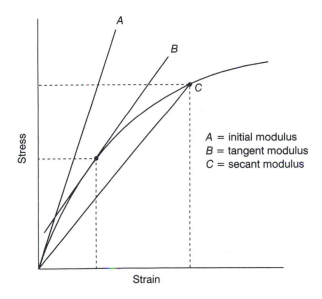

A = initial modulus
B = tangent modulus
C = secant modulus

Table 12–1
Common values for Young's modulus.

Material	Modulus of Elasticity (10^6 lb/in.^{2})	(GPa)
Aluminum	10.4	72
Brass (70–30)	16	110
Cast iron	13.4–21	92–145
Copper	16	110
Glass	10.4	72
Granite	7.3	50
Inconel	31	214
Lead	2.4	16.5
Limestone	8.5	59
Magnesium	6.7	46
Marble	8	55
Molybdenum	50	345
Monel	25	172
Nickel	30	207
Polyethylene	11.5	79
Quartz	10	69
Silver	11	76
Stainless steel	28.5	196.5
Tin	8	55
Titanium	15.5	107
Tungsten	59	407
Zirconium	12.5	86

Some materials' stress-strain curves exhibit no linear proportionality and, therefore, must be handled differently. Concrete and cast iron are two examples. The slope of a line drawn tangent to the stress-strain curve at the origin for a material that exhibits this characteristic is known as the *initial tangent modulus*. The slope of the same curve at some other point along the curve is called the *tangent modulus*. The *secant modulus of elasticity* is the slope of a secant of the curve and is also the ratio of any given stress to its corresponding strain. These concepts are illustrated in Figure 12–5. The modulus of elasticity for a material is commonly given in gigapascals (GPa). The modulus for a material under shear stress is typically two-fifths that for the axial stresses.

The modulus of elasticity is based on the elastic range for a material. There is no measure of stiffness within the plastic range of a material.

Figure 12–6
Strainometer with accessories.

12.7 STRESS-STRAIN DIAGRAMS

A *stress-strain diagram*, or *stress-strain curve*, is a plot of the stress, or load applied, with the corresponding strain produced. Stress is usually plotted as the ordinate (x axis) and strain is given as the abscissa (y axis). Stress-strain diagrams also cover plots of load or applied moments versus extension, compression, deflection, or twist. These diagrams can be plotted manually from data obtained through load readings and the use of a strainometer or can be plotted automatically by the test machine by using an electronic strain gage.

The first thing to do when planning a test is to determine which strainometer to use. *Strainometers* are devices used to measure the dimensional changes that occur during the test: extensometers, deflectometers, compressometers, and detrusion indicators. *Extensometers* and compressometers are used to measure changes in linear dimensions due to tensile and compressive forces, respectively. *Deflectometers* are used to measure the deflection, or bending, of a material under load conditions. *Detrusion indicators* measure the amount of twist of a test specimen under torsion loading. Figure 12–6 shows a strainometer with accessories.

Once a proper strainometer is selected, it is necessary to select the proper load and strainometer increments that will be used for successive readings. These increments can be based on time, changes in applied load, changes in strain, or other relatively constant changes in test conditions. It is desirable to take several readings so that a more accurate representation of test results can be obtained. If there is a long interval between data points, significant test results may be missed, but it may not be practical or feasible to record a large amount of data. The more data points recorded and plotted, the better defined and more accurate the stress-strain diagram will be.

Before recording test data, it is desirable to apply a small load to ensure proper gripping or bearing has been obtained for the test specimen and to make sure that the strainometer is properly placed and seated. The strainometer should then be set to zero, and preparations should be made to begin recording data.

Figure 12–7
Dislocations.

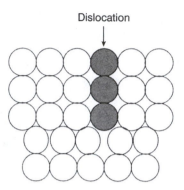

12.8 SLIP

Plastic deformation by slipping takes advantage of defects in the lattice structure of the material. These defects are called *dislocations* and occur in all metals. The bonds between atoms are dislocated, not broken, which accounts for the deformation rather than rupture of the metal. This deformation can occur only above a certain applied load (energy level), which is required to deform the atomic bonds.

A dislocation is one more or one fewer partial row of atoms within a crystal, as shown in Figure 12–7. This incomplete row, caused either by the addition of a partial row or the loss of a partial row of atoms, causes an incomplete bond to be formed and additional skewing of the remaining rows. These dislocations are fixed within the crystalline structure, but under an applied load, they can move through the lattice. They can move only in certain directions within certain planes, called *slip planes*.

Once a dislocation reaches the surface of a crystal, it produces a permanent change in the size of the spaces between atoms. This process is called *slip* because it appears that one part of the crystal has slipped over another. Once at the surface, the original dislocation is finished. Under increasing loads, new dislocations are generated. This action can be observed in the plastic action of real materials under applied loads. Increasing stress is required to drive the dislocations through the material to the surface. Once the load is removed or held constant, the dislocations cease to move through the material.

Slip continues until there is insufficient material left to resist the applied load. For instance, slip will continue in a tensile specimen, with a corresponding reduction in area, until breakage occurs. Slippage in a metal specimen along slip planes produces a reduction in area as the dislocation moves through the specimen. Once these dislocations reduce the material to a certain point, the remaining bonds can no longer support the applied stress. When this happens, a specimen will fail.

12.9 ELASTICITY

Elasticity is defined as the property of a material to recover completely from a deformation once the stress causing the deformation has been removed. No one material can exhibit perfect elasticity throughout the entire range of stress from initial load application to rupture. Some materials, such as steel, are elastic over a considerable range of stress but become inelastic beyond a certain magnitude of stress. Other materials, such as cast iron and concrete, are relatively inelastic regardless of the stress applied.

The elasticity of a material may be altered by elevating the temperature or by prolonged or rapid load applications. For instance, a material (e.g., wood or concrete) may be able to support a load and recover perfectly for a short period of time but will sag or rupture if the load must be sustained for a long period of time.

There are essentially three measures of *elastic strength*: the elastic limit, the proportional limit, and the yield strength. Figure 12–8 illustrates these three concepts.

The *elastic limit* is the greatest amount of stress a material can develop without taking a permanent set. The most accurate method that can be used to determine the elastic limit requires that successive tests be made on a material with increasing loads (very small increments) until a load is found that produces a permanent set. Determining the elastic limit in this manner is a tedious, painstaking process that is not practical or desirable in many cases and is, therefore, not common in practice.

The *proportional limit* is the greatest stress that a material can develop without deviating from linearity between stress and strain. Most materials exhibit linearity within the elastic range. The elastic limit and the proportional limit have often been confused, because the proportional limit can be used as an indication of the elastic limit. The proportional limit is often difficult to determine due to the difficulty in determining exactly where the stress-strain curve deviates from a straight line. Therefore, for many purposes, yield strength is used as a measure of elastic strength.

The *yield strength* is most often determined by the offset method. This is in accordance with ASTM standards, which define yield strength as the stress at which a

Figure 12–8
Elastic limit, proportional limit, and yield strength.

Figure 12–9
Upper and lower yield points.

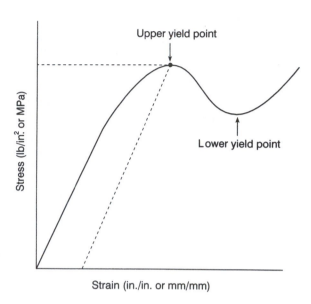

material exhibits a specified limiting permanent set (ASTM E 6). This method is based on an arbitrarily chosen offset or plastic yield but is easier to obtain than the elastic limit or the proportional limit. To determine the yield strength, we must choose an offset figure, or strain. Because the precision of the offset method decreases as the value of the offset decreases, typical strain values for structural materials range from 0.0001 to 0.001.

Two critical points within the yield range for ductile materials are the *upper yield point* (UYP) and the *lower yield point* (LYP). A material is said to have yielded when an increase in strain occurs without a further increase in applied stress. The UYP and the LYP are used to characterize the plastic range of a material and in design criteria. Figure 12–9 illustrates the upper and lower yield points for a material.

12.10 PLASTICITY

Plasticity is the ability of a material to endure a permanent deformation without rupture. Plastic action involves a time element, because a material within its plastic range can undergo a change in strain under a sustained stress. Also involved in plastic action is the concept of deformation before rupture or the limit of strain before rupture. Plasticity is evidenced by the characteristics of *yield, plastic flow*, and *creep*.

Plastic strains are a result of shear stresses developed within the material. The maximum shear stress for tensile loads occurs on a plane 45° from the direction of loading. The sliding of adjacent planes or slippage induced within a material doesn't take place along that 45° plane, because it would require a major rearrangement of the atoms within the crystal structure. Slippage takes the path of least resistance. The plastic strain developed by a material under load is a function of (1) the number of slip planes, (2) the crystal orientation of the material, and (3) the intensity of the applied stress.

Two properties of materials of particular importance to material testing that are related to plasticity are ductility and brittleness. *Ductility* is the ability of a material to be stretched considerably before rupture while maintaining the ability to sustain a reasonable load. An example of a ductile material is mild steel. It can be used to make wire or cable. If a material fractures with little or no elongation, the material is said to be *brittle*. Cast iron and concrete are brittle materials. The tensile strength of brittle materials is significantly lower than their compressive strength. Ductility is often determined by the *percent elongation* and *percent reduction* in area for a given material that has undergone a tensile test. The percent elongation of a test specimen is found by

$$\% \text{ elongation} = [(l_f - l_o)/l_o] \times 100$$

where: l_o = original gage length
l_f = final gage length

Example

If a 1.000-in. test piece elongates to 1.033 in. when loaded, what is the percent elongation of the test piece?

$$\% \text{ elongation} = (1.033 \text{ in.} - 1.000 \text{ in.}) \times 100 = 3.3\%$$

The percent reduction in area is

$$\% \text{ reduction} = [(A_o - A_f)/A_o] \times 100$$

where: A_o = original area of specimen
A_f = final area of specimen

Example

After rupture, a 0.505-in. tensile specimen's cross-sectional area was reduced to 0.492 in. What is the percent reduction in area for the tensile test specimen?

$$A_o = \pi(0.505 \text{ in.})^2 = 0.801 \text{ in.}^2$$
$$A_f = \pi(0.492 \text{ in.})^2 = 0.760 \text{ in.}^2$$
$$\% \text{ reduction} = [(0.801 - 0.760\%.801] \times 100 = 0.0512 \times 100 = 5.12\%$$

When reporting the percent elongation, the *gage length* is always stated. The percent elongation varies, depending on the original gage length. A common test used to determine a material's ductility is the cold-bend test, which is discussed later.

12.11 CATEGORIES OF STRENGTH

The *ultimate strength* (US) of a material is the maximum stress that a material can develop. It is determined by dividing the maximum load applied by the original cross-sectional area. It can also be determined by finding the highest point on the stress-strain curve. The type of loading is also stated—i.e., ultimate tensile strength, ultimate compressive strength, and so on.

The *tensile strength* is the maximum tensile stress that a material is capable of developing during a test. The *rupture*, or *breaking, strength* is the stress applied at rupture or the load applied at rupture divided by the original cross-sectional area.

The *compressive strength* is the maximum compressive strength that a material can develop. In the case of a ductile, malleable, or semiviscous material, the compressive strength is an arbitrary value based on the maximum allowable amount of deformation.

If a material undergoes repeated stress, it eventually fails due to fatigue. The *fatigue strength* of a material is the maximum stress that can be applied over a given number of cycles without causing the material to fail. The *fatigue limit* is the maximum stress below which a material will not fail regardless of the number of cycles.

Example

The test illustrated by Figure 12–10 was conducted on a specimen whose original diameter was 0.507 in. and whose final diameter was 0.493 in. The initial gage length was 2.000 in., and the final gage length was 2.037 in. after rupture. You should be able to determine the following using the given data: (a) modulus of elasticity, (b) proportional limit, (c) ultimate strength, (d) yield strength for a 0.2% offset, (e) percent elongation, (f) percent reduction in area, (g) breaking strength, and (h) the yield point.

Figure 12–10
Example test data.

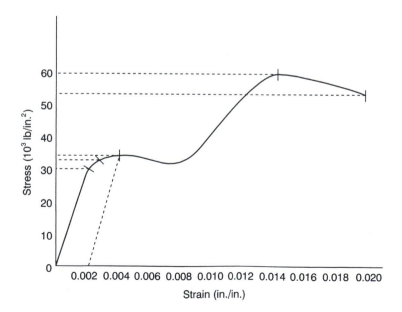

Hardness may be defined as the resistance of a material's surface to indentation or abrasion. It may also be thought of as the stress required to cause an indentation or abrasion on the surface of a material. For example, the Brinell test is a measure of the stress per unit area required to press a ball indenter into the surface of a material. However, most hardness tests values are based on arbitrary figures associated with some scale value or dial reading, which is an indication of the stress required for penetration but not a direct measurement.

12.12 TYPES OF FAILURE

In material testing, the type and characteristics of the failure are important in characterizing the material. Evidence of failure includes *yield, slippage, scaling* or *flaking, necking down* or a *reduction in cross section*, and *cracks*. These are qualitative inspections that can be assigned quantitative values. When conducting a test to failure, it is important to make sure that the material failed as predicted and not due to an improper setup, defects in the material, improper loading, or other test features that may invalidate the test and, therefore, invalidate the results.

When conducting a material test, we should be concerned with the ability of a material to resist deformation and at what point the material can no longer sustain the applied stress and resist such deformation. Many materials do not exhibit a clear rupture or a violent end to the test procedure. Therefore, it is difficult to determine when a material has failed.

Failure can be defined as the change in any characteristic that renders a material unsatisfactory for use. Failure does not necessarily mean rupture, although a rupture would render a material unsatisfactory for use. For example, if a tire picks up a nail that punctures it, the tire loses air and fails. It cannot take the sudden application of a stress. An example of a material that fails but doesn't immediately rupture is plastic wrap. Plastic wrap will exhibit a relatively large strain before rupture. Therefore, manufacturers must arbitrarily specify what is acceptable deformation before failure can be said to have occurred.

Particular types and characteristics of specific test failures are discussed in the sections dealing with those tests. Three general failure indications are slippage, separation, and buckling. However, a combination of these fractures may occur during actual testing.

Slippage is the movement of parallel planes within a material in parallel directions. These material planes slip or slide past each other, causing a fracture along the planes. Thus, the crystalline structure of the material and the arrangement of these crystals is an important factor in a material's predilection toward slippage and toward its ultimate result, shear fracture. Slip takes place more easily in close-packed material planes in which there is a minimum of contact between adjacent rows of atoms.

Creep, also known as *plastic flow*, is a constant slippage at a constant volume without material disintegration. If slippage continues, the molecular forces binding the material together are eventually overcome, and rupture occurs. Shear stresses causing slip can occur under tensile, compressive, torsional, or flexural loadings.

Separation, or *cleavage, fracture* takes place when the applied stress is greater than the internal binding forces of a material under tensile loading. As in slippage, the resistance of a material to separation depends on the material's crystalline structure and arrangement of the crystals within the material. An example of separation is when you twist an empty paper towel roll by grasping the ends and apply a torsional load. The roll tends to separate along the lines where the paper was wrapped.

Buckling or *buckling failure* occurs when a material is unable to resist a compressive stress and a buckle occurs. This type of failure is best shown in slender columns such as beverage cans. A buckling failure occurs when someone steps on a beverage can. The can may also buckle when subjected to a torsional loading, such as when you twist the ends of the can in opposite directions. The compressive stresses on the outside of the can compress the can.

12.13 ENERGY CAPACITY

Energy capacity, the ability of a material to absorb and store energy, is important to material testing. Energy capacity—or, specifically, the energy required to do work—is directly connected to the testing of shock resistance and impact testing. The fundamental behind energy capacity is that work or energy is equal to a force times an associated distance. Both work and energy are measured in joules (J). Work done on a plastic material is converted to heat and is, therefore, lost, but work done on an elastic material can be recovered mechanically.

The *elastic resilience* of a material is the energy or work required to stress a material up to its elastic limit. This energy can be recovered mechanically when the stress is released. Associated with the elastic resilience is the *modulus of resilience*. The modulus of resilience of a material under stress can be defined as the energy stored per unit volume for a material at its elastic limit. This energy is the maximum amount of energy that can be reclaimed. Moduli of resilience are calculated by multiplying the average stress by the strain at the elastic limit:

$$\frac{\sigma^2}{2E}$$

where: σ = stress at the proportional limit
         E = Young's modulus for the material

Referring to the stress-strain diagram, the work done can be shown as the area under the stress-strain curve (Figure 12–11).

The modulus of resilience is a measure of elastic capacity or elastic energy strength of a material. This measure is useful in materials selection for parts that will be subjected to repeated loads while the stresses involved are kept within the elastic limit for the material. Units for the modulus of resilience are joules per cubic meter (J/m^3) or pascals (Pa). For common grades of steel, the modulus of resilience ranges from 100 to 4500 kJ/m^3. The modulus of resilience increases as the percentages of carbon and other alloys in the material increase.

Figure 12–11
Area under curve.

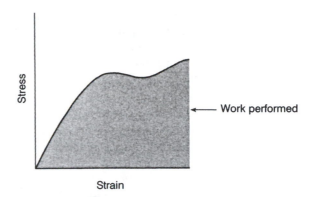

Some energy is always lost during energy transferrals, such as repeated loading and unloading. This is true for all materials and is evidenced in the bouncing of a ball. The initial loading causes the ball to bounce very high, and, even though loading takes place within the elastic range of the material, successive bounces are lower and lower, until the ball loses all of its energy. This phenomenon of losing energy is known as *hysteresis*. Within the elastic range for materials, it is called *elastic hysteresis*. Figure 12–12 shows the stress-strain curves for repeated loadings. The shaded areas illustrate hysteresis losses for a relatively elastic material [Figure 12–12(a)] and for a relatively inelastic material [Figure 12–12(b)]. The most practi-

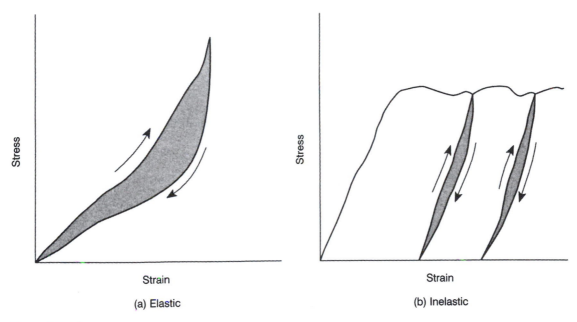

Figure 12–12
Hysteresis.

cal method of measuring this hysteresis is the precise observation of the damping of cyclic vibrations. For example, we could count the number of times a ball bounces after a fixed initial stress was applied.

Toughness is a measure of the energy required to cause a material to rupture. The *modulus of toughness* is the amount of energy per unit volume required to rupture a material under static load. Toughness is represented graphically by the total area under the stress-strain curve. It can also be evaluated qualitatively by examining the overall appearance of the curve. This fact is illustrated in Figure 12–11.

When selecting a material to be used in a situation where impact loads above the yield point are to be endured, the toughness of a material is an important criterion. The units for toughness are joules per cubic meter (J/m^3). Typical values for steel range from 40 to 120 MJ/m^3.

12.14 SUMMARY

When materials are subjected to increasing loads, first they exhibit elastic behavior, then plastic flow, and, finally, rupture or failure. The practical limit of a material's elastic action is referred to as its yield point, or yield strength. The proportional limit is the point on a stress-strain curve just before the material's curve deviates from linearity. The tensile strength, compressive strength, rupture strength, and fatigue strength are other strength values for materials. Measures of elastic ability are the modulus of elasticity, or Young's modulus, and the modulus of resilience. These moduli are ratios of stress developed within the material to the corresponding strain below the elastic limit.

Ductility is a plastic property, or a measure of the ability of a material to deform without rupturing. Toughness is the ability of a material to withstand applied stress and is the amount of energy or work required to cause rupture. Toughness is a function of both the strength and the ductility of a material. Hardness is a measure of the resistance of a material to surface penetration or abrasion and is usually based on arbitrary values. Hardness is particularly useful in that it can be used to indicate other properties.

Elastic properties are based on the molecular forces that exist between atoms. Inelastic, or plastic, properties are based on slip planes that exist between atoms or the type of crystalline structure. Slippage occurs when parallel material planes move in parallel directions. It is based on a shearing stress, which is greatest at 45° to the normal plane.

The ordinary behavior of structural materials at room temperature can be altered by such actions and conditions as raising or lowering the temperature; the rate at which stress is applied; the repeated application of a load, or adjusting various factors within the test environment, such as humidity.

QUESTIONS AND PROBLEMS

1. Define *strength* as it relates to materials.
2. What are mechanical properties?
3. Define the purpose of materials testing.
4. Define the terms *stress* and *strain*.
5. Define the following terms: (a) stiffness, (b) hardness, (c) elasticity, (d) plasticity, and (e) energy capacity.
6. Compare elasticity and plasticity.
7. In designing a product to absorb a sudden impact without breaking, such as safety glass, bulletproof vest, and other such items, would you use an elastic material or a plastic material? Why?
8. Compare the various moduli of elasticity, and discuss the use of the modulus of elasticity in specifying structural materials.
9. Define the terms *nominal stress* and *true stress*, and discuss the difference between the two.
10. What is Poisson's ratio? Describe the test that you would use to determine Poisson's ratio for a rubber band.
11. Define *elastic limit, proportional limit,* and *yield point.*
12. Given the following data, construct a stress-strain curve for the material and determine the proportional limit, the elastic limit, the modulus of elasticity, the yield strength at 0.2% offset, and the ultimate strength.

Stress (lb/in.^{2})	Strain (in./in.)
10,000	0.00027
20,000	0.00053
30,000	0.00105
40,000	0.00142
50,000	0.00202
60,000	0.00237
70,000	0.00307
80,000	0.00381
90,000	0.00433
97,000	0.00691

13. Sketch the stress-strain curve you predict would result from testing a 12-oz aluminum can by compression loading in 10-lb increments. Compare this with actual test data.
14. Sketch the stress-strain curve you predict would result from testing an elastic material such as neoprene rubber under tensile loading.
15. Sketch the stress-strain curve you predict would result from testing a plastic material such as polyethylene under tensile loading.
16. Sketch the stress-strain curve you predict would result from testing a brittle material under tensile loading.
17. Sketch the stress-strain curve you predict would result from testing a ductile material under tensile loading.
18. List the three basic methods of failure, and discuss the significance of each.
19. List and describe the effects of three testing conditions that affect the results of materials testing.
20. Describe the process of slippage.

CHAPTER 13
Introduction to Materials Testing

Objectives
- Define materials testing and recognize what goes into a materials test.
- Differentiate between testing and inspection.
- Define and apply the terms *precision, accuracy,* and *significance* when discussing materials testing.
- Describe the data collection procedure.
- Describe and recognize standard testing procedures.
- Specify correct parameters for selecting mechanical testing machines.
- Recognize and describe various mechanical testing instruments.
- Write a complete test report following recommended reporting procedures.
- Recognize that test selection depends on the properties under consideration and that these are often related.

13.1 INTRODUCTION

Materials testing involves three things:

- Finding out new information concerning known materials
- Discovering the specific properties or characteristics of new materials
- Developing standard methods and procedures for further testing

Materials testing involves both experimentation and testing. These two items are closely related but are not identical. The purpose of an experiment is to gain information when the outcome is uncertain; therefore, we gather as much information about the subject under study and the test conditions as possible. In testing, you follow a well-established, standard procedure and record only the necessary information. The results are expected to fall within established guidelines. The procedures and results of a specific test and experiment may be the same; the only difference would be the objective of the tester/experimenter.

Tests may be conducted in the field or in the laboratory. Tests done in the field are usually less precise than those conducted in the laboratory for the following reasons:

- Hazardous conditions
- Outside interference
- Time constraints or limitations
- Weather conditions
- Economic considerations

Some tests are better suited for the laboratory, and some are better made in the field. For instance, a slump test for a particular concrete batch is best made at the work site. Tests involving delicate equipment or that require highly specialized conditions are best done in the laboratory, where there is a better chance of control.

In selecting a testing procedure, four things to consider are whether to do the following:

- Conduct tests on full-size specimens
- Conduct tests on models or mock-ups
- Test sections or pieces drawn from actual materials
- Test samples of the raw materials

When conducting tests on other than full-size specimens, it is necessary to know the relationship that the sample or model has to the actual full-size article.

Tests may be classified into two distinct groups: destructive and nondestructive. In *destructive testing*, the part is not fit for use after the tests have been conducted. In *nondestructive testing*, the material should suffer no damage and should be able to return to service. An example of destructive testing is to blow up a balloon until it pops. The material is unfit for use after the test. An example of nondestructive testing is to fill the balloon full of water and check it for leaks. After the test, the water is drained, and the balloon is returned to service.

13.2 TESTING VERSUS INSPECTION

Testing is the physical performance of an operation or series of operations aimed at providing quantitative data regarding the properties of a material. *Inspection* is the observance of a material to determine the presence or absence of a desired or desirable attribute. These two procedures also differ in objectives. Testing provides information used to determine the level of quality for a material, whereas inspection is aimed at controlling the quality of a material through the application of established criteria by acceptance/rejection decisions. One is not better than the other. Both are aptly suited for their purposes and intents. It is the responsibility of the individual who is looking for information on a material's properties to decide whether it is more practical, reliable, and economical to conduct tests or to perform an inspection.

13.3 PRECISION, ACCURACY, AND SIGNIFICANCE OF TESTS

The term *precision* is used to indicate the repeatability of a measure or measurement, whereas the term *accuracy* is used to signify how close a measurement or measure is to the "true," or correct, value. This is illustrated in Figure 13–1.

In Figure 13–1(a), five arrows were shot, and all are within a 2-in. circle, which is very precise. However, none of the arrows hit the target, which is inaccurate. In Figure 13–1(b), five more arrows were shot. This time all five arrows hit the target, but they were spread all over the target. This is accurate but not very precise. Finally, in Figure 13–1(c), five arrows were shot, all of which landed within the bull's eye. This shooting is both precise and accurate.

Test significance is a measure of the extent to which the information obtained through the test procedure is a predictor of the performance of the same material in

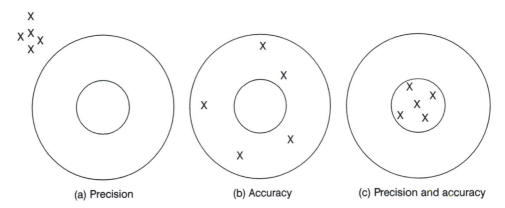

(a) Precision   (b) Accuracy   (c) Precision and accuracy

Figure 13–1
Precision and accuracy.

service. If the information obtained cannot be used to predict actual performance of a material in service, the test has little or no significance. Information gathered through testing may not always be directly applicable to the situation under question. This does not mean that the test was insignificant. After careful study over time, the data obtained by a particular testing method may be shown to be significant in other areas not readily recognizable. For instance, one Charpy test result for a material may not readily give an indication of the fatigue strength of the material, but after many tests and careful study of the data, it may be found that Charpy test data can be used as an indicator of the fatigue limit for that material.

13.4 COLLECTING DATA

Data collection, in regard to materials testing, refers to the operations required to obtain the necessary information on a material or a material specimen. It relies on the ability of the investigator to select the best possible procedure to follow. Even the best procedures can fail if the investigator fails to follow directions properly, skips some of the procedures to save time, or records the information improperly. Attention must be given to data collection if reliable results are to be obtained.

Before beginning to conduct testing on materials, the researcher should ask the following questions:

- Where is the data going to be collected?
- When is the best time to collect the data?
- Who is going to collect the data?
- What data need to be collected?

The first question deals with the location of the test. Is this going to be a field or laboratory test? Does the environment need to be controlled? In what special locations or possible locations can the test be made reliably?

Additionally, you must decide on the best time or times to collect the necessary information. Choose a time that is representative, yet both practical and economical. Select the person who is going to do the test. Is it best done in-house, or should it be contracted out to a qualified materials laboratory? Who is the best qualified to conduct this test? Has anyone had experience in this test procedure or process? Use common sense in deciding what will give the best results.

Finally, make a list of data that should be collected. Carefully scrutinize this list and rationalize each item on the list. If necessary, repeat the questioning process to justify each decision.

The data collected frequently depend on the constraints of the situation (field or laboratory test, time, money, and so on), the variable(s) under investigation, and the measuring instrument(s) available to the investigator. If the proper instrumentation is not available, auxiliary instruments can be used, and the data can be translated into the proper form. Although this may introduce greater error into the test than using the proper instrument in the beginning, sometimes it is the only option available to the investigator.

Details of the data collection that should be included are what type of data was gathered, who gathered the data, and what instrument or instruments were used to gather the data. If special instruments were used or made for a specific testing purpose, include details of this in the report. Give enough information so that someone who is reading it can replicate the procedure and, it is hoped, the results.

Some basic information concerning statistical analysis and manipulation of data is provided in the appendices. That material supplements this discussion and provides a basic understanding of statistical data analysis for the tests described in the following chapters.

13.5 TESTING PROCEDURES

Materials testing deals with the measurement or quantifying of the properties of materials. One area of particular interest is mechanical testing. Mechanical properties are those that deal with the elastic or inelastic behavior of a material under load. The primary measurements involved in mechanical testing are (1) the load applied and (2) the effects of load application. These are also called *stress* and *strain,* respectively.

Tests are generally classified by (1) the method of loading and (2) the condition of the specimen during the test. There are five primary types of loading in mechanical testing: tension, compression, direct shear, torsion, and flexure. Occasionally, a special loading situation may be used in which the stress produced results from (1) load application requiring more than one of the primary loading types, (2) load application in more than one axis involving the same primary loading type, or (3) combining two or more primary loading types to develop a secondary effect related to, but different from, the loading types or to develop localized or uneven stress distribution within the material. In addition to the type of loading that will be used during the test, the time element involved in the actual test procedure should be determined. Tests may be classified into three general groups: static, dynamic, and long-time tests. Decide which will give the most applicable results.

The temperature at which the test is conducted is of primary importance to the condition of the specimen during the test. The most common test condition is normal atmospheric, or room, temperature. Most materials are used at room temperature, and it is both economical and efficient to test at room temperature. The other two remaining options are to test either at elevated temperatures or at reduced temperatures. Both of these conditions require special laboratory equipment or are present in field tests as existing environmental conditions. For example, tests on the heat shields on the space shuttle may be conducted at elevated temperatures in the laboratory, but endurance tests for tires on desert vehicles will probably be conducted in the desert. As a further example, if we were trying to find the insulation value for a particular insulation material, we could do that in the laboratory or in the field at a particular point of application.

Other variables associated with developing and utilizing a testing procedure are humidity; vibration (acoustics); power regulation (electrical); method of holding, gripping, supporting or bedding the specimen; machine calibration, precision, and accuracy, and special atmospheres, such as corrosive, salt spray, or baths, that are required.

When trying to design or understand a test procedure, ask yourself the following questions:

- What answer(s) are to be found?
- What test(s) will help to find the answer(s)?
- Will the test results reflect actual performance?
- What are the limitations of this procedure and investigator?
- What precision should be required to be practical and economical?
- What test specimen(s) are best suited to obtain the required results?
- How many tests should be conducted to get representative results?

Ideally, a test should be purposeful, reliable, replicable, within a specified precision, and economical in both time and money. Accordingly, an ideal test is conducted for a well-defined reason with a specific purpose or purposes in mind. It should produce the same results every time it is conducted. Someone else conducting the same test should get the same results. The precision of the test and testing machines should be known and quantifiable. Finally, the test should be designed so that it can be conducted with a minimum expenditure in both time and money.

13.6 TESTING MACHINES

Many different tests are used to determine the different properties of a material. The most common tests are tension, compression, flexure, torsion, impact, hardness, fatigue, and creep. Although many more tests are available, the preceding are the most common and are described in greater depth in the following chapters. The equipment and methods described in this section are of the most general nature and are given here as an introduction to the concepts involved regarding the construction and utilization of testing machines.

Testing machines are employed to apply measurable loads to test specimens. The universal testing machine is probably the most commonly used. Universal testing machines can, with the proper accessories, be used to conduct a wide variety of tests, including tension, compression, direct shear, and flexure tests. They are made in various load ranges. Sizes range from a few pounds to as much as several million pounds of load capacity. The most common machines used in the average laboratory range from a 10-ton to a 200-ton load capacity.

Universal testing machines contain (1) some way of applying the necessary load to the test specimen and (2) some way of measuring the applied load with the necessary accuracy and precision.

Most universal testing machines use either mechanically driven or hydraulic systems to apply the load. We can picture the two types of loading by imagining using an old screw-type car jack. When we turn the screw body, the jack is raised or lowered, as required by the circumstances. Now, we replace the screw jack with a hydraulic

Figure 13–2
Universal testing machines.

jack. Now all we have to do is pump the handle to raise the jack and unscrew the bleeder valve to release it. The hydraulic jack gives us a considerable mechanical advantage and is easier to operate, yet requires a bit more maintenance. The two jacks (or testing machines) accomplish the same objective but by two different methods. The two basic types of universal testing machines are shown in Figure 13–2.

In order to gain flexibility, universal testing machines are built to adapt to as many situations as possible by employing accessories. These accessories include wedge grips for tension testing, compression blocks, auxiliary beams or platens for flexure testing, or shear blocks for direct shear tests. In any case, whenever a test is being conducted for the first time, make sure the proper accessories are available for the test and that the machine and the accessories, including the strainometer, are capable of conducting the test completely and *safely*.

Selecting the proper testing machine requires asking the following questions:

- What is the purpose of the test?
- What accuracy is desired in the test?
- What equipment is available and most practical for the test?
- Which equipment is most economical?

Depending on the situation and conditions of the test, the final decision regarding test equipment is usually based on the concept of best buy for the least money.

In routine testing, standard equipment will provide the necessary precision and accuracy desired, as long as these are known quantities. A commonly accepted rule for testing machines is that their accuracy falls within 1% (±0.5%) for the applied stress and strain over the entire load range. For better accuracy, the characteristics of various pieces of proposed apparatus should be studied, and actual performance data should be taken to determine which best suits the specific test requirements. Special circumstances dictate special procedures. You may need to talk with specific equipment manufacturers, give them the specific specifications for your testing purposes, and have them locate or manufacture the right testing machine. This is the most expensive option but is the most practical if the situation demands it. Before having a machine designed and built to your specifications, make sure that you have exhausted all other sources.

13.7 TEST INSTRUMENTS

Because there are so many properties of materials that are of interest to us, many different measures and measuring instruments are required to obtain the information needed. After selecting the data or properties that need to be collected, the best instrument with which to collect this data needs to be selected.

Some of the more important quantities that are measured are length, degree of rotation (angle), volume, mass or weight, force, pressure, time, temperature, voltage, current, and resistance. Although many ways of measuring these quantities are available, the one selected should be based on the same four questions previously stated:

- What is the purpose of the test?
- What accuracy is desired in the test?
- What equipment is available and most practical for the test?
- Which equipment is most economical?

To control the accuracy of the test, it is necessary to know the accuracy of all the instruments involved in recording data during the testing. *Error* (the difference between the recorded value and the true value) can be determined through a process called *calibration*. The instrument is used to measure a value of known accuracy, such as a gage block or gage standard.

The measurement error is also a function of the instrument's *sensitivity*. The sensitivity of an instrument is its ability to detect and respond to a change in the characteristic being measured. The better able an instrument is to detect these changes, the more sensitive the instrument is said to be.

Along with the sensitivity, the *least reading*, or smallest identifiable measurement, of an instrument is important in relation to error. The least reading is the smallest value that can be conveniently and reliably obtained from the instrument. Measurement error is dependent on the sensitivity, the least reading of the instrument, and the ability of the investigator to use the instrument. Measurement always involves some error; therefore, it is important to recognize this and be able to predict what this error might be.

There are two general classes of error: methodical and coincidental. *Methodical errors* tend either constantly to inflate or constantly to deflate the measured value. They are illustrated by a scale that constantly weighs 3 lb heavy; a rule that (due to thermal expansion) measures 1/32 in. long; and a dial indicator with a bent needle that measures 0.0005 in. low. Methodical errors may also involve the investigator's ability or experience, such as whether he or she can read an instrument; physical limits, such as whether he or she can stop a stopwatch at the exact moment that an event occurs, or personal prejudice or bias relating to what he or she thinks an experiment's or test's results should be. These errors do occur, but every attempt should be made to minimize and account for them in writing the test results.

Coincidental errors are occurrences that influence the results in a random fashion. They tend to cancel each other out. One coincidental error may inflate the reading, and another may diminish the indicated value. One cause of coincidental errors is the inexperience or inability of the observer to match the precision of the measuring instrument. It would be impossible to measure the lengths of several cars to the nearest 0.001 in. or 0.1 mm with a scale. It is possible, however, if a precision of only ±0.5 in. or 1 cm is required. Thus, coincidental errors are compounded by increasing precision requirements. Coincidental errors cannot be corrected. Instead, they can be dealt with by statistical methods, such as averaging, or other more involved methods, as discussed in statistical texts.

It is often necessary in materials testing to determine the dimensions of a specimen fairly precisely. These measurements are then checked against the specifications to determine whether the specimen is within tolerance. *Tolerances* are the max-

Figure 13–3
Common test instruments.

Figure 13–4
Balanced bridge circuit.

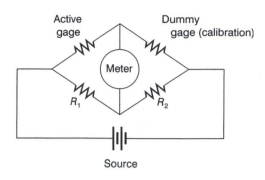

imum and minimum values that can be accepted, often written as a nominal value plus or minus some percentage or value. The tolerance may be unilateral or bilateral, depending on the situation. A *unilateral tolerance* specifies a variance in the dimension in only one direction, such as 2.500 in. + 0.001 in./–0.000 in. or 1.095 in. + 0.000 in./–0.002 in. *Bilateral tolerances* allow variation in both directions, such as 35°C ± 2° or 25 mils + 5 mils/–3 mils.

There are many types of gages or test instruments used in the testing and inspection of materials. These range from very simple devices that can be used successfully with a few minutes' instruction to very complex devices that require many hours of training before individuals can use them properly.

Other types of gages used in materials testing include calipers, micrometers, dial indicators, and dividers. These are just some of the more common measuring instruments used. Each of these instruments is shown in Figure 13–3, and specific instructions on how each one is used are given in the specific test procedures.

In addition to the mechanical and manual instruments already discussed, electric and electronic measuring instruments are also used. These devices allow remote data collection and automatic data recording, which make them more convenient, more adaptable, and less prone to recording errors. One electric gage in particular—the *electric strain gage*—is important to the field of materials testing.

The electrical resistance of a wire fluctuates up or down, depending on whether it is in tension or compression, respectively. In other words, a wire under a tensile load shows a greater resistance than normal, and a wire under a compressive load exhibits a less-than-normal resistance. Thus, this change in resistance can be used to indicate the strain developed in the test specimen. The rate of resistance change or the sensitivity of the wire used in the electric strain gage or load cell depends on the type of wire used in the gage.

The strain gage is connected to a strain indicator or a balanced Wheatstone bridge circuit, as shown in Figure 13–4. The signal from the strain gage is very small and must be amplified before it can be used. After the signal is amplified, it is then fed into a chart recorder, which constantly and consistently logs the strain throughout the test. Electric strain gages are more accurate, reliable, convenient, and sensitive than most other forms of strain measurement, but they are also more expensive. It is cheaper—but not as accurate or reliable—to use dividers and a caliper or scale.

13.8 PRESENTATION OF RESULTS

The presentation of the results and findings of testing and research is very important. You need to share the knowledge gained through these methods with others conducting similar tests and research and also with people involved with similar materials. In addition, this information must be reported accurately and in a manner that is easily read and clear in its meaning. Communication will take place only when both the sender and receiver understand the message.

In preparing a technical report of your findings, a general outline of the contents could be as follows:

1. Statement of the problem
2. Materials, methods, and procedures used during testing
3. Data presentation and results
4. Summary, conclusions, and discussion
5. Appendices to support findings

These are general suggested titles for the test report that indicate what should be included in the different sections.

The statement of the problem describes what the objectives of the test are and what the investigator intends to do about the problem. Also, any specific features of the test, such as multiaxial loading, can be included in the statement of the problem.

The materials, methods, and procedures section should include what materials were tested, the conditions of the test and specimen, important apparatus used during the test procedure, and the procedure used during the actual testing. Reference is usually made to the relevant ASTM or similar standard.

In the data presentation and results section, the data are plotted on graph paper with proper units assigned or listed in tabular form in a clear and meaningful manner. All tables, graphs, or diagrams should be introduced with statements in the text that point out the significance of the material. Don't just stick in some numbers; make sure that the data illustrate what you want to emphasize and make that result clear in the text for the reader. Also, remember that one well-placed graphic may illustrate your findings or some confusing point better than several pages of written material.

The summary, conclusions, and discussion section includes the general findings of the test or experiment and summarizes the important points. Also included in this section should be a discussion of what errors were involved during the testing. You can also to interject your opinions on the material: its use, projected performance, and acceptability for use.

If needed, an appendix section can be included that provides supportive material or material required for clarity that, for one reason or another, was not appropriate for inclusion in the text.

SAMPLE TEST: TENSILE TEST ON STEEL

TITLE: TENSILE TESTING OF METALS: 1020 COLD DRAWN STEEL

INTRODUCTION:

The following is an example of how to construct a test report. Further information on testing and reporting can be found in Section Two. Please use this as a model for preparing your own reports.

OBJECTIVE:

To determine the strength and properties of a specimen; to observe the behavior of the specimen under a tensile load; and to study the fracture developed at rupture.

Equipment Required
1. Universal Testing Machine
2. Extensometer
3. Micrometer or calipers
4. Dividers
5. Safety Glasses and Shields

SPECIMEN:

Smooth-end round tensile test specimen

PROCEDURE:

1. The original cross-sectional area of the specimen was measured and recorded.
2. A 2-in. gage length was marked on the specimen.
3. The proper grippers were selected and mounted on the machine.
4. The crosshead of the machine was adjusted to proper range for the test.
5. The load indicator was reset to zero.
6. The specimen was properly mounted in the grippers, making sure that the specimen was seated properly in the test apparatus.
7. The crosshead was preloaded to take up the slack
8. An extensometer was mounted on the machine and specimen to measure the crosshead movement.

9. Tensile loading was applied to the specimen at a rate of approximately 5000 lb/min.
10. Once the yield point was reached, the extensometer was removed and displacement measured with the dividers. The displacement was measured and recorded.
11. Loading continued until rupture where the ultimate strength and rupture strength were recorded.
12. Measurements of the final gage length and final cross-sectional area were taken and recorded.

DATA:

Material: **1020 cold-drawn steel**   **Date:**

Original diameter: **0.505 in.**   **Gage length:** **2.0000 in.**

Reading	Load (lb)	Stress (lb/in.^{2})	Δ length (in)	Strain (in./in.)
1	1,000	5,100	0.0004	0.0002
2	2,000	10,200	0.0008	0.0004
3	3,000	15,300	0.0010	0.0006
4	4,000	20,400	0.0014	0.0008
5	5,000	25,500	0.0018	0.0009
6	6,000	30,600	0.0022	0.0011
7	7,000	35,700	0.0025	0.0012
8	8,000	40,800	0.0029	0.0014
9	9,000	45,900	0.0033	0.0017
10	9,500	48,000	0.0035	0.0018
11	10,000	50,500	0.0037	0.0019
12	10,500	53,000	0.0042	0.0021
13	11,000	56,000	0.0100	0.005
14	12,000	61,100	0.0160	0.008
15	13,000	66,300	0.0260	0.013
16	14,000	72,000	0.0620	0.031
17	11,800	60,000	0.4000	0.250

Fracture

Yield strength at 0.2% offset: 52,750 psi
Proportional limit: 45,500 psi
Ultimate strength: 68,000 psi
Rupture strength: 60,000 psi

Example of tensile test data.

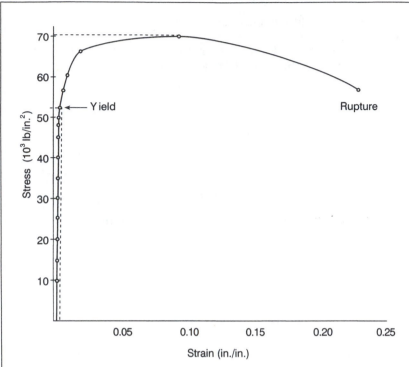

Sample test graph.

From the test graph:

Yield strength $= 53{,}000 \ lb/in^2$
Ultimate tensile strength $= 70{,}500 \ lb/in^2$
Rupture strength $= 60{,}000 \ lb/in^2$
Proportional limit $= 50{,}000 \ lb/in^2$

Calculations:

$$\% \text{ reduction in area} = \frac{A_o - A_f}{A_o} \times 100 = 37\%$$

$$\% \text{ elongation} = \frac{l_f - l_o}{l_o} \times 100 = 15\%$$

$$E = \frac{\sigma}{\varepsilon} = \frac{50{,}100}{0.0018} = 2.8 \times 10^7 \ lb/in^2$$

(modulus of elasticity)

Calculations based on test data.

Data were recorded concerning the applied load and displacement at appropriate intervals, approximately every 1000 lb. From these data, a stress-strain curve was plotted and is provided below. The following features were either taken from the graph or calculated based on results:

RESULTS:

In this section, an attempt is made to answer questions that may have arisen during the test. Sample questions may include the following:

1. Did the test specimen conform to ASTM standards in shape and dimensions?
2. Was the test carried out in accordance with ASTM standards?
3. Did the results match your expectations for this material? Why or why not?
4. What errors may have been introduced during the test? What precautions were taken to eliminate these?
5. Discuss the following items in the results:
   a. Elastic limit
   b. Proportional limit
   c. Yield point
   d. Yield strength
   e. Ultimate strength
   f. Modulus of elasticity
   g. Percent elongation
   h. Percent reduction in area
   i. Nature and type of fracture

13.9 TEST SELECTION

It is important to match the properties of a material to the conditions it will have to sustain in service. It is equally important to find the proper test to measure those properties accurately and precisely. Selecting the proper test and carefully applying standardized testing procedures are important considerations in materials testing. Often, one test will give an indication of another indirect property. For example, hardness is often used as an indication of a material's ductility.

The following chapters deal with the selection and procedures related to the mechanical testing of materials to determine their mechanical properties. This section deals with the remaining physical properties of materials: electrical, magnetic, optical, and thermal. Figure 13–5 represents the entire spectrum of material properties.

Chemical properties are those that result from or relate to the structure of a material and its formation from other elements. Chemical properties are, therefore, not always readily observed and are often tested in chemical labs. Due to their nature, chemical properties often require that the test specimen be destroyed. Chemical properties measured in laboratory facilities include microstructure, chemical composition, crystalline structure, and chemical reactivity. The procedures for these tests can be found in an analytical chemistry text.

Class	Property
General	Density, porosity, moisture content, macrostructure, microstructure
Chemical	Composition, pH, resistance to corrosion, weatherability
Physiochemical	Liquid absorption/repulsion, shrinkage and swell due to hygroscopic action
Mechanical	Strength in: tension, compression, shear, torsion and flexure; static, impact; endurance; stiffness, toughness, elasticity, plasticity, ductility, hardness
Thermal	Specific heat, thermal expansion, thermal conductivity
Electrical/magnetic	Conductivity, permeability, galvanic action
Acoustic	Transmission, reflection, absorption
Optical	Color, transmission, reflection, collimation

Figure 13–5
Material properties.

Physical properties are those that relate to the interaction of materials with various forms of energy and other forms of matter. The physical behavior of materials depends on the atomic structure, atomic arrangement, and crystal structure of the material. Its atomic structure—particularly the valence shells—help determine whether a material is an electrical conductor, semiconductor, or insulator. The atomic arrangement, structure, and crystalline structure all help determine the optical and magnetic properties of a material. Color is a physical property. Density and mass are also physical properties of materials. Many physical properties can be measured without destroying the material.

Physical properties that can be measured are coefficient of linear expansion, density, dielectric strength, electrical resistivity, melting point, refractive index, specific gravity, specific heat, and thermal conductivity. There are various specific tests that have been standardized for applications such as nuclear engineering. These tests are used to measure other physical properties, such as nuclear absorption rates, damage due to excessive radiation, ability to absorb radiation, and other properties associated with nuclear energy. These tests can be found in various nuclear safety and nuclear engineering manuals. Here we deal with the more common and general physical tests.

The coefficient of linear expansion is the rate at which a material will elongate when heated. It is expressed as a unit increase in length per unit rise in temperature (in./in.-°F or cm/cm·°C). This information is important when dissimilar materials are joined and heated. One of the materials will often have a higher coefficient of linear expansion than the other. For example, aluminum has a higher thermal expansion rate than steel. If the two were bolted together and heated, the aluminum would expand twice as much as the steel at the same temperature. This action could lead to disastrous results.

Density is the weight of a material per unit volume. The higher the weight, the denser the material.

Dielectric strength is the highest potential difference or voltage that an insulating material can withstand for a specified time without showing signs of breaking down. This property is important for capacitors.

Thermal conductivity is the rate of heat flow per unit of time in a material under constant conditions, per unit area, per unit temperature in a direction perpendicular to the area. The preceding is a very specific definition for the property of heat transfer. Heat transfer is the ability of a material to conduct heat from one area to another. The units are Btu/h/ft²/°F/ft or W/m·K (watts per meter-Kelvin).

Electrical resistivity is the electrical counterpart of thermal conductivity. It is important when a material is specified to be a conductor or an insulator. It determines the rate at which current is allowed to flow through a given cross section of a material for a given length of that material. Units for electrical resistivity are the microhm-centimeter ($\mu\Omega$·cm) or ohm-meter (Ω·m). Electrical conductivity is simply the reciprocal of (or 1 over) the electrical resistivity. Conductivity/resistivity is used to separate conductors, semiconductors, and insulators.

Melting point is the temperature at which a solid material turns to liquid or a liquid material solidifies when cooled. A range of temperatures is sometimes given for the melting point.

Refractive index is a property that relates the velocity of light in a vacuum to its velocity in another material. This index is expressed as a ratio.

Specific gravity is the ratio of the mass of a solid or liquid material to the mass of an equal volume of water.

Specific heat is the ratio of the amount of heat required to raise the temperature of a unit mass of a material 1° of temperature to the heat required to raise an equal mass of water 1°.

Magnetic classes, such as when a material is permanently magnetic, ferro-magnetic, paramagnetic, or nonmagnetic, depend on the ability of the material to accept magnetism. Properties such as coercivity, hysteresis, permeability, and retentivity depend on a material's ability to be magnetized. A good electrical text or any good physics text will provide an adequate explanation of the magnetic properties and classes of material.

The preceding are some of the common physical properties of materials. These properties depend on a material's atomic structure, atomic arrangement, and crystal structure. They are important in engineering applications but are often overshadowed by the mechanical properties, which have more direct application.

13.10 SUMMARY

The study of materials and materials testing is not a new concept. People have been using materials since the beginning of time. Modern methods of study started with the development of the scientific method of study. Since then, numerous improvements have been made in testing, and the fields of materials science and materials testing have gained in importance.

Materials testing involves the identification of the properties of materials through well-established procedures. These procedures are conducted in a manner that is both accurate and precise. The significance of a test is in its ability to predict the performance of a material in service.

Data collection refers to the procedures or operations that are performed to obtain data on a material or a material specimen. The ability of the investigator to select the best possible procedure to follow determines the effectiveness of the data collection. Even the best procedures can fail if the investigator fails to follow directions properly, skips some of the procedures to save time, or records the information improperly.

Before beginning to conduct testing on materials, the researcher should ask the following questions:

- Where is the data going to be collected?
- When is the best time to collect the data?
- Who is going to collect the data?
- What data need to be collected?

Materials testing deals with the measurement or quantifying of values that indicate the properties of materials. Mechanical properties are those that deal with the elastic or inelastic behavior of a material under load. The primary measurements involved in mechanical testing are the load applied and the effects of load application. These measurements can also be called stress and strain, respectively. Some of the more important related quantities also measured are length, degree of rotation (angle), volume, mass or weight, force, pressure, time, temperature, voltage, current, and resistance.

Tests may be classified as static, dynamic, or long-time tests. Typically, tests are conducted at normal atmospheric, or room, temperature, because such a procedure is both economical and efficient. Tests can also be conducted at elevated temperatures or at reduced temperatures. Both these conditions require special laboratory equipment or occur in field tests as environmental conditions. Other variables associated with testing and testing procedures are humidity; vibration (acoustics); power regulation (electrical); method of holding, gripping, supporting, or bedding the specimen; machine calibration, precision, and accuracy, and special atmospheres, such as corrosive, salt spray, or baths, that are required.

The most common mechanical properties tests are tension, compression, flexure, torsion, impact, hardness, fatigue, and creep. Ideally, a test should be purposeful, reliable, replicable, within a specified precision, and economical in both time and money.

The difference between the reading on the instrument and the true value of the measurement in question is the error. Measurement error can be determined by a process known as calibration. The measurement error is also a function of the instrument's sensitivity. The sensitivity of an instrument is its ability to detect and respond to a change in the characteristic being measured. The better able an instrument is to detect these changes, the more sensitive the instrument is said to be.

There are two general classes of error: methodical and coincidental. Methodical errors tend either constantly to inflate or constantly to deflate the measured value. Coincidental errors are random occurrences that influence the results in a random fashion. They tend to cancel each other out. Coincidental errors cannot be corrected.

Tolerances are the maximum and minimum values that the customer or manufacturer consider acceptable. Tolerances are often written as a nominal value plus or minus some percentage or value.

Gages used in materials testing include snap gages, ring gages, plug gages, calipers, micrometers, dial indicators, and dividers. These are just some of the more common measuring instruments used. One electronic gage in particular—the electric strain gage—is important to the field of materials testing.

It is important to publish test results so that others may share in the findings. Therefore, it is important that the report be clear in its meaning and form. A technical report of findings might follow this general outline:

1. Statement of the problem
2. Materials, methods, and procedures used during testing
3. Data presentation and results
4. Summary, conclusions, and discussion
5. Appendices to support findings

QUESTIONS AND PROBLEMS

1. Measure 10 coins of a particular denomination to the nearest 0.001 in. using a micrometer. Then prepare a written report describing the test and illustrating your findings. Include some recommendations and your conclusions about the variance in coins.
2. Repeat the coin tests using a scale. Were the data similar or different? Why? Explain the variance in data.
3. What other ways can you identify to obtain meaningful results about the different properties of coins?
4. List 10 common items that are tested.
5. How are the items in Question 4 tested? Which tests are used? For what purposes are they used?
6. Describe the purpose of materials testing.
7. Why should materials tests be standardized?
8. Explain the difference between precision and accuracy and how they relate to materials testing.
9. How is the term *significance* used in relation to materials testing?

10. List several conditions and influences that alter the results of materials testing.
11. List five common physical properties of materials.
12. Describe the procedure and data that you would use to determine the height and weight of everyone in your class. Use the procedure described in this chapter.
13. What errors can you anticipate in your chosen methodology (Question 12)?
14. List five factors that you feel would affect the test results in Question 12. [*Example*: time of day (just eaten a meal versus empty stomach).]
15. How can you use the data from Question 12 to predict the height and weight of others?
16. Take a common product or process that is measured (fuel, food product, time, speed, distance, and so on) and describe how the terms *accuracy, reliability,* and *precision* relate to their manufacture or performance.
17. Measure the length and width of a room using three different methods: (1) placing one foot in front of

the other and stepping it off, (2) using a one-foot ruler, and (3) a tape measure. In terms of accuracy, reliability, and precision, describe how these factors were important to your measurements. How could you better plan and execute a measuring plan to reduce the errors in your measurements?

18. Products are generally tested according to a sampling plan: 1 out of 10, 5%, first three out of each new batch, and so forth. What questions or concerns would you have regarding the sampling plan involving medical supplies and equipment, public transportation (aircraft, bus, subway, and so on), and other critical areas?

19. Write a sampling plan for a child safety seat. Defend your plan and describe the probability of an error passing through to the customer. What steps could be taken to prevent this from happening?

20. Often, physical and chemical properties are given numerical values for analysis. List five features, attributes, or characteristics that are important in material selection, but are not quantifiable or best not quantified.

21. Describe five applications where elasticity is a major factor.

22. Describe five applications where plasticity is a major concern.

CHAPTER 14
Tensile Testing

Objectives
- Recognize and define common terms related to tensile testing.
- Describe the terms *stress* and *strain* and the effects of both in tensile loading.
- List the equipment necessary to conduct a tensile test.
- Describe the operation of various equipment related to tensile testing.
- Relate the general procedures used in conducting a tensile test.
- Perform the necessary calculations related to tensile testing.
- Recognize expected tensile test results.
- Describe common variations in standard tensile test procedures.

14.1 INTRODUCTION

One of the most common tests performed on woods, metals, polymers, and other materials is the *tensile test*. Tensile tests are generally conducted on standard specimens but may be conducted on any specimen of known cross-sectional area and original length.

Tensile tests are used to determine the mechanical behavior of materials under static, axial tensile, or stretch loading. Data and calculations for tensile testing usually include elastic limit, percent elongation, modulus of elasticity, proportional limit, percent reduction in area, tensile strength, yield point, yield strength, and other related properties. One special type of tensile test is the creep test, which is discussed in the next chapter. ASTM procedures for tensile tests may be found in sections E 8 (metals), D 638 (plastics), D 2343 (fibers), D 897 (adhesives), D 987 (paper), and D 412 (rubber).

14.2 PRINCIPLES

Tensile loads are those that tend to pull the specimen apart. They put the material in tension. Tensile loads produce deformations. *Deformation* is a change in the form of a specimen that is produced by the applied load. The measurement of a deformation is based on changes in the original dimensions of the specimen—that is, deformation is determined based on an original and final gage length of the material specimen that has undergone testing. A gage length is a standard unit of length used to determine the degree of stretch or deformation produced during the test. The standard length for common tensile tests is 2 in.

In tensile loading, the tensile load is denoted as P. Tensile loads are measured in pounds, kilograms, or newtons. Tensile strength is based on the load that a specimen endures per unit of cross-sectional area. The load equivalent for a 1-in.^{2} cross section is called the *stress*. Stress (σ) is the load per unit of cross-sectional area measured in units of pounds per square inch (lb/in.^{2}) or pascals (Pa):

$$\sigma = \frac{F}{A}$$

where: σ = stress
F = applied force (lb or N)
A = area in (in.^{2} or mm^2)

When using metric units, the load may be recorded in kilograms and converted to newtons. The cross-sectional area is converted to square meters. These calculations are necessary to convert to newtons per square meter, which are pascals. One megapascal (MPa) equals approximately 145 lb/in.^{2} and 1000 lb/in.^{2} equals approximately 6.895 MPa. The general equation for stress is

$$\text{stress} = \frac{\text{load}}{\text{cross-sectional area}}$$

If the tensile specimen has a rectangular cross section, the formula for stress becomes

$$\sigma = \frac{P}{WD}$$

where: σ = stress
P = tensile load or force
W = width of rectangular specimen
D = depth of rectangular specimen

For a round cross section, the formula becomes

$$\sigma = \frac{P}{\pi d^2/4}$$

where: d = diameter of the round specimen

Example

What is the stress developed in a rectangular specimen that is 0.5 × 0.5 in. at 1000-lb tensile load?

$$\sigma = \frac{1000\ \text{lb}}{(0.5\ \text{in.})(0.5\ \text{in.})} = 4000\ \text{lb/in.}^2$$

Example

What is the stress developed in a round specimen with a 0.505-in. diameter at 1000-lb tensile load?

$$\sigma = \frac{1000\ \text{lb}}{\pi(0.505\ \text{in.})^2/4} = 4993\ \text{lb/in.}^2$$

In a standard round test specimen, a 0.505-in. diameter is used so that the original cross-sectional area is 0.200 in². The force applied can be converted to stress by simply multiplying the force by 5. The dimension 0.505 in. was chosen for convenience rather than for any significant mechanical importance.

When a tensile load is increased, the specimen tends to grow longer as it is pulled apart. Therefore, prior to testing, the original cross-sectional area and gage length are measured. A two-point gage punch is often used to mark the original gage length on standard specimens. An extensometer or strain gage may be used to determine the extension of the specimen during the test, or it may be determined by measuring the difference in the gage points. The difference between the final and original gage lengths is called the *elongation*. The units of elongation are inches or millimeters. The elongation divided by the original gage length is called the *strain*.

Strain is typically written as a dimensionless number; however, if units are given, they are inches per inch (in./in.) or millimeters per millimeter (mm/mm).

Example

If the final gage length of a specimen is 1.005 in. (25.53 mm), based on an original gage length of 1.000 in. (25.40 mm), what is the strain?

$$\varepsilon = (l_f - l_o)/l_o = (1.005 \text{ in.} - 1.000 \text{ in.})/1.000 \text{ in.} = 0.005 \text{ (in./in.)}$$
$$\varepsilon = (25.53 \text{ mm} - 25.40 \text{ mm})/25.40 \text{ mm} = 0.005 \text{ (mm/mm)}$$

Deformation in tensile testing occurs in two directions, laterally (at right angles to the applied load axis) and axially (along the axis of loading). The ratio of the lateral deformation to the axial or longitudinal deformation is called *Poisson's ratio*.

Example

If the lateral strain of a specimen is 0.005 and the axial strain is 0.010, what is Poisson's ratio for this material?

$$P.R. = \text{lateral strain/axial strain} = 0.005/0.010 = 0.5$$

For most engineering materials, the values for Poisson's ratio range from 0.25 to 0.7.

Stress is the intensity of internally distributed forces developed in a material during testing. Stresses are calculated by dividing the applied force or load by the original cross-sectional area. Materials that will return to their original form, without permanent deformation, after the applied stress is removed are known as elastic materials. These materials exhibit *elastic behavior*. Hooke's law applies to materials

Figure 14-1
Stress-strain curve.

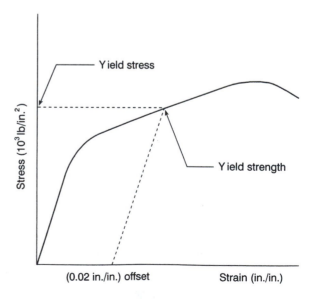

within the elastic region and states that the strain developed is proportional to the applied stress. By examining a stress-strain curve, the elastic region is seen to be the period from the origin of the curve to the elastic limit. The elastic limit is the point beyond which an increase in stress induces a permanent deformation. Figure 14–1 represents a typical stress-strain curve.

An indication of a material's stiffness is gained by calculating the modulus of elasticity. The modulus of elasticity is the ratio of the incremental increase in stress to the coincident incremental increase in strain. The modulus of elasticity can be found by calculating the slope of the stress-strain curve over the period of linearity. The modulus of elasticity is also known as *Young's modulus,* after the English physicist who first proposed its use in materials science. The modulus of elasticity is a measure of a material's stiffness. Stiff materials do not exhibit much deformation under a load.

Example

Calculate the modulus of elasticity for a material that has a strain of 0.0025 for an applied stress of 30,000 lb/in.^{2}

$$E = \text{stress/strain} = 30,000 \text{ lb/in.}^2 / 0.0025 \text{ in.} = 12 \leftrightarrow 10^6 \text{ lb/in.}^2$$

Some standard tensile test specimen recommendations and specifications are shown in Figure 14–2.

Prepared specimens come in a variety of forms and shapes. Typically, the cross section is round, square, or rectangular. In addition, the specimens should all have some form of standard gage length. The ends of the specimen should be suitable for use with standard grippers. Round specimen ends may be smooth, shouldered, pinned, or threaded. The ends should be long enough to be able to be gripped properly.

The primary function of the grippers is to hold the specimen so that the load can be transmitted efficiently from the machine to the specimen. One essential point is

Figure 14–2
Tensile test specimens.

(a) Smooth-end round tensile test specimen

(b) Threaded-end tensile test specimen

1.00 in. (25 mm)   0.75 in. (19 mm)

2.25 in. (57 mm)

8 in. (200 mm)

0.125 in. (3.2 mm)

(c) Flat tensile test specimen

Butt joint

200 in. (51 mm)

(d) Brazed or silver-soldered tensile test specimen

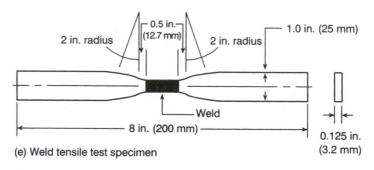

0.5 in. (12.7 mm)

2 in. radius   2 in. radius   1.0 in. (25 mm)

Weld

8 in. (200 mm)

0.125 in. (3.2 mm)

(e) Weld tensile test specimen

3/8 –24 UNF   1 in. (25mm) diameter

0.75 in. (19 mm)

1.75 in. (44.5 mm)

(f) Fixture tensile test of adhesives

Figure 14–2, *continued*

that the load be transmitted axially to the specimen. This requires proper setup and proper gripper selection to help ensure that the specimen is aligned correctly before and during the test procedure.

Wedge grippers are used with ductile materials. Wedge grippers are unsatisfactory for brittle materials, because they tend to pinch the material and cause a failure near the ends of the specimen rather than within the gage length. Most grippers have a range of movement or can be shimmed to align them.

Grippers can be threaded, grooved, beveled, and so on, and some even provide a type of universal joint to help in self-alignment. ASTM standards specify which type of gripper is best for the tensile specimen. Refer to the specific ASTM specification for the material under test to ensure proper gripper design.

14.3 EQUIPMENT

Tensile testing machines vary in size from small, portable, hand-held devices with capacities of approximately 2000 to 5000 lb (908 to 2270 kg) to larger, stationary

Figure 14–3
A. Common tensile testing equipment.

Figure 14–3, *continued*
B. Common accessories.

Part B. Courtesy of Tinius Olsen
Testing Machine Co., Inc.

machines having capacities as high as 300,000 lb (136,200 kg) or more. The proper tensile testing machine, based on individual test requirements, should be selected for the type of material that is being tested. Individual test requirements may vary considerably. Figure 14–3 shows some common tensile testing equipment.

14.4 PROCEDURE

The tensile test is used to determine the tensile properties of a material. Included in the tensile properties of a material is the tensile strength. The *tensile strength* of a material is the maximum *tensile stress* that may be developed in the material.

To conduct a tensile test, you must obtain a specimen of suitable size and shape. This specimen must have a known cross-sectional area and suitable gage length for the test conditions. Once a suitable test specimen has been obtained, the specimen is loaded into the test machine using proper grippers. Then, a steadily increasing tensile load is applied to the specimen. The load is applied in known increments over a specified period of time. The amount of deformation is determined by measuring the gage length directly or by measuring the amount of deflection or deformation that has occurred after each load increment. The amount of deformation and the original gage length can be used to find the strain. When the specimen begins to yield significantly (deformation occurs without any additional increment in load), the material is said to have failed. The strainometer (extensometer) is then removed, and the material may continue to be loaded until failure. The ultimate and rupture strengths may be obtained from this latter loading beyond the *yield stress*. Figure 14–4 illustrates an extensometer and accessories.

To yield significant results, the test specimen must be machined accurately and to the correct specifications. The specimen must be uniform over the gage length. This uniform cross section helps to ensure that the stress developed is uniformly distributed and to eliminate areas of concentrated stress, which would bias the results.

Primary properties determined through tensile testing include yield strength (or yield point for ductile materials), tensile strength, ductility (based on the percent elongation and percent reduction in area), and a description of the type of fracture sustained. In the case of brittle materials, only data concerning the tensile strength and type of fracture are taken. From the data recorded, the relationship between stress and strain developed in the specimen, the modulus of elasticity, and related mechanical properties can be obtained.

Prior to the actual test, the dimensions of the specimen and the test conditions are measured. Linear measurements are made using calipers, scales, or microme-

Figure 14–4
Common tensile test accessories.

ters, depending on the dimension's range and the precision required. Typically, measurements to 0.001 in. (0.01 mm) are of adequate precision.

If the percent elongation is to be calculated, the gage length must be marked off and measured. On ductile specimens of ordinary dimensions, this is done with a center punch. On thin sheets or brittle materials that could be damaged, fine scratches should be made to indicate the gage length. Whichever method is used to mark the gage length, the marks should be light enough so that they don't influence the test results. Heavy or deep marks may cause a localization of stresses and influence the break.

Figure 14–5
Grippers and specimen in place.

Before using the tensile test machine or any testing machine, the inexperienced operator should be aware of the capabilities, controls, and functions of the various dials, levers, and buttons on the machine. Before any test is conducted, the operator should know what works on the machine and how it works. The machine is then adjusted to the proper speed rate for the test. Checks are made to see that the load is removed from the machine. This process is also referred to as *zeroing* the machine. Before applying any load, the extensometer must be checked for free and proper travel limits. A few "dry" runs and some practice will help ensure that the proper sequence of events can be maintained and will give some measure of experience in timing and recording data.

After the equipment has been checked and the operator has become familiarized with the operation of the equipment, the specimen is loaded into the machine. Proper attention should be given to ensure that proper grippers have been selected. The alignment of the specimen is crucial and must be maintained throughout the test procedure. All guards and safety equipment must be in place before conducting any test. The strainometer should be placed so that strain measurements can easily be taken. The slack is then taken out of the testing apparatus by applying a small load. The machine dial is then rezeroed. The test is ready to commence.

Figure 14–5 shows some variations in grippers and test specimens for a variety of materials.

The extensometer is chosen based on the required data. Strainometers, in general, measure the amount of deflection or deformation produced by an applied load. When conducting a tensile test, the proper strainometer must be selected, and its multiplication ratio must be determined. Once these have been determined, the extensometer is placed on the specimen over the previously marked gage length. The extensometer is then set to zero.

Test speed or rate of load application should be moderate and steady. This is a static test, so the speeds are slow to moderate, but the test should not be affected by moderate loading speeds (typically 0.010 to 0.050 in./min.). The speed of the test should not exceed the recorder's ability to take readings accurately and reliably. Specimens can typically accept loads within a wide range as long as the loading does not produce inertial effects. The ASTM standards for the particular test can be consulted if there is a question on loading rate.

The major concern regarding loading rate is the yield point. The yield point is most affected by loading rate. Above the yield point for ductile materials, the loading rate may increase, because there is less effect on the ultimate strength than on the yield strength. The percent elongation, however, may be adversely affected.

The load is applied in a steady, increasing manner, and readings are taken at predetermined intervals. Once the readings are past the yield point, the extensometer should be removed so that it isn't damaged as the material fails. The calipers and scale or micrometer are used to determine elongation. Depending on the purpose and required data for the test, loading is continued until failure or the test is stopped at any point after which the necessary data have been collected.

Once the specimen has failed or ruptured, it can be removed from the machine and inspected. To determine the elongation, the two pieces are put together and the gage

Figure 14–6
Fractures.

Flat   Cup-cone   Star   Irregular

length (final) is measured. Also, the final cross-sectional area of the specimen is measured to help determine the percent reduction in area. The two halves are separated, and the fracture is observed. Tensile fractures are generally characterized according to form, texture, and color. Types of fractures include flat, cup-cone, irregular, and ragged. Descriptions used in conjunction with these types of fractures are silky, fine grain, coarse grain, fibrous, splintery, crystalline, glossy, and dull. Some materials can be easily identified by their fractures. For instance, mild steel usually produces a silky, cup-cone fracture; wrought iron produces an irregular, fibrous fracture, and cast iron produces a grey, flat, coarse-grain fracture. Evaluation of the fracture may give additional clues as to the reason for failure. An attempt should be made to sketch and classify the fracture according to the general characteristics given in Figure 14–6.

14.5 EXPECTED RESULTS

The results of tensile testing can be used to plot a stress-strain curve. In a stress-strain curve, the applied stress is plotted along the vertical axis, and the strain devel-

Figure 14–7
Stress-strain curve for mild steel.

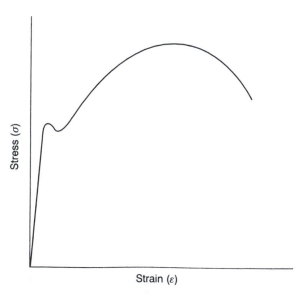

oped is plotted along the horizontal axis. A typical stress-strain curve for mild steel is shown in Figure 14–7.

When testing begins and the stress begins to increase, the strain developed in the material is very slight. During this period of linearity, the strain developed is proportional to the stress applied. This period is known as the *elastic region*. Within the elastic region, a material will resume its original shape when the applied stress is removed. Under elastic conditions, the material acts in a manner described by Hooke's law.

The point where the stress-strain curve first deviates from proportionality, or linearity, is called the *proportional limit*. The material will still exhibit elasticity, but the applied stress and resultant strain are no longer proportional. The point at which a material no longer resumes its original shape after the applied stress is removed is the *elastic limit*. Once beyond the elastic limit, the material is said to exhibit *plasticity*. The material will still exhibit limited elasticity, but it will not return to its former condition upon removal of the load.

The curve continues on an uphill slant until it reaches a peak. This peak is the maximum stress that can be developed in the material. This point is known as the *ultimate tensile stress* of the material. The tensile strength of the material can be found using the applied stress and the resultant strain at this point. Beyond this point, the material starts yielding, and additional strain occurs without developing greater stress. During this period, the specimen begins to "neck down," or develop a ring of reduced cross-sectional area. As long as the stress remains, the material will continue to neck down until failure. Failure occurs within the necked-down region, where the stress at failure is less than the maximum tensile stress.

After determining the ultimate tensile strength, a safety factor is often used to help determine a maximum safe working stress. The ultimate tensile strength is divided by the safety factor (usually 2 to 10). This working stress, or working strength, is used in design calculations, which helps ensure that the material will not fail under normal service conditions.

Consider an example of the use of tensile calculations in materials design.

Example

a. Suppose we want to use a length of cable to pull a cable car weighing 1200 lb plus six passengers weighing 175 lb each. What would the minimum diameter of the cable have to be if the cable's yield strength was 32,000 lb/in.^{2}?

$$\text{area} = \text{force/stress} = [1200 \text{ lb} + (6 \cdot 175 \text{ lb})]/32,000 \text{ lb/in.}^2 = 0.070 \text{ in.}^2$$
$$\text{diameter} = \text{square root of } (4/\pi) \cdot \text{area} = 0.300 \text{ in.}$$

b. If the cable were 0.500 in. in diameter, would it hold if eight people got aboard the cable car?

$$\text{stress} = [1200 \text{ lb} + (8 \cdot 175 \text{ lb})]/\pi(0.250 \text{ in.})^2 = 13,242 \text{ lb/in.}^2$$

Because this value is less than the cable's yield strength, it will hold eight people. To find how many people weighing 175 lb the cable car could hold,

$$\text{force} = \text{area} \cdot \text{stress} = \pi(0.250 \text{ in.})^2 \cdot 32,000 \text{ lb/in.}^2 = 6283 \text{ lb}$$
$$6283 \text{ lb (max. weight)} - 1200 \text{ (cable car)} = 5083 \text{ lb}$$
$$5083 \text{ lb (max. weight of passengers)}/(175 \text{ lb per passenger}) = 29 \text{ people}$$

Figure 14–8
Stress-strain curve for soft steel.

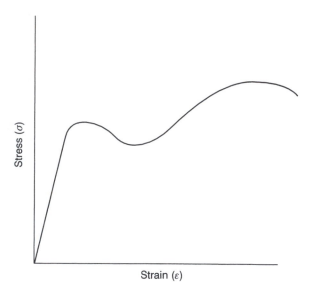

Stress (σ)

Strain (ε)

14.6 VARIATIONS ON STANDARD PROCEDURES

Other materials exhibit different stress-strain relationships. Figure 14–8 shows the stress-strain for a low-carbon (soft) steel. Notice the two yield points. Between the upper and lower yield points is an area where the material continues to develop an increasing strain with little or no coincident increase in the applied stress. Materials that yield in this manner can withstand larger forces before failure. These materials are called *ductile materials*, and they can endure a larger strain before rupture occurs. However, the material develops a permanent set after the upper yield point has been exceeded.

Because stress is the result of dividing the applied force by the cross-sectional area, the actual stress, or *true stress*, developed in the material would have to be calculated based on instantaneous data concerning the stress and strain. This would mean measuring the cross-sectional area between each increment in the applied load. Generally, the true stress-strain curve looks the same as the theoretical one, except that the actual stress as necking down occurs increases rapidly as the cross-sectional area is reduced.

It is not always practical or economical to gather data for true stress-strain calculations. In most cases, this process is very difficult and time-consuming. Therefore, calculations are made based on the original cross-sectional area. These results are called the *nominal*, or *engineering*, stresses. For conditions and applications where the stress developed is well under the maximum tensile stress, there isn't a significant difference between the material's true and nominal stresses. The values are similar, because very little necking down occurs until the stress developed in the material approaches the maximum tensile stress.

Figure 14–9
Offset method.

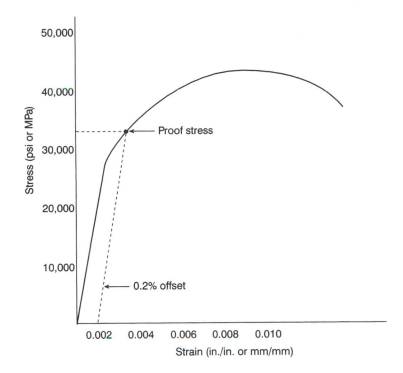

Under conditions where the material has undergone a large plastic deformation, the nominal stress can deviate significantly from the true stress. As stated previously, taking multiple readings at each increment in loading is not an appealing thought. Therefore, an arbitrary offset or deformation is selected. Typically, this offset is 0.2% and is called the 0.2% *offset yield strength*, or *proof stress*. The proof stress is the nominal tensile stress that produces a nonproportional strain equal to the arbitrary offset selected, based on a percentage of the original gage length. This offset method is used for materials that do not exhibit distinct yield points. The determination of the proof stress using the offset method is shown in Figure 14–9.

The tensile strength of the material is an indication of the material's ductility. In addition, the percent elongation and percent reduction in area are also indications of ductility. The percent elongation is 100 times the quotient of the difference in the final and original gage lengths divided by the original gage length. The percent reduction in area is difficult to calculate accurately, due to the difficulty in measuring the final cross section. It is especially difficult if the material's cross section is distorted or damaged due to the method and characteristics of the failure.

Example

a. Calculate the percent elongation of a material that had an original gage length of 2.000 in. and a final gage length of 2.012 in.

Table 14–1
Tensile properties of various
engineering materials.

Material	Tensile Strength (lb/in²)
Metals	
Aluminum alloys	13,000–24,000
Brasses	40,000–120,000
Bronzes	40,000–130,000
Cast iron	18,000–60,000
Cold-rolled steel	84,000
Magnesium alloys	20,000–45,000
Monel	100,000
Stainless steel	85,000–95,000
Titanium	95,000
Zirconium	24,000–40,000
Woods (Perpendicular to Grain)	
Softwoods	200–500
Hardwoods	300–1,000
Thermoplastic Polymers	
Polyamide	9,000–9,500
Polycarbonate	8,000-9,500
Polyethylene	1,000-5,500
Polymethyl methacrylate	7,000-11,000
Polypropylene	4,300-5,500
Polystyrene	5,000–10,000
Polytetrafluoroethylene	2,000–4,500
Polyurethane	5,000–5,500
Polyvinyl chloride	1,500–9,000
Thermosetting Polymers	
Epoxy	4,000
Melamine-formaldehyde	5,500–13,000
Phenol-formaldehyde	6,000–9,000
Urea-formaldehyde	5,500–13,000
Elastomers	
ABS	7,000
Butadiene	3,500
Butadiene-styrene	600–3,500
Silicone	3,000–4,000
Fibers for Composites	
Glass	500,000
Graphite	300,000–410,000
Polyaramid (Kevlar)	410,000
Steel	60,000–320,000

$$\% \text{ elongation} = [(2.012 \text{ in.} - 2.000 \text{ in.})/2.000 \text{ in.}] \cdot 100 = 0.6\%$$

b. Calculate the percent reduction in area of a specimen that had an original diameter of 0.505 in. and a final diameter of 0.487 in.

$$\text{original area} = \pi(0.253 \text{ in.})^2 = 0.200 \text{ in.}^2$$
$$\text{final area} = \pi (0.244 \text{ in.})^2 = 0.187 \text{ in.}^2$$
$$\% \text{ reduction in area} = [(A_o - A_f)/A_o] \cdot 100$$
$$= [(0.200 \text{ in.}^2 - 0.187 \text{ in.}^2)/0.200 \text{ in.}^2] \cdot 100 = 6.5\%$$

Some degree of ductility is sought by design. Materials that show some ductility will bend before breaking or deform before rupture. Ductility also increases the formability of the material.

The area under the stress-strain curve, up to the breaking point, represents the energy required to rupture the material and is, therefore, a representation of the toughness of the material. Actually, the area under the curve represents the modulus of toughness, which is a measure of the energy-absorbing characteristics of the material, including both elastic and plastic deformations. Toughness refers to the difficulty in breaking a material or its ability to withstand shock loading. These results don't necessarily agree with impact test results, which are also representative of toughness. However, results are similar for the true stress-strain results.

The results of tensile testing are significantly altered by temperature. Generally, the yield strength, tensile strength, and modulus of elasticity for a material decrease in magnitude as the temperature increases, and the ductility of the material increases. In certain instances, this phenomenon is used as an advantage. For instance, hot working takes advantage of the higher ductility and lower stress required to work and form these materials.

The tensile properties of various engineering materials are shown in Table 14–1.

14.7 SUMMARY

Tensile tests provide important data, which is often used to specify the physical and mechanical properties of materials. The tensile test was one of the first tests conducted on materials. It can be used to provide data concerning the tensile strength of the material being examined. This information is often important in designing and constructing engineering products.

Tensile test data can be used to construct a stress-strain diagram, or curve, for the test specimen. The results of the test represent the action of the material in service. Two important points that occur on the stress-strain curve are the yield point and the ultimate tensile stress, or tensile strength, of the material. These values provide information on the load-bearing capabilities of the material.

Other important information gathered through tensile testing are the percent elongation, the percent reduction in area, and the strain developed during testing. Data concerning these characteristics are used to provide an indication of the deformation of the material during the test. The amount of deformation that the material exhibits before rupture is often used to classify the material as ductile or brittle.

During tensile testing, materials display both elastic and plastic deformation. Within the elastic limit, materials are able to recover fully from the applied stress and return to their original dimensions. Plastic behavior is the ability of a material to deform without rupture. It is the "stretching" that occurs during tensile testing. The shape of the stress-strain curve gives an indication of the elastic and plastic properties of the material. Once a material begins to yield, changing from elastic to plastic behavior, a specified offset can be used to determine the material's yield strength at that arbitrary offset.

Young's modulus, or the modulus of elasticity, represents the resilience or elastic stiffness of the material. It is calculated by dividing the stress within the elastic limit by the corresponding strain developed at that stress. The modulus of elasticity provides a figure of merit for comparison between materials.

QUESTIONS AND PROBLEMS

1. What changes in the results would you anticipate if a material's cross-sectional area were 50% of the height? 75%? 100%?

2. Discuss the relationship between the length and the height in tensile specimens.

3. Why does a tensile test specimen neck down?

4. What is the relationship between the nominal and true stress?

5. What are the effects on the test results if the axis of loading and the axis of the specimen don't coincide? (In other words, what will the effects be if the test specimen is improperly aligned?)

6. List the types of grippers and their purposes.

7. Explain the terms *proportional limit, elastic limit, yield point, ultimate strength, breaking strength, modulus of elasticity, toughness, stiffness*, and *ductility*.

8. Explain the difference between yield point and yield strength. Draw a graph clearly showing this difference.

9. What additional information is collected in a tensile test, other than the applied load and extension (deformation)?

10. What is the stress developed in a 0.505 in.–diameter round test specimen at an applied load of 2500 lb?

11. What is the stress developed in a 1.000 in.–diameter round test specimen at an applied load of 5000 lb?

12. What is the strain produced by a 4.5-in. specimen that elongates 0.050 in.? What is the percent elongation?

13. What is the percent reduction in area of a specimen with an original cross section of 0.505 in. and a final cross section of 0.502 in.?

14. What is the stress developed in a rectangular specimen 3 ↔ 5 in. at an applied load of 8000 lb? Convert your answer to megapascals, where 1 MPa equals 145 lb/in.^{2}.

15. What is the stress developed in a 3 ↔ 5 cm rectangular test specimen at an applied load of 12,000 kg?

16. Would a steel beam measuring 2 ↔ 3 in. designed to withstand a 100,000 lb/in.^{2} load fail if it were used to support a 50-ton load? What is the maximum load it could support?

17. A 10 ↔ 10 cm steel bar is used to support a continuous load of 100 MPa. What would be the maximum weight (in kg) it could support? Would the bar be able to withstand a stress of 10,000 lb/in.^{2}?

18. A steel bar 18 in. long should not be strained more than 0.010 in./in. What would the elongation of the bar be at the maximum strain?

19. If a steel bar were designed to withstand a maximum stress of 100,000 lb/in.^{2}, what would be the maximum load it could support if the beam diameter were 0.750 in.?

20. Based on the following data for a 0.505-in. diameter round test specimen with a 2.000-in. gage length, construct a stress-strain curve. If the specimen has a final cross section of 0.503 in., label the following points and make the appropriate calculations.

Load (lb)	Elongation (in.)
1,000	0.0010
2,000	0.0015
3,000	0.0020
4,000	0.0025
5,000	0.0030
6,000	0.0040
7,000	0.0050
8,000	0.0065
9,000	0.0080
10,000	0.0100
10,800	Rupture

Original length
Final length
Percent elongation
Original cross-sectional area
Final cross-sectional area
Percent reduction in area

Yield strength
Tensile strength
Modulus of elasticity

21. Provide five applications each where an object is under (1) tensile stress, (2) compressive stress, (3) direct shear stress, (4) torsional stress, and (5) flexural stress.
22. Describe three applications where tensile strength is the primary factor in material selection. What material would you select for that application and why?
23. What design considerations are important to applications involving tensile loading?
24. Describe the shape of a tensile stress-strain curve for (1) a ductile material and (2) a brittle material.
25. Develop and describe your own tensile test procedure for testing shoelaces. Describe the test setup, grippers, extensometer, loading concerns, data collection, and expected results.

CHAPTER 15
Creep Testing

Objectives • Recognize and define common terms related to creep testing.
• Describe the terms *stress* and *strain* and the effects of both in creep loading.
• List the equipment necessary to conduct a creep test.
• Describe the operation of various equipment related to creep testing.
• Relate the general procedures used in conducting a creep test.
• Perform the necessary calculations related to creep testing.
• Recognize expected creep test results.
• Describe common variations in standard creep test procedures.

15.1 INTRODUCTION

Creep tests are designed to measure the effects of long-term applications of loads that are below the elastic limit of the material being tested. *Creep* is the plastic deformation resulting from the application of a long-term load. Even though the magnitudes of the applied load are less than the materials' elastic limits, materials tend to deform over long periods. Elevated temperatures, severe service conditions, and other factors may tend to accelerate creep in applications.

Creep is a design concern for engineers and technologists and a service concern for technicians. It is especially important in such structures as bridges, buildings, and other load-bearing structures in which the members are subjected to long-term static loads.

15.2 PRINCIPLES

Environmental conditions may also play a role in creep. Temperature is a major concern in creep testing. The higher the temperature, the more accelerated the creep under the same load. Creep tests at elevated temperatures serve as a satisfactory guide to the effect of creep in service. Many of the tests previously studied are applied in acceptance testing when inspecting incoming materials. Creep tests are inherently long-time tests and are, therefore, unsatisfactory for acceptance sampling.

At elevated temperatures, materials behave somewhat like a very viscous fluid. Therefore, an applied load, much smaller than would normally cause deformation, may cause appreciable results if allowed sufficient time. This slow, but steadily increasing strain is termed *creep*. Most materials are subject to creep. The degree to which they are affected depends on their internal structure and variations in composition. For example, concrete exhibits a different creep rate than do plastics, fabrics, and other materials. Environmental temperatures and temperature cycles may also affect the creep rate. For example, some plastics may exhibit marked creep at normal temperature while steel may require several hundred degrees before any appreciable effect.

During a creep test, data are collected on the duration of the test, the applied load or stress, and the resultant strain. The strain may be either elastic or plastic, depending on the material, the temperature, the applied stress, and the duration of the test.

The creep test itself is comprised of three stages. In the initial stage, creep continues at a decreasing rate, levels off to a more constant rate in the secondary stage, and rapidly accelerates in the final stage until failure. Typical creep test curves as shown in Figure 15–1. Under high stress, failure in metals may occur at low temperatures by fracturing through the crystals. Failure at high temperatures generally occurs through the crystal boundaries. Therefore, it is important to characterize the type and details of the fracture. The temperature that characterizes the difference between fracture within the crystal and between crystals is called the *equicohesive temperature*.

The strain mechanism also varies according to temperature. Below the equicohesive temperature, strain may occur as a result of elastic deformation of the crystals themselves. At elevated temperatures, strain may result from crystal movement

Figure 15–1
Typical creep test results.
A. Stress versus temperature.
B. Creep versus time.

(a) Stress versus temperature.

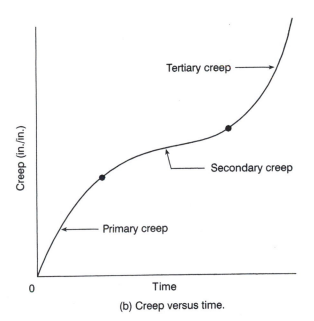

(b) Creep versus time.

within the matrix. The strain mechanism varies with the material being subjected to the stress and the other factors previously mentioned. During the creep test, removing the stress results in some recovery. However, if the stress is removed in the second stage, some permanent plastic strain will remain.

There are many applications in which creep is an important consideration: internal combustion engines, for example. Creep can occur at any temperature; its rate depends on the amount of stress applied. A material may act differently if subjected to normal stress at low temperature, then placed in an application, and subjected to high temperatures and/or high stresses over a period of time. Therefore, adequate testing of materials in environments where they are expected to perform is essential to predicting their performance in service.

15.3 EQUIPMENT

During any creep test, four variables are concerned: time, temperature, applied stress, and resultant strain. Typically, creep tests are performed on specimens with varying loads and temperatures. A composite profile is then constructed that illustrates the limiting stress for a given percentage of creep over periods of time.

To determine the creep characteristics for a given material at elevated temperatures requires (1) a means for elevating the temperature and maintaining it, (2) a

Figure 15–2
Creep apparatus.

Fixed support

Furnace

Specimen

Applied load

Strain indicator

Creep test apparatus

method of measuring the resultant strain (extensometer), and (3) a device for applying stress. Figure 15–2 illustrates a typical setup.

Temperature variation should be controlled according to ASTM standards to prevent thermal stresses and strains in the specimen that will adversely affect results.

15.4 PROCEDURE

In creep testing, the applied load and temperature remain constant throughout the test. The resulting strain and time are taken as data points and plotted on a curve. Specimens resemble typical tensile test specimens.

To conduct the test, the specimen is first heated to the required temperature. When the temperature is reached and maintained, the gage length is measured and the load applied. Measurements are then taken at predetermined intervals to collect strain-time data. Initially, there will be an elastic strain that forms the initial stage of the creep test.

Over the duration of the test, at least 50 data points should be collected to help ensure that significant results are obtained. Because creep tests are long-time tests, it may be several weeks between data collection. Typical creep tests may run for 1000 hours for plastics up to 100,000 hours with perhaps 1% elongation for steels.

15.5 EXPECTED RESULTS

Creep results are dependent on the composition of the material. For example, alloying elements such as nickel, chromium, molybdenum, and others increase creep resistance. Other factors that influence creep results include manufacturing process, heat treatment, and grain size. Typical test reports follow the following pattern:

Objective: To determine the creep resistance for metals at room temperature
ASTM Standard: ASTM E139 and #150
Specimen: 1 in. × 6 in. round, treaded
Procedure:

1. A tensile test was performed to determine the ultimate strength of the material.
2. The specimen was measured and inserted in the testing machine.
3. A dial indicator was used to measure strain and zeroed to start the test.
4. Sufficient time was allowed for the specimen to reach test temperature.
5. A load was selected which was about two-thirds of the ultimate strength of the specimen. This load was applied, rapidly and smoothly.
6. Readings were taken at 1-min intervals to start and at increasing intervals during the test.
7. Readings continued until rupture.

Results: The applied stress was calculated and the temperature recorded. Environmental conditions were recorded. All known properties of the specimen were recorded. The results were presented as a creep curve with strain versus time on a logarithmic scale. Information was gathered concerning the shape and condition of the fracture.

15.6 SUMMARY

Creep tests are designed to measure the effects of long-term applications of loads that are below the elastic limit of the material being tested. Creep is the plastic deformation resulting from the application of a long-term load. Even though the magnitudes of the applied load are less than the material's elastic limit, materials tend to deform over long periods. Elevated temperatures, severe service conditions, and other factors may tend to accelerate creep in applications.

A creep test is a long-term test conducted at elevated temperatures. The results of a creep test aid in design and in predicting the effects of temperature and sustained loads on a material over long periods of time. Structures such as bridges and buildings are expected to deform less than 1% over a period of 100,000 hours.

Creep is a design consideration for engineers and technologists and a service concern for technicians. It is especially important in such structures as bridges, buildings, and other load-bearing structures in which the members are subjected to long-term static loads.

QUESTIONS AND PROBLEMS

1. What is creep?
2. How does temperature affect creep rate?
3. Explain how elastic and plastic strain develop during a creep test.
4. What happens to a shelf after weight sits on it for long periods?
5. What are the four basic variables in creep tests?
6. Given a strain versus time curve for a material, how might you use this data to predict the performance of the material in service?
7. How could you design a shelf to reduce the effect of creep?
8. Explain how you would design a creep test on a sheet of paper.
9. Give some examples where creep would be a concern in an internal combustion engine.
10. Explain the three stages of creep and the shape of the curve during each stage.
11. Describe three applications where creep is the primary factor in material selection. What material would you select for that application and why?
12. What design considerations are important to applications involving creep?
13. Describe the expected creep rates for various materials. Which would you expect to have the lowest creep rate? The highest?
14. Develop and describe your own creep test procedure for testing glass. Describe the test setup, grippers, extensometer, loading concerns, data collection, and expected results.

CHAPTER 16
Compression Testing

Objectives
- Recognize and define common terms related to compression testing.
- List the equipment necessary to conduct a compression test.
- Describe the operation of equipment related to compression testing.
- Relate the general procedures used in conducting a compression test.
- Perform the necessary calculations related to compression testing.
- Recognize expected compression test results.
- Describe common variations in standard compression test procedures.

16.1 INTRODUCTION

In general, a compression test can be thought of as the opposite of a tensile test. Where a tensile test tends to pull a specimen apart, a compression test tends to squeeze, or compact, the specimen. The decision to select a compression test over other types of testing is largely determined by the types of service to which the material is likely to be subjected. Metals, for example, are more efficient at resisting tensile loads than other materials with relatively low tensile strengths, such as concrete. They are, therefore, more commonly tested through tensile loads. The same is true for most drawn and extruded plastics. Brittle materials, such as concrete, brick, and most ceramic products, have relatively low tensile strengths compared to their compressive strengths and are, therefore, generally employed to resist compressive loads. Thus, the compression test is more significant and useful for testing these types of materials.

16.2 PRINCIPLES

Compression results from forces that push toward each other. This action tends to shrink, or compress, the specimen that is under compression loading. Concrete is one material nearly always tested in compression.

The specimens required for compression testing tend to be short and of larger diameter. The longer the compression specimen is, the greater the risk of buckling and of *column action*, which is an instability that develops in the specimen. This column action leads to a buckling of the material in which the material bulges out to the side due to an elastic instability. Column action is illustrated in Figure 16–1.

Figure 16–1
Column action.

Figure 16–2
Metallic compression specimen specifications.

Compression specimen
(Cast iron, cast aluminum and ductile metals)

A circular cross section is recommended for compression testing specimens. However, some materials (such as bricks) are not easily obtained or produced in circular sections. In these cases, the material is tested in a convenient shape.

The *length-to-diameter ratio* of the specimen must be carefully chosen, depending on the material and test conditions. The specimen's stability decreases as the height or length is increased. There is a greater tendency for the piece to buckle as it gets longer. This is due, in part, to the uneven distribution of the internal stresses developed during the test. Therefore, there is a practical limit of 2:1 for the length-to-diameter ratio. The length of the sample should not exceed twice the smallest cross-sectional diameter.

Conversely, if the length-to-diameter ratio is reduced significantly, the specimen's apparent strength is increased. This is a result of the intersection of the slippage planes and the test fixture. This intersection tends not to allow the specimen to slip along normal channels and, therefore, apparently to increase its compressive strength. This figure varies according to the specimen material. Another factor to take into consideration is the space required to mount the compressometer. Specimen dimensions may need to be changed to accommodate the compressometer, but the length should not exceed twice the smallest cross-sectional diameter.

The ends of the compression test specimen should be flat, parallel, and perpendicular. Specimen parameters for metallic materials are given in ASTM E 9. A sampling of these specifications is given in Figure 16–2.

In addition to metallic materials, it is often desired to determine the compression strength of concrete. The standard specimen requires a cylinder 6 × 12 in. (15 × 30 cm). An optional size for the compression testing of concrete is 3 × 6 in. (7.5 × 15 cm), which satisfies the 2:1 ratio.

For testing the compression strength of wood, a 2 × 2 × 8 in. (5 × 5 × 20 cm) specimen is used (see ASTM D 143). Additional compression tests can be made on construction bricks and tiles (ASTM C 67), rubber (ASTM D 395, 575), and stone (ASTM C 170). Sheet metals can be tested for compressive strength by mounting them in special fixtures to prevent them from buckling. Specifications for selected compression test specimens are given in Figure 16–3.

(a) Compression specimen
(concrete)

(b) Wood compression specimen
(parallel to the grain)

Figure 16–3
Specifications for various materials.

16.3 EQUIPMENT

The most common machine used is the universal testing machine (UTM), although testing machines are built especially for compression testing. The capacity of the testing machine is determined by the size and type of material to be tested. Any testing machine that can provide a compressive load at reasonable determinable increments can be used, as long as it exceeds the compressive strength of the material under test.

A compressometer is used in much the same way as the extensometer was used in tensile testing. The compressometer measures the amount of compression sustained during testing.

Figure 16–4
Eccentric loading.

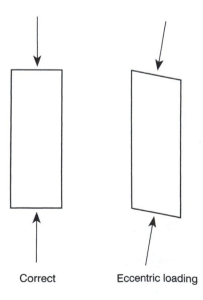

Correct   Eccentric loading

The ends of the specimen should be flat and parallel to each other and the load applied perpendicular to the ends. Otherwise, stresses will be concentrated at the ends. *Eccentric loading* occurs when the load is applied in a direction that is not perpendicular to the axis of the specimen. Eccentric loading is shown in Figure 16–4.

Materials such as concrete, stone, and brick require additional precautions and apparatus. These materials often require end caps or bedding material to satisfy the loading conditions. These bedments are typically made of plaster, cements, or sulfur compounds. The bedment is spread over the end of the specimen and then covered with an end cap. Some additional fixture or jig may be used to help ensure that the cap is held perpendicular to the specimen's central axis. The bedment should at least have the same strength properties as the specimen, if not greater. Soft materials such as foam and rubber tend to flow outward in a lateral direction under load, causing the specimen to split. Softer materials should, therefore, be avoided.

16.4 PROCEDURE

The primary objective of the compression test is to determine the compressive strength of the material. Therefore, data concerning the compressive strength, applied load, and deformation are taken. In brittle materials, the compressive strength is easily obtained. However (as in the tensile test), the compressive strength for ductile materials must be based on some arbitrary deformation. Ductile materials do not exhibit the sudden (and often explosive) fractures that brittle materials do. Because ductile materials tend to "barrel" out, deformation data concerning this deformation should be taken and used to determine the compressive strength.

Prior to the test, the dimensions of the specimen should be measured with adequate precision using proper instruments. After the proper dimensions have been obtained, the specimen should be loaded into the testing machine. Proper care should be taken to ensure that the axis of the specimen is aligned and centered in the machine. It is difficult to gain a uniform stress distribution during a compression test. Therefore, strain measurements are usually taken 120° apart around the diameter of the specimen. An averaging-type compressometer may also be used.

Specific loading rates for each test material are given in the ASTM standards. An average test pace is typically a 0.125 mm/min or 0.005 in./min strain rate. Typically, the applied stress rate ranges from 500 to 1000 lb/in.^{2}/min (3 to 7 MPa/min). Testing rate will affect the results of the test. In general, the more rapid the rate, the higher the indicated strength. For example, the indicated strength of a specimen loaded at 1000 lb/in.^{2}/min may be as high as 5% greater than one loaded at 100 lb/in.^{2}/min, for instance. In many cases, any convenient load, such as every 1000 lb/in.^{2}, is allowed, up to one-half the expected maximum. After that point, ASTM specifies the rates (maximum and minimum, time, and loading) for the last half of the data. If the readings are taken manually, the test pace can be adjusted to accommodate the accurate recording of data. Proper safety precautions should be taken, particularly when conducting compression tests on materials such as concrete, which explode upon failure.

Figure 16–5
Compression stress-strain curve
for concrete.

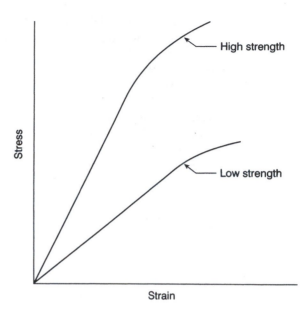

16.5 EXPECTED RESULTS

Data concerning compression and applied load are taken during compression test-ing. These data can then be plotted on a compression stress-strain curve, such as the curve for concrete in Figure 16–5.

Compressive strength, like tensile strength, is given in pounds per square inch or in Pascals. The applied load divided by the cross-sectional area provides the stress at that particular point on the curve.

Example

What is the compressive strength of a piece of 2 × 2 × 6 in. wood that fails under a 7500-lb load?

$$\text{compressive strength} = \frac{7500 \text{ lb}}{(2 \text{ in.})^2} = 1875 \text{ lb/in.}^2$$

Example

A 6 × 12 in. concrete specimen failed at a load of 135,000 lb. What is the compres-sive strength of the concrete?

$$\text{area} = \pi r^2 = \pi \cdot 3^2 = 28.3 \text{ in.}^2 \quad \text{compressive strength} = \frac{135,000 \text{ lb}}{28.3 \text{ in.}^2} = 4770 \text{ lb/in.}^2$$

Wood, bricks, metals, polymers, ceramics, and other materials are tested in the same basic manner.

16.6 SUMMARY

Compression testing is much the same as tensile testing, with the exception that the load tends to compress the specimen rather than pull it apart. Data concerning the applied load and compression are taken to construct a compressive stress-strain curve. The compressive strength is calculated by dividing the applied compressive force (in pounds or newtons) by the cross-sectional area (in square inches or square millimeters). Concrete is nearly always tested in compression. Wood, metals, polymers, bricks, ceramics, and other materials may also be tested in compression, depending on their intended use.

QUESTIONS AND PROBLEMS

1. Discuss the similarities and differences between the tensile test and the compression test.
2. Why is the compression test the most commonly conducted test for concrete?
3. List five materials that are dependent upon their compressive strength.
4. Describe how test pace, length-to-diameter ratio, and end conditions affect the compression test results.
5. If a 2 × 2 × 6 in. piece of wood failed at a load of 5000 lb, what would its compressive strength be?
6. A 6 × 12 in. concrete specimen failed at a load of 105,000 lb. What is its compressive strength?
7. If an engineer has specified that the concrete to be used in a new parking garage must have a minimum compressive strength of 3500 lb/in.^{2} and you tested three 6 × 12 in. specimens that failed under loads of 85,000, 87,500, and 93,000 lb, would you recommend the concrete as acceptable for use?
8. If a table with four legs is loaded with ten 50-lb bags of dog food, what is the compressive load and stress on each leg if each measures 1.5 in. in diameter?
9. Using concrete with a compressive strength of 3200 lb/in.^{2}, what would the dimensions of a square pedestal have to be to support a 100-ton statue? 200-ton?
10. Calculate the diameters required for a round pedestal in Problem 9.
11. A load of 60,000 lb has to be supported by four square concrete columns. If the columns have a compressive strength of 2750 lb/in.^{2}, what size columns are required?
12. A load of 60,000 lb has to be supported by a number of concrete columns, each with a cross section that is 1.5 in. in diameter. If the compressive strength of the concrete is 1500 lb/in.^{2}, how many columns are required to support the load?
13. If you wanted to test a 6 × 12 in. concrete specimen with an expected compressive strength of 4500 lb/in.^{2}, could you go to failure with a 60,000-lb testing machine?
14. What is the maximum concrete specimen with a length-to-diameter ratio of 2:1 and a compressive strength of 3,500 lb/in.^{2} that could be tested to failure on a 60,000-lb testing machine?
15. What is the minimum compressive load capacity testing machine required to break a 6 × 12 in. concrete specimen with a 3200 lb/in.^{2} compressive strength?
16. Develop and describe your own compression test procedure for a tennis ball. Describe the test setup, grippers, extensometer, loading concerns, data collection, and expected results.
17. Describe three applications where compressive strength is the primary factor in the application. What material would you select for that application and why?

18. What design considerations are important to applications involving compression loading?

19. Describe the shape of a compressive stress-strain curve for (1) a ductile material and (2) a brittle material.

20. Compare the similarities and differences between tensile and compressive testing.

CHAPTER 17
Shear Testing

Objectives
- Recognize and define common terms related to shear testing.
- List the equipment necessary to conduct a shear test.
- Describe the operation of equipment related to shear testing.
- Relate the general procedures used in conducting a shear test.
- Perform the necessary calculations related to shear testing.
- Recognize expected shear test results.
- Describe common variations in standard shear test procedures.

17.1 INTRODUCTION

A *shearing stress* is an applied force that acts in a direction parallel to the plane in which the load is applied. This shearing action is different from tensile and compressive forces, which act normal, or perpendicular, to the axis of loading. Two shearing forces are of particular interest in determining mechanical properties. These forces are direct shear and torsional shear.

17.2 PRINCIPLES

Direct shear occurs when parallel forces are applied in opposite directions. Direct shear forces are further categorized as *single* or *double shear*. Single-shear forces occur along a single plane, whereas double-shear forces occur between two planes simultaneously. Theoretically, the shear strengths found through single- and double-shear testing should be equal, but due to bending or flexure errors, they are not always equal. Figure 17–1 illustrates the principles of single and double shearing forces.

Torsional shearing forces occur when the forces applied lie in parallel but opposite directions, where the planes do not coincide with the longitudinal axis of the specimen. Torsional forces can be thought of as twisting motions, which tend to rotate a body. Torsional shear occurs when these twisting motions act in opposite directions, as shown in Figure 17–2.

The torsional forces developed in a material are the result of an applied *torque*. Torque is a force (f) applied through some distance (d). The applied torque can be calculated by multiplying the applied force by the distance through which it is applied: $T = f \cdot d$. In SI units, the torque is calculated by the formula $T = L \cdot a \cdot d$, where L is the applied load in kilograms, a is the acceleration (9.8 m/s^2), and d is the

Figure 17–1
Single- and double-shear action.

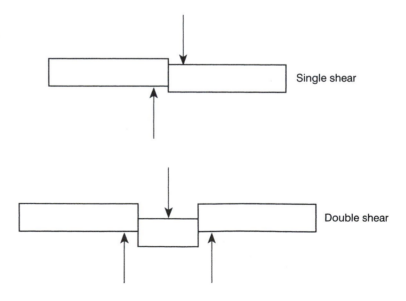

Single shear

Double shear

Figure 17–2
Torsional shear action.

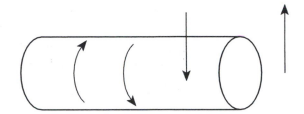

distance in meters over which the load is applied. This formula yields units of new-ton-meters, or joules. The common units for torque are inch-pounds, foot-pounds, and newton-meters, or joules. The principle of torque is illustrated in Figure 17–3.

Example

Calculate the torque developed by applying a 100-lb force over a distance of 18 in.

$$T = f \cdot d = 100 \text{ lb} \cdot 18 \text{ in.} = 1800 \text{ in.-lb}$$

or

$$T = f \cdot d = 100 \text{ lb} \cdot 18 \text{ in.}/(12 \text{ in./ft}) = 150 \text{ ft-lb}$$

The stress developed in a solid cylindrical test specimen is calculated by using the formula:

$$\text{stress} = \frac{16T}{\pi d^3}$$

where: stress = the maximum torsion strength (in lb/in.^{2} or Pa)
$\quad\quad\quad T$ = the torque (in in.-lb or N·m)
$\quad\quad\quad d$ = the diameter of the object (in in. or m)

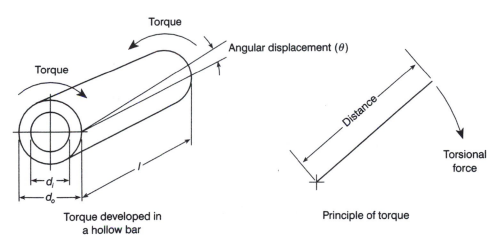

Torque developed in
a hollow bar

Principle of torque

Figure 17–3
Torque.

Example

Calculate the maximum torsion strength of a 0.500-in.–diameter solid bar that requires a torque of 500 in.-lb to break.

$$\text{stress} = \frac{16\,T}{\pi d^3} = \frac{16(500 \text{ in.-lb})}{\pi \cdot 0.5 \text{ in.}^3} = 20{,}372 \text{ lb/in.}^2, \text{ or } 140 \text{ MPa}$$

If the test specimen is hollow tubing or pipe, the formula for the maximum torsion strength becomes

$$\text{stress} = \frac{16 \cdot T \cdot d_o}{\pi(d_o^4 - d_i^4)}$$

where: stress = maximum torsion strength (in lb/in.^{2} or MPa)
   T = torque (in in.-lb or N.m)
   d_o = outside diameter of the specimen
   d_i = inside diameter of the specimen

Example

Calculate the maximum torsion strength of a 0.500-in. hollow specimen with an inside diameter of 0.450 in. that requires a torque of 500 in.-lb to break.

$$\text{stress} = \frac{16 \cdot T \cdot d_o}{\pi(d_o^4 - d_i^4)} = \frac{16 \cdot 500 \cdot 0.5 \text{ in.}}{\pi[(0.5 \text{ in.})^4 - (0.45 \text{ in.})^4]} = 59{,}238 \text{ lb/in.}^2, \text{ or } 408 \text{ MPa}$$

The strain that results from an applied shearing force is due to the effort of thin parallel slices of the specimen trying to slide over one another. This shearing strain is called *detrusion*. Detrusion is based on (1) the change in the angle between adjacent sides of the specimen or (2) the angular deformation of the specimen under torsional loading. The shearing strain within the elastic range is very small. The *angular displacement* is, therefore, given in radians (there are 2π rad in 360°). Shearing forces and detrusion are illustrated in Figure 17–4.

The torsion strain for a specimen, whether it is a solid or a hollow specimen, is calculated using the formula

$$\varepsilon = (\theta \cdot d \cdot \pi)/360l$$

where: ε = torsion strain (in in./in. or m/m)
   θ = angular displacement (in degrees)
   d = diameter (in in. or m)
   l = gage length over which the strain is measured (in in. or m)

Figure 17–4
Detrusion.

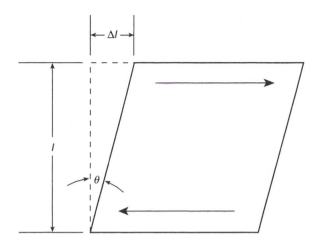

Example

Calculate the torsion strain for a 0.500-in. solid specimen whose displacement is 3°
over a gage length of 6 in.

$$\varepsilon = (\theta \cdot d \cdot \pi)/360l$$
$$= (3° \cdot 0.500 \text{ in.} \cdot \pi)/360 \cdot 6 \text{ in.}$$
$$= 0.0022 \text{ in./in.}$$

The direct shear test and the torsional test are the most commonly conducted shear
tests. In the direct, or *transverse shear*, test, the specimen is transversely mounted, as
illustrated in Figure 17–5. The transverse shear test is a close approximation of the shear
strength of the material. The accuracy of the results depends on the hardness and sharpness of the plates used in the test fixture. A further limitation of the transverse shear test
is the inability to collect strain data. Without data concerning the strain developed
throughout the test, there can be no calculation of the elastic strength or the modulus of
rigidity. The only indication of mechanical properties gained through a transverse shear
test concerns the force required to shear the specimen. Based on the original cross section of the specimen, an approximation of the shear strength can be calculated.

Another form of the direct shear test is the punching shear test. The punching
shear test is limited to flat specimens through which a hole can be punched. The
hole is punched by a slicing motion around a thin ring adjacent to the outside cutting edge of the punch. Care must be taken regarding the clearance between the
punch-and-die combination. If the clearance is too great, the punch tends to pull or
bend the material while attempting to shear it. Data from the punching shear test
should not be used in calculating the shear strength of the material. The punching
shear test is used to provide data concerning the force required to punch a hole in
the material specimen. These data can be used for comparing the relative punching
resistance of different materials.

Figure 17–5
Transverse shear test.

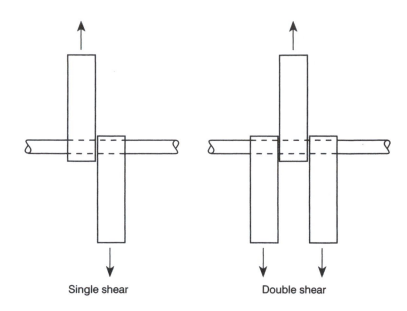

Single shear   Double shear

To gain a more complete understanding of the shear properties of a material, the torsional shear test is used. The specimen can be solid or hollow and must be of sufficient length to allow strain measurements to be taken. A *tropometer* is used to take strain measurements during the test. The tropometer readings can then be used in determining the proportional limit, the yield strength, resilience, and stiffness, or modulus of rigidity, of the material. These calculations are based on the observations of applied torque and angular displacement of the specimen during testing. In addition, the ductility of the material is indicated by the angular displacement up to rupture. The angular displacement and the shear strength indicate a material's toughness.

Torsional tests have a wide range of application. Torsion tests can be conducted to determine the shearing strength of cast iron, plastics, wire and cable, and a variety of other materials with varying cross sections.

17.3 EQUIPMENT

Shear testing is usually conducted on the universal testing machine. Depending on the type of test (single or double shear) and the material being tested, some special apparatus is used to hold the material specimen. The machine should be of sufficient capacity to complete the test so that the material fails.

Torsional testing is usually conducted on a machine specifically designed for that purpose. One end of the specimen is placed in a fixture that applies a torsional load. The other end of the specimen is connected to a tropometer, which measures the detrusion. Some simpler machines use an indicator and graduated scale (in radians) in place of the tropometer.

Torsional testing machines typically range from 1 kN·m/rad to as high as 200 kN·m/rad. An average machine has a capacity of 10 kN·m/rad.

Figure 17–6
Direct shear apparatus.

Direct shear apparatus for plates

Direct shear apparatus for round stock

17.4 PROCEDURE

Before testing, the specimen is accurately measured, and the gage length is marked. The tropometer or a suitable replacement is attached and zeroed. Proper precautions should be taken to center the specimen in the machine or fixture. The grippers are tightened so that they are not allowed to slip, yet not so tight as to cause a deformation or to alter the test results.

The procedure for direct shear testing involves supporting one end of a specimen in a fixture while applying a shear force through the use of a die. This procedure can be altered to include a double shear. If the specimen extends through the fixture, supported on both ends, the specimen is then exposed to a double shear. In testing involving metal plate, a round punch is often used. However, slotted punches and square punches can be used for specific specimens, although the round punch is recommended. Once the specimen is mounted in the test fixture, a tensile or compressive load is applied until rupture. The apparatus required for various direct shear tests are shown in Figure 17–6.

Shear testing of wood is made in a special fixture and involves a special specimen. A wood shear specimen and fixture, as shown in Figure 17–7, are used. Failure should occur along the plane indicated. In addition, two pieces of wood can be glued together along the line marked and shear tested.

Shear test specimen
for wood

Figure 17–7
Wood shear apparatus.

In general, the test fixture should hold the material firmly to ensure proper alignment throughout the test. The shear load should be applied evenly and perpendicular to the longitudinal axis of the piece. The ends of the test specimens must protrude through one or both dies far enough to avoid bending stresses.

The loading rate should be slow enough to allow a sufficient number of accurate readings to be taken. Once the material has yielded, the testing rate may be increased. Torque and angular displacement readings are taken until rupture or until the material has failed to the limit of the test's objectives. The rate at which the load is applied should not exceed 1.25 mm/min (0.05 in./min) for most materials. In the case of wood specimens, 0.6 mm/min (0.025 in./min) should not be exceeded.

Torsional specimens should be long enough to facilitate strain measurement and to reduce the stress effects developed at the ends as a result of the grippers. The ends of the specimen should be of sufficient proportion and configuration that they can be easily and securely gripped. Both ends of the specimen should be centered in the fixture. To help ensure that the ends are centered, the fixture typically has guides (small points) that help align the specimen in the fixture.

If data concerning the proportional limit of a specimen in torsional loading are desired, a thin-walled tubular specimen should be employed. The tubular specimen should have a short reduced section with a length-to-diameter ratio of 1:2 and a diameter-to-thickness ratio of approximately 10:1. For larger diameter-to-thickness ratios, the specimen tends to buckle, which would affect the accuracy of the test. Dimensions can be altered to conform to the test machine and material used in the actual testing.

To conduct a torsional shear test on a piece of tubing (hollow specimen), the ends must be plugged to avoid crushing them during the test. The plugs should be only as long as the grippers and should not extend beyond the grippers, which would adversely affect the test results. Torsional tests for plastic materials are conducted on square specimens that have holes drilled in the ends to facilitate pins for gripping (see ASTM D 1043).

17.5 EXPECTED RESULTS

The shear strength of a material is found by dividing the applied shearing force (in lb or N) by the cross-sectional area of the specimen (in in.^{2} or m^2).

Example

a. Calculate the shear strength of a round test specimen whose diameter is 0.25 in. and that requires a single shear load of 5000 lb to shear.

$$\text{shear strength} = \text{applied load/cross-sectional area}$$

where the cross-sectional area for the round specimen is calculated using the formula $a = \pi r^2$.

$$
\begin{aligned}
\text{shear strength} \quad &= 5000 \text{ lb}/[\pi(0.125 \text{ in.})^2] \\
&= 5000 \text{ lb}/0.049 \text{ in.}^2 \\
&= 102{,}041 \text{ lb/in.}^2, \text{ or } 704 \text{ MPa}
\end{aligned}
$$

b. Calculate the shear strength of the same specimen if the test is conducted in double shear with the same results.

$$
\begin{aligned}
\text{shear strength} \quad &= 5000 \text{ lb}/[\pi \cdot (0.125 \text{ in.})^2 \cdot 2] \\
&= 5000 \text{ lb}/0.098 \text{ in.}^2 \\
&= 51{,}020 \text{ lb/in.}^2, \text{ or } 352 \text{ MPa}
\end{aligned}
$$

c. What would be the diameter of a specimen that requires a shear load of 5000 lb to break in double shear and whose shear strength is 100,000 lb/in.?

$$
\begin{aligned}
\text{shear strength} \quad &= \text{applied load/cross-sectional area} \\
101{,}859 \text{ lb/in.}^2 \quad &= 5000 \text{ lb/CSA} \\
\text{CSA} \quad &= 5000 \text{ lb}/100{,}000 \text{ lb/in.}^2 = 0.050 \text{ in.}^2
\end{aligned}
$$

For tensile tests, the modulus of elasticity, or Young's modulus, was calculated based on the stress and corresponding strain developed during the elastic range for the material under test. In shear testing, a shear modulus is calculated for the material using the shear stress and shear strain within the material's elastic range. The formula $E_s = \tau/\gamma$ is used, where E_s is the shear modulus for the material, the lowercase Greek letter tau (τ) represents the shear stress, and the lowercase Greek letter gamma (γ) represents the shear strain. For an angular displacement, γ can be calculated by taking the tangent of the resulting angle, so $\gamma = \tan \theta$. The units for the shear modulus are pounds per square inch (lb/in.^{2}) or pascals (Pa).

Example

Calculate the shearing modulus of elasticity for a specimen that exhibits 0.006 in. strain for a corresponding stress of 60,000 lb.

$$E_s = S_s/\varepsilon_s = 60,000 \text{ lb}/0.006 \text{ in.}^2 = 10^7 \text{ lb/in.}^2, \text{ or } 68,930 \text{ MPa}$$

Example

Calculate the shear modulus for a material that shows a 3° displacement under a stress of 2500 lb.

$$E_s = \tau/\gamma = 2500 \text{ lb/in.}^2/\tan 3° = 44,703 \text{ lb/in.}^2, \text{ or } 329 \text{ Pa}$$

The shear modulus can also be calculated by dividing the shear stress by the shear strain or by the formula $E_s = (F \cdot h)/(A \cdot x)$, where F is the applied force, h is the length over which the strain is measured, A is the area of the cross section, and x is the displacement.

Example

Calculate the shear modulus for a 2 × 2 in. square specimen that has a 1500-lb load applied and a strain of 0.020 in.

$$E = \frac{Fh}{Ax} = \frac{1500 \text{ lb} \cdot 2 \text{ in.}}{(2 \text{ in.} \cdot 2 \text{ in.}) \cdot 0.020 \text{ in.}} = 37,500 \text{ lb/in.}^2, \text{ or } 259 \text{ MPa}$$

The shear modulus is related to the material's modulus of elasticity. The relationship is given as

$$E = 2E_s (1 + n)$$

where: E = modulus of elasticity for the material (Young's modulus)
E_s = shear modulus for the material
n = Poisson's ratio for the material

Because Poisson's ratio for most engineering material falls between 0.25 and 0.50, the shear modulus is typically 30% to 40% of the modulus of elasticity for a given material.

The modulus of rigidity can be calculated using the formula $E = (T \cdot L)/(J \cdot \theta)$, where T is the applied force (in lb or kg), L is the distance over which the strain is measured (distance between fixed points of the tropometer, in in. or m, J is the polar moment of inertia of the cross section ($\pi r^4/2$ for a circular cross section, in in.^{4} or m^4), and θ is the angular displacement measured over distance L (in rad). The units for the modulus of rigidity are pounds per square inch (lb/in.^{2}) or megapascals (MPa).

Example

Find the torsional modulus of rigidity for a specimen at an applied torque of 5000 in.-lb, whose outside diameter is 1 in. The angular displacement of the specimen was 0.2 rad over a distance of 18 in.

$$E = (T \cdot L)/(J \cdot \theta)$$

where: T = 5000 ft-lb
  L = 1.5 ft
  J = $0.5 \cdot \pi r^4$ = 0.5(3.14159)(0.5)^4 = 0.0982 in.^{4}
  θ = 0.2 rad
  E = $(T \cdot L)/(J \cdot \theta)$ = (5000 in.-lb $\cdot$ 18 in.)/(0.0982 in.^{4} $\cdot$ 0.2 rad)
    = 4.6 × 10^6 lb/in.^{2}, or 3.2 × 10^4 MPa

The modulus of rigidity should be calculated within the proportional limit. Beyond the proportional limit, the modulus of rigidity becomes inflated. For materials that tend to fail due to tension under torsional loading, the modulus of rigidity is approximately equal to the tensile strength of the material.

The ductility of a material under torsional shear loading can be found by using the original and final fiber lengths of the torsion specimen. The original distance over which the angular displacement was measured (L) and the final gage length (L') are used in the formula $[(L' - L)/L] \cdot 100$. The percentage of elongation for the outside fibers of the specimen is obtained and used as an indication of the ductility for the material.

Example

Find the percentage of elongation for a specimen whose original gage length was 18 in. and whose final gage length is 19.2 in.

$$\% \text{ Elongation} = [(L' - L)/L] \cdot 100 = [(19.2 - 18/18] \cdot 100 = 6.7\%$$

The results of a shear test are dependent on the type of test conducted. Efforts should be made to select the test that most closely approximates the conditions to which the part or specimen is subjected in service. For example, the direct shear test is appropriate for rivets, nails, and various pins and fasteners. The torsion test is better for drive shafts, couplings, and levers. Direct shear test conditions must further be specified as being single or double shear.

The stress-strain relationship is primarily connected with torsional testing. Theoretically, when a specimen is torsionally loaded, the planes remain planar after displacement. This is true only for circular cross sections, however. If other than circular cross sections are tested, some correction factor must be used in the calculations.

In solid cylindrical specimens, the internal fibers are less stressed than the surface fibers under test. This fact can be shown by the angular displacement of the central fibers as opposed to the surface fibers. As the surface fibers are stressed beyond their proportional limit, the integrity of the specimen is dependent on the internal, less highly stressed fibers. The core of the specimen then resists the load after the surface fibers have yielded. Testing hollow specimens or tubing can minimize this effect. All of the fibers of the tubing realize the same amount of stress. There is a more uniform distribution of forces, and a more accurate representation of the material's shear strength is obtained. However, if the tubing wall thickness drops below some critical value, the material will buckle due to the compressive stresses developed before yielding to the torsional stresses developed. Thus, torsional results for tube specimens depend on the wall thickness chosen for the specimen.

17.6 SUMMARY

Shear is the result of parallel forces acting slightly off-axis from each other. The result is a cutting action on one or more surfaces. Shear forces are also present in torsion. This twisting action tends to separate the specimen. Rivets, bolts, nails, and pins are common materials that undergo shearing forces. These materials must be designed to withstand the shearing forces present for them to function properly.

QUESTIONS AND PROBLEMS

1. What forces, other than shear, are present during the two types of shear tests? What effects do these extraneous forces have on the results of the test?
2. Conduct a torsional shear test on an empty paper towel roll (twist it with your hands). Describe and explain the results of the test. What forces acted upon the roll?
3. What is the shear strength of a 0.500 in.–diameter rivet that failed at a single-shear load of 1500 lb?
4. What is the shear strength of a 0.750 in.–diameter bar that failed under a double-shear load of 3500 lb?
5. What is the shear strength of a 5-cm–diameter bar that failed under a single-shear load of 5000 kg?
6. What is the shear strength of the 5-cm–diameter bar if it failed under a double-shear load of 5000 kg?
7. If Poisson's ratio for a material is 0.35 and its modulus of elasticity is 2×10^7, what is the shear modulus for the material?
8. Based on a shear modulus of 10×10^6 and a Poisson's ratio of 0.30, what is the modulus of elasticity of this material, in megapascals?
9. If a 5000-lb load causes a 5° angular displacement in a 0.25 in. shaft, what is the shear modulus for the material?
10. If a 3500-kg load causes an angular displacement of 3° in a 2-cm-diameter shaft, what is the shear modulus for the material? If the material has a Poisson's ratio of 0.28, what is its modulus of elasticity?
11. A force of 35 lb is applied to the end of a 12-in. adjustable wrench. What is the torque?
12. A load of 25 kg is applied to the end of a 25-cm lever. What is the torque?
13. A solid round bar has a 0.75-in. diameter and an applied torque of 250 ft-lb. What is the torsional stress developed in the bar?
14. A piece of pipe with an inside diameter of 3.5 cm and an outside diameter of 5 cm has an applied torque of 25 N·m. What is the stress developed in the pipe in pascals?
15. What is the torque stress developed in a 1/4-in.–diameter bolt when it is tightened to 50 ft-lb?

16. A solid shaft that can endure a torsion stress of 100,000 lb/in.^{2} is needed to transmit a torque of 125 ft-lb. What is the minimum diameter that the shaft could be?

17. A piece of 1/2-in. tubing 3 ft long is twisted through an angle of 5°. What is the strain developed in the tubing?

18. Which has the greatest torsional strength, a solid bar 0.500 in. in diameter or a 0.500-in.–diameter piece of pipe with a 0.010-in. wall thickness? What would the wall thickness of the pipe need to be to equal the torsional strength of the solid bar?

19. Which would exhibit the greater torsional strain, a 0.750-in.–diameter piece of pipe 18 in. long with an inside diameter of 0.650 in. or a solid bar 1 in. in diameter 24 in. long, if both specimens were displaced 7° by an applied torque of 130 ft-lb?

20. Construct a torsion stress-strain curve for a 1-inch diameter bar with the following data:

Applied Torque (ft-lb)	Angular Displacement (degrees)
20	1
40	2
60	3
80	5
100	7.5
120	10
140	13
160	17
180	22
200	Rupture

21. Describe three applications each where direct and torsional shear strength is the primary factor in the application. What materials would you select for these applications and why?

22. What design considerations are important in applications involving shear testing?

23. What efforts can be made to increase shear strength using the same material in an application?

24. Develop and describe your own direct shear test procedure for testing paper and cardboard. Describe the test setup, grippers, extensometer, loading concerns, data collection, and expected results.

CHAPTER 18
Bend or Flexure Testing

Objectives • Recognize and define common terms related to flexure testing.
  • List the equipment necessary to conduct a flexure test.
  • Describe the operation of equipment related to flexure testing.
  • Relate the general procedures used in conducting a flexure test.
  • Perform the necessary calculations related to flexure testing.
  • Recognize expected flexure test results.
  • Describe common variations in standard flexure test procedures.

18.1 INTRODUCTION

When a force is applied that induces a compressive stress over a portion of the cross section and a tensile stress over the remaining portion, the material is *bending*. Bending can be accompanied by direct stress, transverse shear, or torsional shear. Bending action in beams is often referred to as flexure. The term *flexure* refers to bending tests of beams subject to transverse loading. The *deflection* of a beam is the displacement of a point on the neutral surface of a beam from its original position under the action of applied loads. Deflection is a measure of the overall stiffness of a given beam and can be seen to be a function of the stiffness of the material and the proportions of the piece.

18.2 PRINCIPLES

Flexure is the bending of a material specimen under load. The strength that the specimen exhibits is a function of the material of which the specimen is made and the cross-sectional geometry of the specimen. For example, a rectangular beam 1 × 4 in. exhibits higher flexural strength than a 2 × 2 in. square beam of the same material. Therefore, the mathematics involved in calculating the flexural strength of a material is quite involved.

The properties involved in bend or flexure testing are the same as those addressed in tensile testing: ultimate strength, yield point, modulus of elasticity, and others. Because these properties are the same for flexure as for tensile testing, the flexure test is more commonly used to determine the best cross-sectional geometry for a given application.

When a specimen is loaded while being supported at both ends and the load is applied in the middle, the bottom of the specimen goes into tension and the top of the specimen goes into compression. *Tension* is the tendency to be pulled apart, and *compression* is the squeezing, or compressing, force. Somewhere between these two opposing forces is a neutral line, or axis, separating the forces. Along the neutral axis, the specimen is neither in tension nor in compression.

It is appropriate to record data for applied load and deflection that can be used to plot a flexure stress-strain curve describing the characteristics of the flexure test for the material. In addition to load and deflection, the characteristics and type of failure should be reported. For example, describe the type of fracture: Did it fail in tension, compression, or shear? Did it buckle before failure? What was the condition of the fracture: rough, jagged, or smooth? At this point, a sketch, along with a short paragraph, is often the better way to describe the type and characteristics of the fracture.

18.3 EQUIPMENT AND PROCEDURES

The flexure test is often performed on a universal testing machine. Figure 18–1 illustrates a simple test setup for flexure testing. Some precautions should be taken when setting up for a flexure test:

Figure 18–1
Setup for flexure testing.

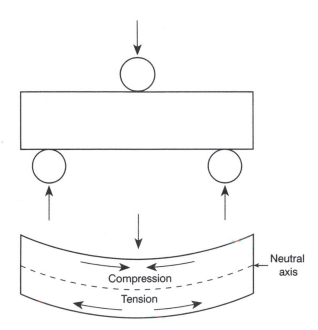

- The shape of the test specimen should be such that it permits use of a definite and known length of span. Specimen length should be between 6 and 12 times the width to avoid shear failure or buckling.
- The areas of contact with the material under test should be such that unduly high stress concentrations do not occur.
- Longitudinal adjustment of the position of the supports should be provided for so that longitudinal restraint will not be developed as loading progresses.
- Some lateral rotational adjustment should be provided for to accommodate beams having a slight twist from end to end, so that torsional stresses will not be induced.
- The arrangement of parts should be stable under load.

To satisfy these conditions, many flexure tests are conducted in universal testing machines with the supports placed on the platen or an extension of it and with the loading block fastened to or placed under the moving head (Figure 18–2).

Data concerning applied load and deflection should be recorded and plotted. The speed of applying the load is important and should not produce failure too rapidly. Test specifications for specific materials and specimen characteristics should be referenced. Flexure testing provides a crude estimation of the ductility of the material tested.

In brittle materials, such as ceramics and composite materials, the normal tensile test cannot be performed. When appropriate grippers are fastened to the specimen, flaws on the surface tend to promote cracking, which invalidates the test. In these situations, the flexure test is used to determine the flexural strength or modulus of rupture. By loading the specimen in three points, the tensile force acts from the center load point outward toward the ends. The flexural strength is calculated using the formula

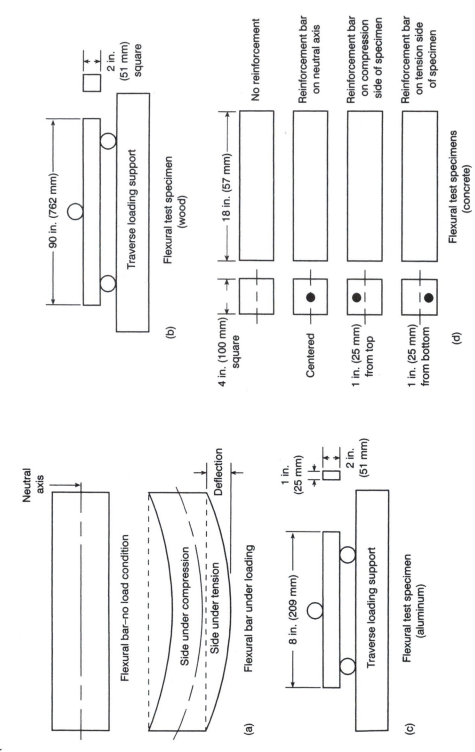

Neutral axis

Flexural bar–no load condition

Side under compression

Side under tension

Deflection

Flexural bar under loading

(a)

1 in. (25 mm)

2 in. (51 mm)

8 in. (209 mm)

Traverse loading support

Flexural test specimen (aluminum)

(c)

2 in. (51 mm) square

90 in. (762 mm)

Traverse loading support

Flexural test specimen (wood)

(b)

4 in. (100 mm) square

18 in. (57 mm)

No reinforcement

Centered

Reinforcement bar on neutral axis

1 in. (25 mm) from top

Reinforcement bar on compression side of specimen

1 in. (25 mm) from bottom

Reinforcement bar on tension side of specimen

Flexural test specimens (concrete)

(d)

Figure 18–2
Flexure test.

376

$$\text{flexural strength (lb/in.}^2) = \frac{3\,FL}{2\,wh^2}$$

where: F = applied load (in lb or kg)
L = distance between two outer points (in in. or cm)
w = width of specimen (in in. or cm)
h = height of specimen (in in. or cm)

Example

What is the flexural strength of a specimen 6 in. wide and 0.5 in. high that is loaded between two points 6 in. apart, if it takes 10,000 lb to break it?

$$\text{FS} = \frac{3 \cdot 10{,}000 \text{ lb} \cdot 6 \text{ in.}}{2 \cdot 6 \text{ in.} \cdot (0.5 \text{ in.})^2} = 60{,}000 \text{ lb/in.}^2$$

In compression, cracks and flaws tend to remain closed. Therefore, many brittle materials are used in applications where only compressive stresses are present. One such material is concrete.

One variation on the standard flexure test is the weld bend test. A good weld joint must be stronger than the material surrounding it. A weld bend tester is used to determine the strength of weld joints. Figure 18–3 shows a typical weld bend tester. This tester will accept material up to 3/8 in. thick.

Figure 18–3
Weld bend tester.

When conducting a weld bend test, the two parent pieces are welded together, generally with a butt weld. The excess material is removed by grinding or machining it smooth. The specimen is then placed in the tester so that the weld is placed over the center mandrel. The load is applied so that the specimen is bent through an angle of 180°, and the specimen is examined for defects in the weld. Welders must be able to pass this test to become certified welders.

18.4 SUMMARY

The flexure test or bend test is used to place the specimen in tension and compression simultaneously, the results of which can describe the modulus of elasticity, ultimate strength, and other similar properties gained by the tensile test. The flexure test is often conducted on brittle specimens that cannot accurately be tested in tension. These include many ceramics and composites.

QUESTIONS AND PROBLEMS

1. What is the flexural strength of a specimen 3 in. wide and 0.25 in. high that is supported at two points 4 in. apart, if it failed at an applied load of 1500 lb?

2. Convert the answer in Problem 1 to megapascals.

3. What load would be required to break a specimen 6 in. wide and 0.5 in. high, supported at two points 12 in. apart, if the flexural strength of the material is 50,000 lb/in.^{2}?

4. How high would a specimen need to be to support a load of 25,000 lb if the specimen is 12 in. wide, is supported at both ends 24 in. apart, and has a flexural strength of 55,000 lb/in.^{2}?

5. What is the maximum load that a concrete platform can support if the platform is 36 in. wide and 12 in. high and is supported at both ends 72 in. apart, if the flexural strength of the concrete is 3000 lb/in.^{2}?

6. Describe three applications where flexure strength is the primary factor in the application. What material would you select for that application? Why?

7. What design considerations are important in applications involving flexure loading?

8. Describe the shape of a flexural stress-strain curve for (1) a ductile material and (2) a brittle material.

9. Develop and describe your own flexure test procedure for testing bookshelves. Describe the test setup, grippers, extensometer, loading concerns, data collection, and expected results.

CHAPTER 19
Hardness Testing

Objectives • Recognize and define common terms related to hardness testing.
• List the equipment necessary to conduct a hardness test.
• Describe the operation of equipment related to hardness testing.
• Relate the general procedures used in conducting a hardness test.
• Perform the necessary calculations related to hardness testing.
• Recognize expected hardness test results.
• Describe common variations in standard hardness test procedures.

19.1 INTRODUCTION

Hardness is generally referred to as a material's property that indicates resistance to surface penetration. Although no single hardness test is applicable to all materials, some of the more common are presented in this chapter, including indentation hardness, rebound hardness, scratch hardness, wear hardness, and machinability.

The most common hardness tests involve driving a penetrator or indenter into the material and either recording the force required or measuring the resultant indentation from a given force. This is referred to as *indentation hardness.* If a known mass is dropped onto the surface of a material and the amount of rebound is measured, this is termed *rebound hardness.* The easiest scratch hardness test is attempting to scratch the surface of a specimen with a file. The quantity and quality of the resultant marks are used as an arbitrary indication of hardness. The amount of surface wear under test conditions is used to determine wear hardness and resistance to abrasion. Finally, machinability is an indication of the ease or difficulty that a material exhibits.

19.2 PRINCIPLES

The principle behind the hardness test involves the idea that hardness is measured by resistance to indentation, which serves as the basis for a variety of instruments. The *indenter*, either a ball or a plain or truncated cone or pyramid, is usually made of hard steel or diamond and is ordinarily used under a static load. Either the load that would produce a given depth of indentation or the indentation produced under a given load could be measured. In the rebound test, as in the scleroscope, a dynamic or impact load is dropped onto the surface of the test specimen. The amount of rebound determines the hardness of the specimen.

Probably the most commonly used hardness tests for metals in this country are the Brinell and Rockwell tests. However, a number of other tests are now being used as the result of the increase in harder steels and hardened steel surfaces. These other tests include the Shore scleroscope and the Vickers, Monotron, Rockwell superficial, and Herbert machines. Also, the need for determining the hardness of very thin materials, very small parts, and the hardness gradients over very small distances has led to the development of the so-called microhardness testers, such as the Knoop indenter.

A few of the more common hardness tests and machines are presented here as examples of the principles, procedures, and equipment used in testing a material's hardness.

19.3 EQUIPMENT AND PROCEDURES

19.3.a Brinell Test

One of the oldest tests for hardness is the *Brinell hardness test.* The Brinell test is a static hardness test that involves pressing a hardened steel ball into a test specimen. It is customary to use a 10-mm case-hardened steel or tungsten carbide ball under a

Figure 19–1
Brinell hardness testers.

Part B. Courtesy of King Tester Corporation.

(a)

(b)

3000-kg load for hard metals, a 1500-kg load for metals of intermediate hardness, and a 500-kg or lower load for soft materials.

Various types of machines for making the Brinell test are available. They differ according to (1) the method of applying load (e.g., oil pressure, gear-driven screw, or weights with lever); (2) the method of operation (e.g., hand or motive power); (3) the method of measuring the load (e.g., piston with weights, bourdon gage, dynamometer, or weights with lever), and (4) the size [e.g., large or small (portable)]. Figure 19–1 shows two typical Brinell hardness testers.

The Brinell test can be made in a universal testing machine by using an adapter for holding the ball. Tests of sheet metal can be made using a hand-held device much like a pair of pliers, which has a 3/64-in. ball and a 22-lb spring.

To conduct a Brinell test, the specimen is placed on the anvil and raised to contact with the ball. Load is applied by pumping oil into the main cylinder, which forces the main piston down and presses the ball into the specimen. The plunger has a ground fit, so frictional losses are negligible. A bourdon gage is used to give a rough indication of the load. When the desired load is applied, the balance weight on top of the machine is lifted by action of the small piston; this action ensures that an overload is not applied to the ball.

In the standard test, the diameter of the ball indentation is measured by use of a micrometer microscope or Brinell microscope, which has a transparent engraved scale in the field of view. The Brinell test is a good measure of hardness, but it does have limitations. One of these is that it is not well adapted for very hard materials. The ball deforms too much. It also isn't very appropriate for very thin pieces, where the indentation may be much greater than the thickness of the material. Another limitation is that it is not suited for case-hardened materials, where the indentation may have a greater depth than the thickness of the case, so that the yielding of the soft core invalidates the results. In the standard test, the full load is applied for a minimum of 30 seconds for ferrous metals and 60 seconds for softer metals.

The Brinell hardness number is nominally the pressure per unit area, in kilograms per square millimeter, of the indentation that remains after the load is removed. It is obtained by dividing the applied load by the area of the surface indentation, which is assumed to be spherical. The applied load and the diameter of the indentation are entered into the formula

$$BHN = \frac{2L}{\pi D \left(D - \sqrt{D^2 - d^2} \right)}$$

where: BHN = Brinell hardness number (in kg/mm^2)
  L = applied load (in kg)
  D = diameter of ball penetrator (10 mm)
  d = diameter of indentation (in mm)

Example

What is the Brinell hardness number for a specimen in which an indentation of 5 mm is produced with an applied load of 3000 kg?

$$\text{BHN} = \frac{2L}{\pi D\left(D - \sqrt{D^2 - d^2}\right)} = \frac{2(3000 \text{ kg})}{\pi(10 \text{ mm})[10 \text{ mm} - \sqrt{(10 \text{ mm})^2 - (5 \text{ mm})^2}]} = 142.6 \text{ kg/mm}^2$$

The Brinell hardness number usually falls within a range of 90 to 630, with higher numbers indicating greater hardness. The deeper the penetration, the larger the diameter of the indentation and the lower the hardness number. If the penetrator leaves an indentation with a diameter greater than 6 mm, a lighter load should be used. When the penetrator is driven into the surface of the material, a ridge develops around the indentation. This ridge tends to give a false reading on the diameter of the indentation.

On the upper end of the scale, Brinell numbers over 650 should not be trusted for two reasons: (1) The diameter of the indentation is so small that very slight inaccuracies in readings will greatly affect the BHN, and (2) the ball penetrator tends to flatten out and give inaccurately high diameters. Not many common materials have BHNs over 650.

As a rule of thumb, a 3000-kg load should be used for a BHN of 150 and above; a 1500-kg load, for BHNs between 75 and 300, and a 500-kg load, for materials with BHNs below 100. These figures overlap to give a reasonable latitude in testing.

The material's thickness should be no less than 10 times the depth of the indentation. When thinner specimens are tested, the hardness of the anvil underneath the specimen is what is really being tested.

The Brinell test has some limitations and disadvantages. It is a destructive test because it leaves an indentation in the material. In many cases, this renders the specimen unfit for use. In addition, many Brinell testing machines are heavy (more than 200 lb), which makes them cumbersome and unfit for field service. They also tend to be more expensive than other machines. For example, a basic Brinell testing machine with accessories and microscope generally costs more than $5000. The test itself tends to be subjective, in that the training, experience, and attitude of the technician measuring the indentation may vary the results. This variation is generally low, but two technicians performing the same test on the same specimen may have results that vary up to an average of 10%. The necessity of calculating the BHN rather than obtaining the number from direct reading is also considered a limitation and disadvantage.

The Brinell test has some advantages. Because it is older and is well established, most people are acquainted with the test, and the test results are generally accepted throughout industry. The test can be performed quickly, generally in about 2 minutes. After the initial cost of equipment, the test is generally inexpensive to run. Finally, the Brinell test is not greatly affected by imperfections in the material. A hard spot or crater in the material will not significantly affect the overall results.

19.3.b Rockwell Test

The *Rockwell hardness test* is similar to the Brinell test in that the hardness number found is the degree of indentation of the test piece caused by an indenter under a given static load. The Rockwell test varies from the Brinell test in principle because the Rockwell test is conducted with a choice of three different loads and three different indenters. It differs from the Brinell test in that the indenters and the loads are smaller, and the indentation made by the load is smaller and shallower. It is applicable to testing of materials beyond the scope of the Brinell test, and it is faster because it gives arbitrary direct readings (ASTM E 18). Figure 19–2 shows a typical Rockwell hardness tester and indenters.

The Rockwell test is conducted in a specially designed machine that applies load through a system of weights and levers. The indenter or penetrator may be a 1/16-in.

Figure 19–2
Rockwell hardness tester.

Figure 19–3
Rockwell scales.

Scale	Indenter	Applied Load (kg)
A	Brale	60
B	1⁄16 in.	100
C	Brale	150
D	Brale	100
E	1⁄8 in.	100
F	1⁄16 in.	60
G	1⁄16 in.	150

hardened steel ball, a 1/8-in. hardened steel ball, or a 120° diamond cone with a somewhat rounded point, called a *brale*. The hardness number indicated on the dial is an arbitrary value that is inversely related to the depth of indentation. The scale used for the test is a function of the load applied and the indenter used. For example, if the 1/16-in. ball penetrator is used, the 100-kg load is applied, and the reading is taken from the B scale. The brale indenter is used in conjunction with the 150-kg load, which produces readings for the C scale.

In the operation of the machine, a minor load of 10 kg is first applied, which causes an initial indentation that sets the indenter on the material and holds it in position. The dial is then placed on the "set" mark on the scale, and the major load is applied. The major load typically ranges from 60 to 100 kg when the steel ball is used, although other loads may be necessary, and is 150 kg when the diamond brale is used. The ball indenter is normally 1/16 in. in diameter, but 1/8-, 1/4-, and 1/2-in. diameters are available for softer materials. After the major load is applied and removed, the hardness reading is taken from the dial while the minor load is in position. Figure 19–3 shows the scale, indenter used, and the applied load for the Rockwell hardness tests.

There is no Rockwell hardness value consisting of a number alone, because the value depends on the indenter and load employed in the test. Therefore, a letter designation is also required for the test value to have any meaning. The dial on the machine has two sets of figures—one red and the other black—which differ by 30 hardness numbers. The dial was designed to accommodate the B and C scales, which were the first ones standardized and are the most commonly used. The red scale is used for readings obtained through the use of ball indenters, regardless of the size of the ball or magnitude of the major load, and the black figures are used only for the diamond brale.

The B scale is for testing materials of medium hardness, and the working range for the B scale is 0 to 100. If the ball indenter is used for values over 100, there is a chance it will be flattened, thus giving a false reading.

The C scale is the one most commonly used for materials with hardness greater than B100. The hardest steels have values of about C70. The useful range of the C scale is from C20 upward. Variations in the grinding of the diamond brale make readings below C20 unreliable for smaller indentations. Choose a scale to employ the smallest ball that can be properly used, because sensitivity is lost as the indenter size increases.

Rockwell scales are divided into 100 divisions, and each division, or point of hardness, equals 0.002 mm in indentation; thus, the difference in the readings B53 and B56 is 3 × 0.002 mm, or 0.006 mm. Because the scales are reversed, the higher the number, the harder the material.

The test piece should be flat and free of scale, oxide, pits, and foreign material on top and bottom. The thickness of the piece should be such that the piece will not bulge or deform on the anvil side when undergoing the test. Such deformation will give a false reading. On round stock, a flat place can be filed or ground into the material. All tests should be conducted on a single thickness of material. To avoid error induced by hitting a hard spot, impurity, small hole, or other surface imperfection, three tests are generally conducted on three different spots, and the average of the three readings is recorded.

It is possible to convert from Rockwell to Brinell, and vice versa. For Rockwell C values (HR_C) between −20 and 40, the Brinell hardness number is calculated by

$$BHN = \frac{1.42 \times 10^6}{(100 - HR_C)^2}$$

For HR_C values greater than 40, use the following formula:

$$BHN = \frac{2.5 \times 10^4}{100 - HR_C}$$

For HR_B values that fall between 35 and 100, use the following:

$$BHN = \frac{7.3 \times 10^3}{130 - HR_B}$$

Example

Convert the Rockwell hardness number HR_C 60 to a BHN number.

$$BHN = \frac{2.5 \times 10^4}{100 - HR_C} = \frac{2.5 \times 10^4}{100 - 60} = 625$$

19.3.c Rockwell Superficial Hardness Test

Some testers are fitted for *Rockwell superficial hardness tests*. This test is designed to test the surface hardness, where only shallow indentations are possible or desired. The superficial tester operates on the same principle as the regular Rockwell hardness tester, although it employs lighter minor and major loads and has a more sensitive depth-measuring system. The superficial tester employs a minor load of 3 kg and major loads of 15, 30, or 45 kg. One point of hardness on the superficial machine corresponds to a difference in indentation depth of 0.001 mm. This test uses the same indenters as other Rockwell tests.

The superficial tester uses the N and T scales. The W, X, and Y scales are used for very soft materials. The superficial tests correspond to the 15T, 30T, 45T, 15N, 30N, and 45N scales, depending on the indenter and the load used for the test. The T scale uses the indicated load and the 1/16-in. ball indenter, whereas the N scale uses the load indicated and the brale.

Like the Brinell test, the Rockwell test has its advantages, disadvantages, and limitations. The Rockwell test is accurate and precise. It is also a fast test to run, and the readings are taken directly from the machine, without calculation. It is universally accepted, as is the Brinell. The Rockwell test is more of an objective test in that anyone should be able to repeat the results of a test on the same material, regardless of the technician performing the test. The range of the Rockwell test covers a wide variety of material hardness.

The disadvantages of the Rockwell test closely resemble those of the Brinell test. The testing machine is relatively expensive and is not suitable for field service. The test is generally considered to be destructive.

19.3.d Vickers Hardness Test

The principle of using the ratio of the applied load to the area of indentation (such as in the Brinell test) for determining hardness is applied to other tests as well. One such test is the *Vickers hardness test*, which uses a small, diamond, square-pyramid-shaped indenter that has a 136° tip and applied loads between 5 and 120 kg in 5-kg increments.

In conducting the test, a specimen is placed on the anvil and raised by a screw until it is close to the point of the indenter. By tripping a starting lever, a 20:1-ratio loading beam is unlocked, and the load is slowly applied to the indenter and then released. A foot lever is used to reset the machine. After the anvil is lowered, a microscope is swung over the specimen, and the diagonal of the square indentation measured to 0.001 mm. The machine may also employ 1- and 2-mm ball indenters.

The Vickers test is used primarily in research applications. One advantage claimed by some Vickers machine operators is in the measurement of the indentation: A much more accurate measure can be made of the diagonal of a square than can be made on a circle, where the measurement is made between two tangents to the circle. It is a fairly rapid method and can be used on metal as thin as 0.006 in. It is claimed to be accurate for hardness as high as 1300 (about 850 Brinell). The indenter does not exhibit the tendency to flatten out as much as with the Brinell test.

Disadvantages are that the Vickers test is a destructive test. The test is much slower than either the Brinell or the Rockwell tests. To use the Vickers test, the surface of the specimen must be polished, which takes considerable time. Although the Vickers test is more accurate than either the Brinell or Rockwell tests, the cost of the equipment is much higher, although the results are as widely accepted.

The Vickers diamond pyramid hardness number, abbreviated DPH, is calculated using the following formula.

$$DPH = \frac{1.8544\,P}{d^2}$$

where: DPH = Vickers diamond pyramid hardness number (in kg/mm$^{2)}$
$\quad\quad\quad$ P = load (in kg)
$\quad\quad\quad$ d = length of the measured diagonal (in mm)

Example

What is the DPH for a specimen in which an indentation of 0.75 mm was produced using a load of 100 kg?

$$DPH = \frac{1.8544\,P}{d^2} = \frac{(1.8544)\cdot(100\text{ kg})}{(0.75\text{ mm})^2} = 330\text{ kg/mm}^2$$

19.3.e Monotron Hardness Test

The *Monotron hardness tester* also operates on the indentation principle; however, it is essentially a constant-depth indicator. The hardness is the pressure in kilograms per square millimeter necessary to give an indentation depth of 0.0018 in. This corresponds to a depth of indentation of 6% of the diameter of the 0.75-mm spherical-tipped diamond indenter and yields an indentation 0.36 mm in diameter.

The load is applied by a hand lever, which makes it difficult to control precisely. The load and depth are read from separate dials. Because the dial reading for depth lacks adequate sensitivity, it is difficult to produce reliable indentations. The depth is measured under load from the original surface of the specimen. This machine is well adapted to very thin materials or case-hardened surfaces. It is useful over the entire range of hardness and is rapid in operation.

19.3.f Microhardness Testers

The *Knoop indenter* is made of diamond and is ground so that it makes a diamond-shaped indentation, the ratio of the long to the short diagonals being 7:1. The *Tukon tester*, with which the Knoop indenter is used, can apply loads of 25 to 3600 g. It is fully automatic in making the indentation. The operator places the specimen under a microscope and selects an area for testing. The test is conducted, and the indentation is measured under a microscope. The Knoop hardness number is the ratio of the applied load (in kg) to the unrecovered projected area (in mm^2):

$$KHN = \frac{1.43\,P}{L^2}$$

where: KHN = Knoop hardness number (in km/mm^2)
$\quad\quad\quad$ P = load (in kg)
$\quad\quad\quad$ L = length of longest diagonal (in mm)

Figure 19–4
Knoop indenter.

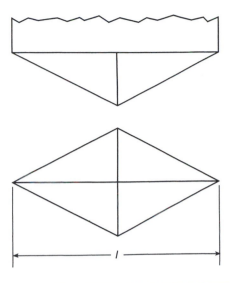

Example

What is the KHN for a material in which a 1000-g load produces an indentation with length of 0.123 mm?

$$KHN = \frac{1.43\,P}{L^2} = \frac{1.43 \cdot 1 \text{ kg}}{(0.123 \text{ mm})^2} = 95 \text{ kg/mm}^2$$

The Knoop hardness number is between 60 and 1000. It is a measure of the resistance of the material's surface to penetration, or a measure of the surface plasticity of the material.

The Tukon-Knoop device and the similar Wilson-Knoop device are useful for hardness testing on small parts or for testing hardness over a small area. Although the Tukon tester is often supplied with a Knoop indenter, it can also be adapted for use with a Vickers 136° diamond-pyramid indenter. Figure 19–4 shows the Knoop indenter.

19.3.g Durometers

Durometers come in various types, depending on the test material. All durometers are basically the same, differing only in the sharpness of the indenter point and the magnitude of the load applied to the indenter by a calibrated spring. The durometer hardness is also a measure of indentation depth; it varies from 100 at zero indentation to 0 at an indentation of 0.100 in. The load acting on the indenter varies inversely with the depth of penetration, being a maximum load at zero penetration and minimum load at full indentation (0.100 in.). Test specimens should be at least 1/4-in. thick, with no tests made within 1/2 in. of the edge of the material. Results made on one type of durometer cannot be correlated to those made on other types of durometers.

Figure 19–5
Shore durometer.

   The *Shore durometer* is a hand-held tester that is specifically designed to test softer materials, such as rubber, plastics, and many composites. Figure 19–5 shows a hand-held Shore durometer. There are two basic models of durometer: the A and D. Both of these models operate in basically the same way: The spring-loaded indenter is pressed against the surface of the material specimen. The resistance of the surface of the material to penetration is read directly from the dial face. The type A durometer has a spring loading between 56 and 822 g. The type D durometer has a spring loading between 0 and 10 lb. Both of these models read from 0 to 100, with the higher number being a harder material with greater surface resistance to penetration. In reporting the results of a durometer test, the reading along with the type is given—i.e., 50A or A50.

19.3.h Hardness Test of Wood

The only standardized hardness test for wood is an indentation type (ASTM D 143). The hardness is determined by measuring the load required to embed a 0.444-in. steel ball to one-half its diameter into the wood. The wood hardness value is for comparative purposes only. The approximate range for wood hardness values is from 400 lb for poplar to 4000 lb for persimmon. The hardness for Douglas fir is about 900 lb.

19.3.i Dynamic Hardness Tests

Most *dynamic hardness testers* are indentation tests and depend on the calculation of energy absorbed by the specimen during the test. Therefore, they are questionable for use in a fixed test procedure and yield arbitrary results.

Among the first dynamic hardness tests were those of Rodman, who experimented with a pyramidal punch in 1861. Later experiments used a hammer with a spherical end; they were made to verify and substantiate Rodman's results. These tests proved that the work of the falling hammer is proportional to the volume of the indentation. The hardness is, therefore, the work required to produce a unit volume of indentation. This method is very useful in determining the hardness of specimens at high temperatures, because the hammer is not in contact with the specimen for any considerable length of time and is, therefore, not affected appreciably by the heat.

There are a number of foreign machines that use a dynamic load. Typically, the hardness number is calculated by dividing the net energy of the blow by the volume of the indentation. One of the more important is the *Pellin hardness tester,* in which the indentation is produced by a falling rod of known mass, which has at its lower end a steel ball 2.5 mm in diameter.

Another dynamic tester is the *Whitworth autopunch,* which is a hand-held Brinell machine actuated by release of a spring in the handle that supplies a standard striking energy to the ball indenter in the bottom of the punch. The diameter of the indentation is measured in a manner similar to the Brinell test.

The *Waldo hardness tester* utilizes a conical-point steel indenter weighing 0.1 lb and is dropped from a height of 12 in. Hardness is determined based on the diameter of the indentation. The *duroskop hardness tester* depends on the rebound of a pendulum hammer. Finally, the *Avery hardness tester* is a modification of the Izod impact machine for dynamic hardness testing.

At present, the *Shore scleroscope* is probably the most widely used dynamic hardness tester. The hardness test conducted using a scleroscope is often referred to as *rebound hardness*. Scleroscope hardness is a number representing the height of rebound of a small pointed hammer that falls within a glass tube from a height of 10 in. against the surface of the specimen. The standard hammer is approximately 1/4 in. in diameter, is 3/4 in. long, weighs 1/12 oz, and has a diamond tip with radius rounded to 0.01 in.

The scale is graduated into 140 divisions, a rebound of 100 being equivalent to the hardness of martensitic high-carbon steel. For this material, the area of contact between the hammer and specimen is only about 0.0004 in.^{2}, and the stress developed is in excess of 400,000 lb/in.^{2}. These hardness numbers are arbitrary and comparable only when obtained from similar material.

19.3.j Herbert Pendulum Device

The *Herbert hardness tester* operates on a principle different from other testers. An arched metal frame (pendulum) is supported on a 1-mm steel or diamond ball. The center of gravity of the pendulum is brought to a predetermined height below the center of the ball by means of an adjustable weight. A graduated tube (from 0 to 100) is mounted on the top of the frame.

The ball is located on top of the specimen surface where the hardness is to be tested. The pendulum is swung, and various measures of hardness are obtained. In a scale test, the angular oscillation during one swing (starting with the bubble at zero) gives a scale hardness number. The scale hardness number is said to be a measure of the resistance of the metal to flow. In the "time test," the number of seconds it takes for the pendulum to swing 10 times is measured. The time test is an indentation hardness test similar to the Brinell test.

19.3.k Scratch Hardness

An arbitrary scale of hardness is the ability of one material to scratch another material with a lower hardness number. The first material can also be scratched by material of higher hardness number, and so it continues, using increasingly harder materials to scratch those with lower hardness numbers. One well-known scale is the Mohs' scale used by mineralogists. Figure 19–6 illustrates the *Mohs' hardness scale.*

Using Mohs' scale, if a material will scratch one material but not the next-highest material, the two numbers are averaged to get the hardness number for the material.

Example

If a material will scratch quartz (7) but not topaz (8), the material is said to have a Mohs' hardness number of 7.5.

A *sclerometer* is a device that attempts to quantify either the pressure required to make a given scratch or the size of scratch produced by a stylus drawn across a test surface under a fixed load. Sclerometer tests are hard to standardize and interpret.

The file test is widely used as a qualitative, or inspection, test for hardened steel. The inspector or tester runs a file over the test specimen. The file passes over mate-

Figure 19–6
Mohs' hardness scale.

Mineral	Hardness Number
Talc	1
Gypsum	2
Calcite	3
Fluorite	4
Apatite	5
Feldspar	6
Quartz	7
Topaz	8
Corundum	9
Diamond	10

rials of the proper hardness and digs into soft materials. Thus, acceptance or rejection can be determined by the file test.

19.3.1 Abrasion Testing

One example of abrasion hardness testing is the *Deval abrasion test*, where 50 test specimens are tumbled in a cylinder. A standardized machine is used to determine the loss in weight (given as a percentage) due to the tumbling process. The primary measurement associated with the Deval test is the *French coefficient*. The French coefficient is determined by dividing 400 by the wear in grams per kilogram of rock used.

A standard abrasion test for coarse aggregate is called the *Los Angeles rattler test*. It is made by tumbling a charge of aggregate together with a charge of steel balls in a standard period of time. The percentage of wear is then determined.

19.4 SUMMARY

There are many different testing methods and testing machines used to determine the hardness of a material. This chapter includes some of the more common: Brinell, Rockwell, Vickers, Tukon, and Monotron tests and durometers. These and

Figure 19–7
Hardness number comparison.

HR$_C$	HR$_B$	BHN 3000 kg	Vickers 10 kg	Knoop 500 g
65	—	700	820	850
60	—	627	765	732
55	120	555	632	630
50	117	495	540	542
45	115	430	455	465
40	112	372	390	402
35	108	340	351	365
30	105	283	286	312
25	103	260	263	284
20	97	222	223	250
15	94	206	207	225
10	90	185	186	200
5	87	167	167	183
—	80	146	146	164
—	75	133	133	150
—	70	121	121	140
—	65	111	111	130
—	60	108	108	120
—	55	100	—	112
—	50	—	—	107

other testing devices measure the resistance of a material's surface to the penetration of the indenter. Scratch hardness is another common test, in which one material is used to scratch another. By determining which materials will scratch other materials, we can arrive at a figure of merit for the material under scrutiny. Hardness is one of the principal properties of a material and is a widely used tool to understanding the condition and applications of a material. Figure 19–7 is a comparison of the various hardness numbers discussed in this chapter.

QUESTIONS AND PROBLEMS

1. A Brinell hardness test produced an indentation of 4 mm using a 10-mm ball indenter and a load of 3000 kg. Calculate the BHN for the material.
2. If a 500-kg load produced an indentation of 2.5 mm using a 10-mm ball indenter, what is the BHN for the material?
3. What is the Vickers hardness number (DPH), for a material where a diagonal length of 0.2 mm was produced using a 10-kg load?
4. If a 10-kg load produced an indentation with a 0.33-mm diagonal, what is the DPH for the material?
5. What is the Knoop hardness number for a material that exhibited an indentation with a length of 0.050 mm after being subjected to a 500-g load?
6. If a 10-kg load produced an indentation with a length of 0.075 mm, what is the Knoop hardness number for the material?
7. Convert an HR_C value of 30 to a BHN.
8. Convert an HR_B value of 50 to a BHN.
9. What Rockwell scale would be used with a 1/16-in. indenter and a 100-kg load?
10. What Rockwell scale would be used with a brale indenter and a 150-kg load?
11. What scale would use a brale indenter and a 60-kg load?
12. What scale would use a 1/16-in. indenter and a 60-kg load?
13. If you wanted to test the hardness for a plastic specimen, what tester or testers would be appropriate?
14. If you wanted to test a rubber specimen for hardness, what tester or testers could you use?
15. What is the Mohs' hardness number for a specimen that would scratch feldspar but would not scratch quartz?
16. Describe three applications where hardness is the primary factor in material selection. What material would you select for that application? Why?
17. What design considerations are important in applications relying on material hardness?
18. Develop and describe your own hardness test procedure for testing coins. Describe the test setup, grippers, extensometer, loading concerns, data collection, and expected results.

CHAPTER 20
Impact Testing

Objectives
- Recognize and define common terms related to impact testing.
- List the equipment necessary to conduct an impact test.
- Describe the operation of various equipment related to impact testing.
- Relate the general procedures used in conducting an impact test.
- Perform the necessary calculations related to impact testing.
- Recognize expected impact test results.
- Describe common variations in standard impact test procedures.

20.1 INTRODUCTION

Materials often exhibit different properties, depending on the rate at which a load is applied and the resulting strain that occurs. Most materials can withstand greater loads before failure if the load is applied gently over a longer period of time (static testing). If a smaller load is applied suddenly (dynamic testing), the material may fail and appear to have less strength. Because the properties of many materials are so strain-rate-dependent, tests have been standardized to determine the energy required to break materials under sudden blows. These tests are classified as *impact tests*. The general outcome of an impact test is the determination of the energy required to break the specimen.

20.2 PRINCIPLES AND EQUIPMENT

Impact testing is used to measure the transfer of energy required to break a given volume of material. Impact strength is then an indication of how well a material can withstand shock loading. Because energy cannot be created or destroyed, the energy in the impact must be released through different channels. For example, energy can be consumed in the elastic deformation of the specimen, the plastic deformation of the specimen, friction between the moving parts, and other such actions.

In the design of structures and machines, an attempt is made to provide for the absorption of as much energy as possible through elastic action and rely secondly on some form of damping to dissipate it. In impact testing, the object is to use the energy of the blow to rupture the test specimen.

Energy is defined as the ability to do work. *Work* is defined as a force operating through a distance. These two factors together give us the equation

$$W = FD$$

where: W = the work done (in ft-lb or N·m)
F = force applied (in lb or N)
D = distance through which the force was applied (in ft or m)

The object of the impact test is to determine the work done in breaking the object under a sudden blow. This can be accomplished by dropping a known weight on the material specimen from a known height. The equation involved in dropping a known weight from a measured height is

$$E = wh \text{ or } E = mgh \text{ (metric)}$$

where: E = the energy of the falling weight (in ft-lb or N·m)
w = weight of the object (in lb)
m = mass of the object (in kg)
h = the height of the object before falling (in ft or m)
g = 9.8 m/s^2

Example

What is the energy of a 12-oz (0.34-kg) can full of soda when it hits the ground after falling off the top of a 100-ft (30.5-m) building?

$$E = 0.75 \text{ lb} \cdot 100 \text{ ft} = 75 \text{ ft-lb}$$
$$E \text{ (metric)} = 0.34 \text{ kg} \cdot 9.8 \text{ m/s}^2 \cdot 30.5 \text{ m} = 102 \text{ N} \cdot \text{m}$$

The property of a material associated with rupture is called *toughness*. The causes of rupture are impact and force. Toughness depends chiefly on the ductility and strength of a material. Impact testing is an adequate measure of a material's toughness.

In making an impact test, the load can be applied in flexure, tension, compression, or torsion. Flexure loading is the most common, and tensile is less common. Compression and torsional loadings are used only in special circumstances. The impact blow can be delivered through dropping weights, swinging a pendulum, or rotating a flywheel. Some tests involve rupturing the test specimen in a single blow; others use repeated blows. In tests involving repeated blows, some are multiple blows of the same magnitude, whereas others, such as the *increment-drop test*, gradually increase the height from which the weight is dropped until rupture is induced.

Perhaps the most common tests are the Izod and Charpy impact tests. Both employ a swinging pendulum and are conducted on small, notched specimens broken in flexure. The two tests differ in the design of the test specimen and the velocity at which the pendulum strikes the specimen. In the *Charpy test,* the specimen is supported as a single beam, but in the *Izod test,* the specimen is supported as a cantilever. In these tests, a large part of the energy absorbed is taken up in a region immediately adjacent to the notch. A brittle type of fracture is often induced. Figure 20–1 shows the machine used to conduct the Izod and Charpy tests.

Figure 20–1
Testing machine for Izod and Charpy tests.

Charpy test specimen   Izod test specimen

Figure 20–2
A. Charpy test specimen. B. Izod test specimen.

In the Izod test, the specimen shown in Figure 20–2(b) is supported vertically, being held in a vise at the base. The standard test velocity for the Izod test is 11.5 ft/s. In the Charpy test, the specimen is mounted horizontally, being held at both ends as the pendulum strikes the specimen. The standard test velocity for the Charpy test is 17.5 ft/s. The Charpy specimen is shown in Figure 20–2(a). The purpose of the notch in both specimens is to concentrate the stress, thus allowing the energy to be absorbed at a single point in the specimen. This facilitates the breaking of the specimen in a known region. Without the notch, the stress would tend to be evenly distributed throughout the specimen. The specimen would then plastically deform by bending rather than breaking. This would render the results invalid, because the purpose of the impact test is to determine the amount of energy required to break the specimen.

For wood, the Hatt-Turner test is used. This test is a flexural-impact test of the increment-drop type. The height of drop at which failure occurs is taken as a measure of toughness. From the data obtained from the Hatt-Turner test, the modulus of elasticity, the proportional limit, and the average elastic resilience can be found.

20.3 PROCEDURES AND EXPECTED RESULTS

Items that require standardization in impact testing are the foundation, anvil, specimen supports, specimen, striking mass, and velocity of the striking mass.

Principal features of a single-blow testing machine are as follows:

- A moving mass whose kinetic energy is great enough to cause the rupture of the test specimen placed in its path

- An anvil and a support on which the specimen is placed to receive the blow
- A means for measuring the residual energy of the moving mass after the specimen has been ruptured

The kinetic energy is determined from, and controlled by, the mass of the pendulum and the height of free fall measured from the center of the mass. The pendulum should be supported to reduce the lateral play, or restrain, and friction that may be felt as it swings in an arc toward the test specimen. The release mechanism should be constructed to reduce any binding, acceleration, or vibratory effects.

The anvil should be heavy enough in relation to the energy of the blow so that impact energy is not lost due to deformation or vibration. The specimen should be supported firmly and in the correct position throughout the test.

20.3.a Charpy Test

A Charpy test machine usually has a capacity of 220 ft-lb for metals and 4 ft-lb for plastics (ASTM E 23). The pendulum consists of a relatively light, although rigid, rod or piece of channel, at the end of which is a heavy disk. This pendulum swings between two upright supports and has a rounded knife-blade edge at the end aligned so that it contacts the specimen over its full depth at the time of impact.

The standard test specimen is $10 \times 10 \times 55$ mm, notched on one side in the center. Some tests require keyhole notches; others require U-shaped notches. The specimen is supported between two anvils so that the knife strikes opposite the notch at the midswing point. The pendulum is lifted to the initial release angle. It is then released and allowed to swing and strike the specimen.

The pendulum is set at a known angle, α, shown in Figure 20–3. Theoretically, if the pendulum does not encounter any resistance, it should swing to angle α on the opposite side. This theory discounts the effects of friction, which should be taken into account when calculating the energy required for rupture. When conducting the

Figure 20–3
Impact test conditions.

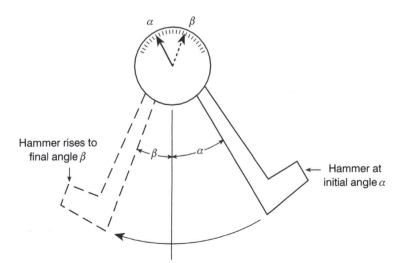

test, the pendulum is set to angle α. With the test specimen properly placed, the pendulum is released and allowed to swing freely. The pendulum gains momentum as it swings through its arc toward the specimen. When the pendulum strikes the specimen, breaking the specimen, it transfers some of its energy to the specimen. The pendulum then continues on to the opposite side of the machine at an angle ß.

Having recorded the weight of the pendulum, the length of the pendulum arm, and the initial and final angles (α and ß), the following calculation can be made:

$$E = wr(\cos ß - \cos \alpha) \text{ or } E = mgr(\cos ß - \cos \alpha) \text{ (metric)}$$

where: E = energy required to break the specimen (in ft-lb or N·m)
 w = weight of pendulum (in lb)
 m = mass of pendulum (in kg)
 g = 9.8 m/s^2
 r = length of pendulum (in ft or m)
 α = initial angle (angle of fall)
 ß = final angle (angle of rise)

Example

A standard Charpy test is conducted on a specimen. Using a 50-lb pendulum that is 36 in. long, with an initial angle of 76° and a final angle of 32°, what is the energy absorbed by the specimen before breakage?

$$E = (50 \text{ lb})(3 \text{ ft})(\cos 32° - \cos 76°) = 91 \text{ ft-lb}$$

20.3.b Izod Test

The common Izod test machine is made with a 120 ft-lb capacity. The test is similar to the Charpy test, although the specimen placement and features are different. In Izod impact testing, the pendulum strikes the front face of the test specimen, just the reverse of the Charpy impact test. In the Izod test, the specimen is $10 \times 10 \times 75$ mm, having a 45° notch cut 2 mm deep. The impact strength of the specimen is based on the angle of rise after rupture occurs. The energy in foot-pounds can be read directly from the dial on the tester.

Example

What is the energy absorbed by the specimen if a 30-kg pendulum 75 cm long is used at an initial angle of 76°, resulting in a final angle of 22°?

$$E = 30 \cdot 9.8 \text{ m/s}^2 \cdot \frac{75 \text{ cm}}{100 \text{ cm/m}} \cdot (\cos 22° - \cos 76°) = 151 \text{ N} \cdot \text{m}$$

20.3.c Hatt-Turner Test

The Hatt-Turner test is used primarily for flexure-impact tests of wood in which the height of drop is increased by increments until failure occurs (ASTM D 143). A tup weighing 50 lb is held by an electromagnet, which is raised by a motor. The tup drops through vertical guides when the current is stopped by opening a relay.

The specimen is a clear piece of wood having nominal dimensions of 2 × 2 × 30 in. The piece is supported over a 28-in. span so that the tup drops directly over the center of the span. The first drop is 1 in., and each succeeding drop is increased by 1-in. increments. If the piece has not ruptured after 10 in. is reached, 2-in. increments are maintained until either rupture or a 6-in. deflection occurs.

20.4 SUMMARY

In a tensile test, a tensile load is slowly applied to obtain data on the tensile strength and related properties of a material. When a material specimen is subjected to a sudden, sharp blow, the material behaves in a more brittle manner. The impact test, of which the Charpy and Izod are two common types, is used to evaluate the behavior of materials under sudden, sharp blows. In the test, a pendulum of known weight is suspended at a known angle above the specimen. The pendulum is then released and swings through an arc through the specimen. The angle through which the pendulum swings after breaking the specimen gives the energy absorbed by the specimen in breaking it. The energy in foot-pounds can be read directly from the dial on the tester. The initial and final angles are used to calculate the impact strength of the material. The specimen is generally notched to concentrate the stress developed by the specimen and to facilitate breakage. There are variations on this procedure, but the purpose of the impact test is to determine the force required to break a specimen of known dimensions.

QUESTIONS AND PROBLEMS

1. If you drop a 20-oz hammer through a distance of 3 ft onto the floor, what is the impact energy?

2. What is the impact energy of a box of marbles weighing 10 kg if it is dropped from a table 1 m high onto your foot?

3. A standard impact testing machine is used to conduct a test on a standard Charpy test specimen. If the 31.5-in.–long arm with a 60-lb hammer is initially set to 76° and rises to 27°, what is the energy absorbed by the specimen? What is the energy in newton-meters?

4. If a hardened steel bar 0.5 in. in diameter is loaded into a standard impact testing machine, what is the impact energy for a 31.5-in.–long arm with a 60-lb hammer if the initial angle of release is 76° and the rise is 0°? What is the energy in newton-meters?

5. A standard test specimen is listed as being able to absorb 75 ft-lb of energy. What is the angle of rise if the standard 31.5-in.–long arm and 60-lb hammer are used at an initial angle of 76°?

6. If a material is said to be able to absorb 200 N·m of energy, what is the angle of rise for a 30-kg

hammer on an 80-cm arm that has an initial angle of 76°?

7. Some low-impact-strength materials, such as plastics, are tested with lower initial angles. What is the impact energy of a 60-lb hammer with a 31.5-in. arm that has an initial angle of 30° and an angle of rise of 12°? What is the energy in newton-meters?

8. In testing some low-impact-strength materials, a lighter hammer is used. What is the impact energy for a 30-lb hammer that is 31.5 in. long if the starting angle is 25° and the angle of rise is 11°? What is the energy in newton-meters?

9. If a 60-lb hammer on the end of a 31.5-in. arm is released from 76° and the energy absorbed by the specimen is 150 ft-lb, what is the angle of rise?

10. Suppose a 16-lb sledgehammer is held by a shaft through the handle, giving it an arc swing of 30 in. If the sledge is allowed to swing freely toward a brick held in place directly below the shaft, what is the impact energy if the hammer is dropped from an angle of 90° and has an angle of rise of 15°? What is the energy in newton-meters?

11. Describe three applications where impact strength is the primary factor in material selection. What material would you select for that application and why?

12. What design considerations are important in applications involving shock loading or impact strength?

13. Develop and describe your own impact test procedure for testing eyeglasses. Describe the test setup, grippers, extensometer, loading concerns, data collection, and expected results.

CHAPTER 21
Fatigue Testing

Objectives
- Recognize and define common terms related to fatigue testing.
- List the equipment necessary to conduct a fatigue test.
- Describe the operation of equipment related to fatigue testing.
- Relate the general procedures used in conducting a fatigue test.
- Perform the necessary calculations related to fatigue testing.
- Recognize expected fatigue test results.
- Describe common variations in standard fatigue test procedures.

21.1 INTRODUCTION

Most structural assemblies are subject to variation in applied loads, causing fluctuations in the applied stresses in these parts. If these fluctuations are of sufficient magnitude, even though they may be considerably less than the static strength of the material, failure can occur when the stress is repeated enough times. For example, if you bend a wire back and forth enough times, it will break. Bending the wire once will not break it, but doing it repeatedly will. The material is fatigued. Terms related to fatigue testing are contained in ASTM E 616.

21.2 PRINCIPLES AND EQUIPMENT

Fatigue is defined as the failure of a material due to repeated or cyclic stresses being applied to it. The number of cycles required to break a specimen depends on the magnitude of the applied stress and the conditions through which it is applied. For example, bend a piece of wire gently a few degrees in both directions. The piece of wire will last a long time before breaking. Bend the same wire double in both directions quickly, and it will last only a few cycles. The applied force was the same in both tests, but the conditions of the test (speed and angle) were changed. These changes caused a subsequent change in the test results.

Parts such as crane hooks, machine parts, and aircraft skins fracture because they are subjected to cyclic stresses. These localized stresses produce permanent structural changes in products, usually through cracking or complete fracture after a number of fluctuations. Once a crack has been initiated, localized stresses around the crack may produce fracture, especially in brittle materials. Stresses tend to concentrate around the edges of the crack, and the crack continues to grow until fracture occurs under tensile stress. As the crack grows, the material exhibits plastic strain. Most fatigue failures originate as microscopic cracks, which grow and spread due to concentrated stresses until the point of fracture occurs. These cracks are commonly formed from slip lines, which form in the material from repeated stresses and microscopic dislocations. Imperfections such as surface scratches, notches, inclusions, and other defects can help initiate cracking.

The stress required for fatigue failure should be designated by degree of stress variation and type of stress. The stresses may be axial, shearing, torsional, or flexural. The stress at which a material fails by fatigue is termed *fatigue strength*. The *limiting stress*, or *critical stress*, is the stress below which a load may be repeated indefinitely without failure; it is also called the *endurance limit*.

Fatigue tests are conducted over long periods of time, sometimes months or even years. Consequently, fatigue tests are generally not used for quality control or inspection because of the time and effort required to collect the necessary data.

There are many types of fatigue-testing machines, but they all have two things in common: (1) They must maintain an accurate count of the number of cycles until failure, and (2) they must accurately record the applied stress required to cause fail-

Figure 21–1
Fatigue curve.

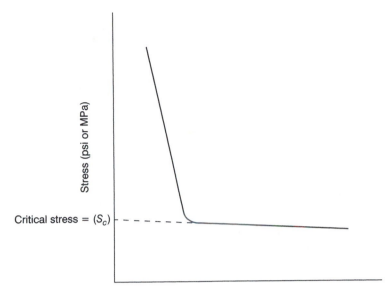

Critical stress = (S_c)

Number of cycles to failure

ure. If many specimens of a material were tested at different stresses and the number of cycles it took to cause failure were plotted against the applied stress, a fatigue curve like that shown in Figure 21–1 would result.

Notice that this curve consists of two nearly straight lines. If we were to extend these two lines until they met, we would arrive at the limiting, or critical, stress of the material.

Temperature also affects the fatigue strength of materials. Generally, as the temperature increases, the strength decreases, and the fatigue strength and endurance limit also decrease. In addition, the fatigue strength of metals is affected by composition and structure of the material. For example, certain alloys are added to increase the fatigue strength of metals. Manufacturing processes such as heat treatment and cold working increase the fatigue strength of metals. Smoothing and polishing metal surfaces also help improve the fatigue strength of these materials. Any alloying material or manufacturing process that helps reduce the occurrence of dislocation and reduces the tendency of crack initiation or propagation will increase the fatigue strength of the material.

21.3 PROCEDURES

Many different machines are employed for fatigue testing. Among these are the following:

- Machines for axial stresses

- Machines for flexural stresses
- Machines for torsional shearing stresses
- Universal machines for axial, flexural, or torsional shearing stresses or a combination of these

The general procedure for a fatigue test is to prepare several representative specimens. The first specimen is treated with a high amount of stress so that it fails rapidly. The second specimen is subjected to less stress until it fails. This procedure continues, decreasing stress and increasing repetitions, until the applied stress falls below the endurance limit of the specimen and it will not fail.

Another common procedure is the rotating cantilever beam test, where one end of a machined, cylindrical specimen is mounted in a rotating chuck. A weight is suspended from the other end of the specimen. This design initially puts a tensile force on the top surface of the specimen and a compressive force on the bottom. As the specimen rotates through one complete cycle, the forces acting on any one point on the specimen reverse and then return to their original condition, completing one cycle. The maximum stress acting on the specimen can be calculated by the formula

$$\sigma = 10.18 \frac{LF}{d^3}$$

where: σ = maximum stress (in lb/in.^{2} or MPa)
        L = length of specimen (in in. or m)
        F = applied force (in lb or N)
        d = diameter of specimen (in in. or m)

Example

What is the maximum stress that a 0.5-in.–diameter bar that is 12 in. long can endure with an applied load of 50 lb?

$$\sigma = \frac{10.18 \cdot 12 \text{ in.} \cdot 50 \text{ lb}}{(0.5 \text{ in.})^3} = 48{,}864 \text{ lb/in.}^2$$

Example

What is the minimum diameter of a 12-in.–long bar with a fatigue strength (maximum stress) of 80,000 lb/in.^{2} that can withstand a 2000-lb load?

$$d^3 = \frac{10.18 \cdot 12 \text{ in.} \cdot 2000 \text{ lb}}{80{,}000 \text{ lb/in.}^2} = \sqrt[3]{3.054 \text{ in.}^3} = 1.45 \text{ in.}$$

21.4 SUMMARY

In many applications, materials are subjected to repeated cycles of stress below their yield strengths. In these applications, the material can develop fatigue. Fatigue can result from repeated rotational, flexural, or vibrational loadings. Therefore, it is beneficial to test materials under controlled conditions in order to estimate what will happen in service. If the expected fatigue strength of the material is known, based on fatigue tests, the expected life of a material can be calculated and adjusted for.

QUESTIONS AND PROBLEMS

1. List several common applications where fatigue strength is important.
2. Describe an apparatus and testing procedure, using commonly available parts, that you could use to conduct a fatigue test on a piece of wire.
3. List other factors, in addition to temperature, that might influence the results of fatigue tests.
4. What is the maximum stress of a 1.375-in.–diameter bar, 3 in. long, with an applied load of 1500 lb?
5. If the maximum stress allowed for a 1.5-in.–diameter bar is 55,000 lb/in.^{2} and it is subjected to a load of 5000 lb, what must be the minimum length of the bar for it to withstand the stress?

6. Describe three applications where fatigue strength is the primary factor in material selection. What material would you select for that application? Why?
7. Describe the shape of the fatigue curve for (1) a ductile material and (2) a brittle material. What would be your expectations in curve shapes for plastics, ceramics, wood, and composites?
8. Develop and describe your own fatigue test procedure for testing metal wire. Describe the test setup, grippers, extensometer, loading concerns, data collection, and expected results.

CHAPTER 22
Nondestructive Testing

Objectives
- Recognize and define common terms related to nondestructive testing.
- List the equipment necessary to conduct various nondestructive tests.
- Describe the operation of equipment related to nondestructive testing.
- Relate the general procedures used in conducting a variety of nondestructive tests.
- Recognize expected test results.
- Describe common variations in standard test procedures.

22.1 INTRODUCTION

In many circumstances, it is better to analyze the properties of a specimen without destroying it in the process. Any method of testing that does not render the part unusable for its intended purpose is considered to be a *nondestructive test*. In this chapter, material is presented on various types of nondestructive tests (NDT), including radiographic, electrical, magnetic, visual, acoustic, photoelastic, and other tests that do not destroy the test part. In many cases, nondestructive tests are conducted on full-sized parts and assemblies. However, the procedures that are described may be used on test specimens as well as full-sized parts.

22.2 RADIOGRAPHY AND DIFFRACTION

Radiographic tests include those tests that use short electromagnetic waves, such as X or gamma rays. Using these waves, it is possible to record features in the interior of solid objects and obtain information about their size and other aspects. Radiography can be used for a wide range of applications by varying the voltage, current, film type, and other parameters of the test.

Figure 22–1
Radiography effects.

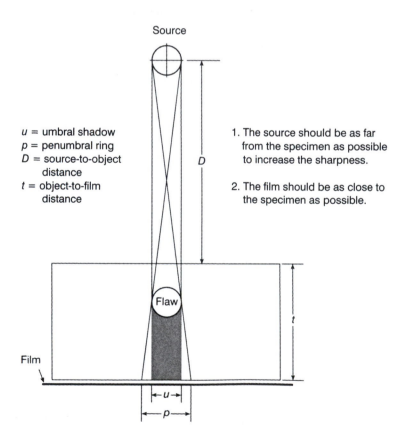

u = umbral shadow
p = penumbral ring
D = source-to-object distance
t = object-to-film distance

1. The source should be as far from the specimen as possible to increase the sharpness.

2. The film should be as close to the specimen as possible.

Metallic parts can be X-rayed or gamma-rayed to find hidden flaws. This procedure involves placing a film behind the part and beaming rays through the part. Flaws in the part show up on the film. Dense areas of the material absorb more of the rays, whereas less dense areas, such as flaws and cavities, permit a greater number of rays to pass through the material. The greater the number of rays per unit area on the film, the darker the radiograph at that point. Voids or inclusions also appear darker on a radiograph.

Because X-ray machines (such as in hospitals) are large and bulky, high-intensity gamma rays produced from radioactive elements such as radium and its salts are used in the field. Gamma rays are generally confined to heavier metals, such as steel and lead.

Scattering and lack of parallelism in the rays lead to a fuzziness in the radiograph called the *umbra-penumbra effect*. This effect results in a halo effect, or a fuzziness, in the outline of the part on the radiograph. Figure 22–1 illustrates the causes of the umbra-penumbra effect.

In addition to their use in radiography, X rays are applied in diffraction methods to determine the crystalline structure of atoms. X rays that strike parallel layers of a material's atoms cause those atoms to vibrate at a particular wavelength. These atoms emit energy of the same wavelength as that of the X ray. The source is adjusted so that the angle of incidence is equal to the angle of reflection for the various layers. This gives information concerning the spacing of the atomic planes, the crystal lattice structure, the orientation of the crystal, and other important information characteristic of the particular atom. Figure 22–2 illustrates this process.

The two most common applications of radiography are in testing welds and castings. Radiography using X rays is much faster than other methods. An X-ray radiograph requires only seconds or minutes, whereas other methods may require hours to develop.

Fluoroscopy involves the conversion of X rays to light on a fluorescent screen. The screen is placed behind the specimen and bombarded with X rays. This process is often used for nonmetallic specimens, such as letters, packages, and luggage.

In fluoroscopy, the image can be transmitted to a remote location, if desired. A closed-circuit television (CCTV) camera is focused on the fluorescent screen and piped through cable or coax to a TV monitor. Using this method, technicians can inspect specimens as they pass in front of the screen in "real time." This capability

Figure 22–2
X-ray diffraction.

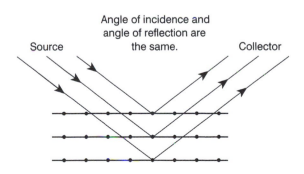

Angle of incidence and
angle of reflection are
the same.

Source   Collector yields savings in inspection costs by allowing the technician to watch for gross defects in critical areas as parts pass in front of the screen.

Another type of radiography is neutron radiography. A neutron is an uncharged particle found in the nucleus of atoms. Because they are uncharged, neutrons can be used to inspect specimens with negligible interaction with other charged particles. Unlike X and gamma rays, neutron radiography can be used to inspect a wide range of materials and thicknesses safely and discriminate more efficiently between elements and features of the specimen.

Fast neutrons [(higher than 10,000 eV (electron volts)] produced by accelerators, radioactive sources, reactors, and other such sources are used in inspection. When these fast neutrons hit a nucleus, they are deflected and lose some of their energy. Americium and californium are two popular radioactive sources of neutrons.

When neutrons strike an object, that object becomes radioactive and emits gamma radiation. In neutron radiography, either screens are used to capture the "image" of the test object and to transfer that image to radiographic film, or the image is placed directly on film. Dysprosium and indium are often used for indirect filming, whereas gadolinium is used for direct transfer of images. Fluorescent screens can also be used to capture an electronic image of the specimen.

Neutron sources are expensive and bulky and require very special handling considerations. Despite these disadvantages, neutron radiography is used to detect flaws in radioactive objects, such as nuclear fuel pellets, fuel rods, reactor vessels, and

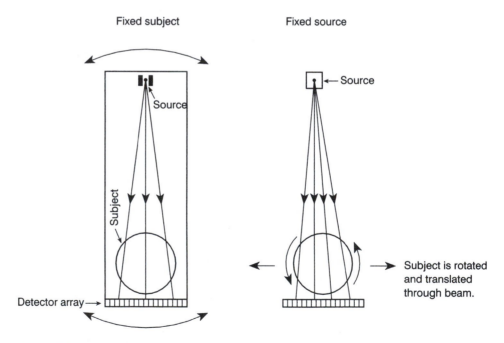

Figure 22–3
Computerized axial tomography.

other nuclear energy–related systems, turbine blades, and aerospace materials, such as seals, gaskets, and shields. Neutron radiography is also used to inspect explosive charges such as bombs, missile cones, and similar materials.

Computerized axial tomography (CAT) uses the concept of measuring the radiation intensity across an X-ray image, digitizing that image, and analyzing the data to reconstruct an image based on the data. CAT scans are widely used in medical applications but are finding wider uses in industrial applications. Either the object is held firmly in place and the source and detectors are rotated around the specimen, or the object is rotated between source and detector. These two methods are shown in Figure 22–3.

CAT scanning techniques are used in a wide variety of situations, such as studying the inner construction of trees, explosive shells, and multistrand cable connections. Figure 22–4 shows an actual CAT scan of a human brain.

Each of these methods described has its particular advantages, uses, and disadvantages as well as hazards in nondestructive testing. Intense radiation is harmful to living tissue. Radiation often affects not only the initial receiver, but also later generations as well. Before working with radioactive components, it is important to become aware of and follow recommended safety procedures, such as protective clothing, dosimeters, badges, detectors, and similar safety devices to be protected from dangerous radiation.

22.3 MAGNETIC AND ELECTROMAGNETIC METHODS

Magnetic and electromagnetic tests use the magnetic characteristics of a material in order to test the material. Materials containing iron, nickel, cobalt, and similar elements are strongly attracted to themselves and to each other when magnetized. These materials are called *ferromagnetic materials*. Materials such as oxygen are only semiattracted to each other. These materials are called *paramagnetic materials*. Some materials even repel each other slightly in the presence of a magnetic field. These materials are known as *diamagnetic materials*.

The principles of magnetism are attributed to the Chinese in approximately 2700 B.C. Because of magnetism, the needle of a compass always points north. Like charges repel, whereas unlike or opposite charges attract. The forces that are in effect when an object is magnetized are invisible, but they can be seen by using iron filings. When iron filings are placed on a piece of paper and a small bar magnet is also placed on the paper, the filings align themselves with the invisible lines of magnetism, which are called *lines of magnetic flux*. The larger the area of the magnet, the more lines of flux or the larger the magnetic force available to the magnet.

Magnets have poles, just as the earth does. Large magnets are made up of many tiny magnets, each with a north and south pole, which are called *domains*. When these domains are aligned so that all of the poles of the same polarity point in the same direction, the material is magnetized. When these domains point in random directions, the material is demagnetized. The *permeability* of a magnetic material is the property of the material that enables it to be magnetized. It is a measure of how easily that material can be magnetized. Low-permeability pieces are hard to magne-

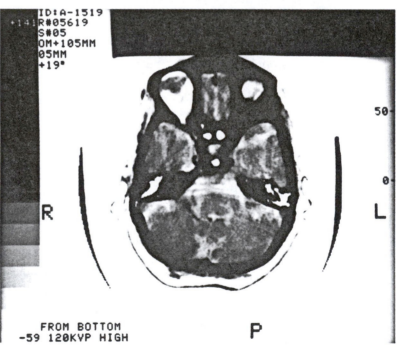

Figure 22–4
CAT scan.

Figure 22–5
Magnetic particle inspection device.

Courtesy of Econospect Corp.

tize. Cracks, flaws, and other defects set up local poles, which change the permeability of the specimen. When magnetic powder particles are applied to the surface of a test specimen, they tend to collect around a flaw or discontinuity in the magnetic flux. This result provides quick visual evidence that a discontinuity exists in the specimen at that point. The flaw must be perpendicular to the lines of flux for it to appear. Once the inspection is completed, the material is demagnetized.

This technique is used to detect flaws in castings, forgings, welds, and other similar parts. The magnetic powder can be applied in either wet or dry form. The dry form is spread on the area to be tested from a shaker, blower, bulb, or other container. The wet form can be applied by immersion, spreading a paste, or by aerosol. A fluorescent light is used to detect unusual concentrations of magnetic particles. The wet forms are available in several colors, depending on the application. The dry form is generally preferred for field work. Figure 22–5 shows a magnetic particle inspection device.

Electromagnetic testing is often used to sort materials, but it can be used to determine the microstructure and heat treatment of a material. Properties are analyzed based on the hysteresis loop characteristics for ferromagnetic materials. These loops are used to sort steels based on their composition, hardness, heat treatment, or amount of ferrite present in austenitic steels. Figure 22–6 represents hysteresis loop analysis.

Eddy current and leakage flux techniques are two common methods of nondestructive electromagnetic inspection. Eddy current inspection uses alternating magnetic fields and can be applied to any electrical conductor. Leakage flux inspection uses a permanent magnetic or dc-generated electromagnetic field and is used for ferromagnetic materials only. In eddy current testing, the alternating fields set up eddy currents in the specimen around flaws, which can be detected. Any flaw that produces lines of leakage flux can be detected using the leakage flux technique. These techniques are often combined with other nondestructive techniques, such as ultrasonic inspection.

Figure 22–6
Hysteresis loop analysis.

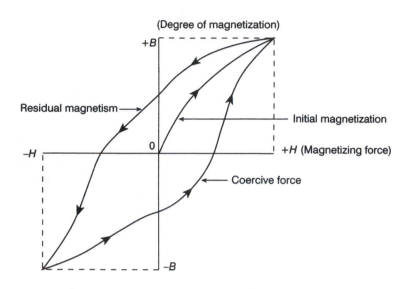

Eddy current inspection uses ac current to excite a test coil, which produces alternating magnetic fields and concentrates lines of flux at the center of the test coil, as shown in Figure 22–7.

As the test coil is placed near the test specimen, eddy currents are produced. Larger eddy currents are produced near the surface of the specimen. Cracks in the

Figure 22–7
Eddy current principle.

Figure 22–8
Eddy current tester.

specimen obstruct the flow of the eddy current and lengthen the flow of the current path, which has the effect of reducing the magnetic field reflected back to the coil and increases the impedance (total resistance to current flow) of the coil. The impedance rises as the severity of the crack or flaw increases. This change in imped-ance can be detected, measured, and output to various devices. Thus, the technician has feedback as to the location and severity of the flaw. Figure 22–8 is a simplified schematic of the eddy current tester.

The eddy current tester generally has two coils: (1) the primary field coil, which generates the eddy current, and (2) the secondary pickup coil, which senses and

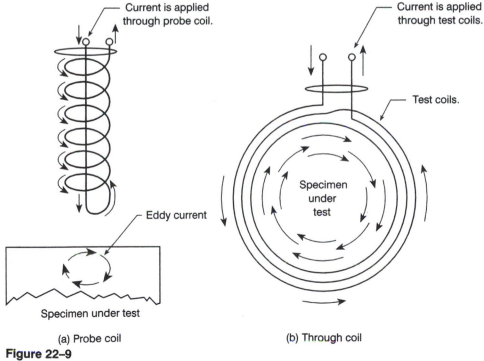

Figure 22–9
Eddy current pickup coils.

(a)   (b)   (c)

(d)

(e)

Figure 22–10
Hall effect sensors.

Parts A. and B. Courtesy of Honeywell's Microswitch Division. Parts C., D., and E. Courtesy of
F. W. Bell Inc.

detects the changes in coil impedance caused by changes in the specimen's permeability. Figure 22–9 illustrates two types of pickup coils.

The second form of electromagnetic inspection is leakage flux inspection, which involves ferromagnetic materials. This type of inspection is based on the concept that magnetic lines of force (flux) flow from one pole to another around the specimen. A flaw on the surface of the specimen will cause a change in the permeability of the piece at that point, and leakage flux will flow from the specimen at the flaw. By measuring the magnitude of the leakage flux at that point, the severity of the flaw can be measured.

Hall effect sensors are popular magnetic flux–measuring devices, due to their long life and wide operating parameters. A Hall effect sensor is a solid-state device whose output can be ac or dc, depending on the current and magnetic field inputs. They can act as high-speed, bounceless, contactless switches and are widely applied in areas such as mechanical testing and electronics.

E. H. Hall discovered what is known as the *Hall effect* in 1879. Hall was researching the effects of magnetic forces on current-carrying devices. He noticed that strong

Figure 22–11
The Hall effect.

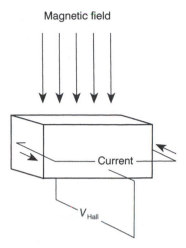

Magnetic field

Current

V_{Hall}

By measuring V_{Hall}, the
strength of the magnetic
field can be calculated.

magnetic fields influenced the conductor so that a small voltage perpendicular to the current direction was developed. This effect went unexplained until the early 1960s, when the first commercially available indium arsenide Hall generators became available. Availability of these generators coincided with increases in the semiconductor industry. Figure 22–10 shows Hall effect sensors.

When a conductor carrying electrical current is placed in a magnetic field, a Hall voltage is produced perpendicular to the direction of current and the magnetic field. The Hall voltage produced is very small (approximately 30 μV for a 1-G magnetic field) and therefore must be amplified and conditioned to be practical for inspection use. Figure 22–11 illustrates the Hall effect.

22.4 ULTRASONIC AND ACOUSTIC EMISSIONS METHODS

Ultrasonic testing is widely used in industry for detecting flaws, inclusions, and other internal defects and for measuring wall thicknesses on hollow pieces. Ultrasonic testing is able to detect these internal defects and allows a trained technician to determine the size, shape, and location of the defect(s). In conjunction with ultrasonic transducers, ultrasonic methods make it possible to measure the wall thicknesses of pipe, tubing, and other hollow shapes. This ability is important when material corrodes or builds up on the inside of pipes, for example.

Ultrasonic testing can also be used to determine differences in material structures and physical properties rather than the structures and properties themselves. For example, the ultrasonic signal is affected by heat treatment, grain structure and size, and the modulus of elasticity of samples.

Figure 22–12
Contact testing technique.

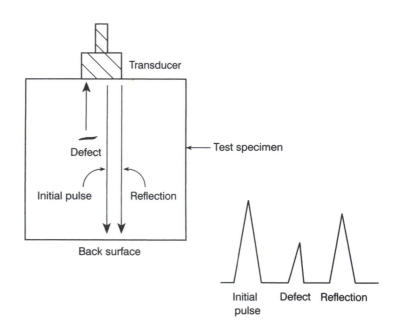

Piezoelectric crystals are used in ultrasonic detection. When high-voltage electrical pulses contact a piezoelectric crystal, the crystal will resonate, or "ring," at its resonant frequency. This ringing produces a burst of high-frequency vibrations. A common use of the piezoelectric crystal is in high-frequency audio speakers, or tweeters. In ultrasonic testing, vibrations produced by the transducer are transmitted into the material being tested. When a flaw is detected, an acoustic mismatch occurs, which reflects some or all of the ultrasonic energy back to the receiver. The crystal in the receiver converts this reflected sound energy (echo) into electrical pulses. The amplitude of the pulses helps determine the characteristics of the flaw. In addition, the time it takes for the reflection helps determine the distance of the flaw from the surface of the material. The back or bottom of a material will also reflect a signal. If the two signals are compared, the depth of the flaw can be determined. Because a signal is reflected from the back or bottom of the material, the physical size of the flaw—i.e., width, length, and thickness—can be determined. These signals are typically displayed on a cathode-ray tube (CRT) screen or oscilloscope. Figure 22–12 illustrates the technique used in ultrasonic testing and a typical display of initial pulse, flaw (defect) reflection, and back reflection.

The process described earlier is called *contact testing,* where the receiver is in direct contact with the material sample. *Immersion testing* is another type of ultrasonic testing. In immersion testing, the material sample and the test equipment are both submerged in a liquid, usually water. The liquid eliminates air and provides a better medium for transmission, known as a *coupling agent.* The liquid provides a sonic path between the test unit and the sample. Figure 22–13 illustrates the immersion testing process and a typical display of the initial pulse, flaw reflection, and back reflection.

Figure 22–13
Immersion testing technique.

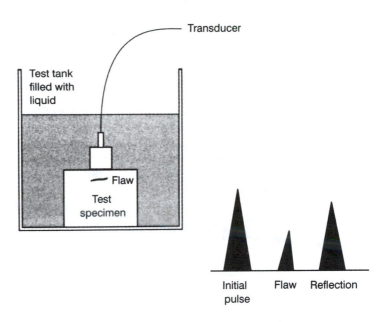

The frequencies used in ultrasonic testing range from 200 kHz to 50 MHz [(1 cycle per second equals 1 hertz (Hz)]. Typically, any frequency above 20 kHz is thought to be ultrasonic, or above the normal hearing range of 20 to 20 kHz. These sound waves travel through the vibration of particles. They travel with ease in uniform solids and low-viscosity liquids but are attenuated (diminished) by flaws or gases, such as air. The circuitry in the test unit is designed so that one wave train (series of waves) dissipates before the next train is introduced. Wave collisions and overlapping signals alter test results.

Ultrasonic signals are produced at the crystal's or crystals' fundamental resonant frequencies in bursts between 1 and 3 μs in duration. The waves then travel through the material sample and are reflected by the back surface or flaw. After sending the waves, the sender then acts as a receiver and waits for the reflected waves or echo pulses. The testing unit then waits for the wave train to decay and sends the next wave train to the sender. The speed at which the waves travel through the sample depends on the frequency and wavelength used:

velocity (ft/s or cm/s) = wavelength (ft or cm)·frequency (cycles/s)

The velocity also depends on the properties of the material sample. The elasticity and density of the sample affect the velocity of the wave trains.

The test frequency used is based on the sensitivity of the test equipment and penetration required in the material sample. As a general rule, the higher-frequency crystals tend to be more sensitive to flaws than the lower-frequency crystals, but lower-frequency crystals provide greater penetration. All frequencies work well with materials of smaller grain size and in immersion testing, where a liquid acts as a coupling agent. If higher-frequency crystals are used with larger-grain-size materials, the material tends to scatter the waves.

Ultrasonic waves are also affected by the acoustical impedance (Z) of the material. The acoustical impedance is the product of the density of the material (d) and the longitudinal wave velocity (v):

$$Z \text{ (g/cm}^2\text{·s)} = d \text{ (g/cm}^3) \, v \text{ (cm/s)}$$

The acoustical impedance or resistance of the material to acoustic waves determines the reflection and transmission characteristics of the material. For example, as the acoustical impedance of a joint formed by two dissimilar metals increases, the amount of sound that will travel through the joint decreases.

Ultrasonic testing relies on the principles of converting electrical pulses to mechanical vibrations (wave trains) and converting returned (echoed) mechanical vibrations back into electrical pulses for display and analysis. This is accomplished through the use of the piezoelectric crystal, which acts as a transducer. The piezoelectric element is a plate of polarized ceramic or crystalline material with electrodes on opposite surfaces. This material may be composed of barium titanate, lead metaniobate, lead zirconate titanate, lithium niobate, lithium sulfate, quartz, or tourmaline. All these materials exhibit piezoelectric properties. In other words, when struck by mechanical vibrations, these materials produce electrical pulses. Typically, these pulses range from 0.001 to 1.0 V and are produced at the same frequency as the applied mechanical vibrations.

The three most common materials used in ultrasonic testing are lithium sulfate, quartz, and polarized ceramics. All of these materials exhibit good wear resistance and have high mechanical strength. Quartz is known for its electrical and thermal stability. Quartz crystals suffer low electromechanical conversion efficiency and are, therefore, the least effective of the three in ultrasonic testing. Lithium sulfate crystals offer the best resolution and moderate conversion efficiency; therefore, they are generally considered to be the best material for receivers. Lithium sulfate receivers, however, cannot be used at temperatures greater than 165°F (75°C). Polarized ceramic transducers, such as barium titanate, offer the highest efficiency and sensitivity. However, they frequently offer lower mechanical strength and higher electrical capacitance, which reduces the upper operating frequency of the transducer by altering its impedance.

The simplest setup for ultrasonic testing is the single-test unit, although multiple-test units may be used to provide data on irregular objects, to increase the speed of testing, and in a sender/receiver configuration for special test conditions. The principles that apply to single-test units also apply to multiple-test-unit situations.

Modern ultrasonic equipment, also called *pulse-echo equipment,* includes a pulse generator, a variable attenuator, an amplifier, filters, and a spectrum analyzer, or oscilloscope. The sensitivity of this equipment depends on the search unit, the pulse generator, and the amplifier characteristics.

The pulse generator provides the electrical pulses for the sending unit and can also be used to trigger the oscilloscope. The generator may also be equipped with functions to control the repetition rate, pulse dampening, and pulse energy. These factors affect the resolution of the received echo. Too great a pulse energy will increase the

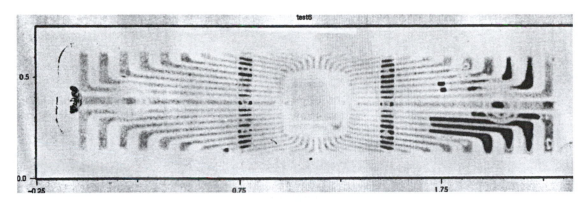

Figure 22–14
Ultrasonic testing display.

Courtesy of Panametrics.

pulse strength and decrease the resolving power of the display. In addition, too fast a pulse rate may not allow the first wave train to die out before the next wave train is sent. Pulse repetition rate also affects the scanning rate. If the repetition rate is too slow or the search unit is moving too fast, flaws will be overlooked. Pulse dampening improves transducer response and helps prevent damage to the transducer.

The amplifier and variable attenuator in combination receive the small electrical pulses produced by the receiver and help eliminate noise, random pulses, and other errors. They also modify the return pulses for display. The filters are used to reduce noise.

After being filtered and amplified, the pulses are sent to a spectrum analyzer or oscilloscope for display and analysis. The display will typically show the initial pulse, any flaws, and the return pulse reflected from the back of the material. Because the acoustical impedance of air is approximately 1.5×10^5 times greater than most metals, the wave train will travel approximately 18 times faster in a metal object than it will in air. Therefore, when the wave train strikes a flaw, it hits a wall and is reflected back to the sender/receiver unit. This is the middle spike in the display in Figure 22–14.

Ultrasonic test units range from portable, hand-held units to larger, stationary units. Depending on the size of the test sample, manual testing may be impractical. The size and shape of the material sample, the required resolution of the test unit, the type of flaw, probe size, material grain size, test conditions, and other factors should be used in determining the appropriate test unit. Figure 22–15 illustrates some common types of ultrasonic testing equipment.

Ultrasonic testing is the application of sound waves (wave trains) for the purpose of finding flaws in a material sample. In ultrasonic testing, sound is applied to the sample. Other acoustic methods involve listening for acoustic emissions produced by the sample at or near the point of failure. Most solid materials emit sound (acoustic emissions) when they are stressed to the point of deformation and, ultimately, fail-

Figure 22–15
Common ultrasonic testing equipment.
Courtesy of Panametrics.

ure. Almost all materials emit some sound when stressed to failure—for example, the "snapping" sound produced by dry branches when crushed beneath people's feet; the sound that a can makes when crushed; and glass breaking when struck by a rock or other object with force. These sounds can be heard with the ear. Other materials produce softer emissions, which must be amplified to be heard. Many materials produce sounds outside the range of human hearing that require ultrasonic equipment for detection.

Acoustic emission testing is often used to detect cracking in structures such as pipes, storage tanks, circuit boards, and welded joints. It is also used to detect leaks in valves and bearing failures in pumps, motors, and compressors. To the trained technician, these acoustic emissions are warning signs of impending or actual failure of components and structures.

Acoustic emission transducers are also piezoelectric sensors, which typically operate in the 20 kHz to 1.5 MHz range. These sensors are placed in contact with the sample. Testing equipment generally consists of a sensor, preamplifiers, signal-processing equipment, and monitors or displays. This equipment is controlled by state-

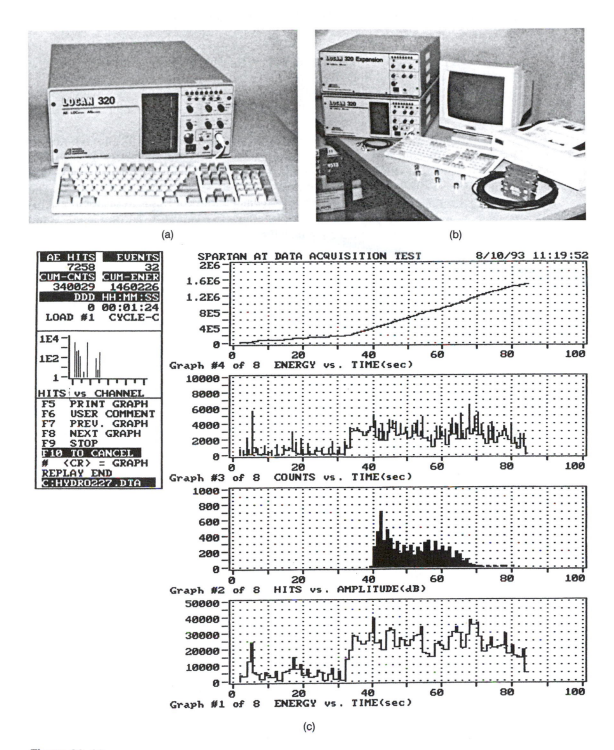

Figure 22–16

A. and B. Acoustic emissions test apparatus. C. Acoustic emission test result.

Courtesy of Physical Acoustics Corp.

of-the-art microprocessor-based systems that collect, condition, synthesize, and analyze emissions for real-time display of test data. Single or multiple sensors may be used; typically, up to eight channels are used in multiple systems. The system counts and measures acoustic emissions. Figure 22–16 illustrates the testing apparatus for acoustic emissions testing.

The sensors and test equipment are set up and calibrated to eliminate background noise, spurious emission, and other environmental factors that may influence the test. The sample is then strained in tension or compression within its elastic limit. The sensors pick up any acoustic emissions by the sample during testing. The sensors can also be applied to samples at their point of application for in-service analysis. Acoustic emission testing can be applied to any material that emits sound under stress. Almost all materials, including metals, nonmetals, and composites, are included in this category.

22.5 VISUAL METHODS AND HOLOGRAPHY

Visual inspection can be performed with or without the aid of other devices and either directly or from a remote site. The first technique often used in nondestructive testing is visual inspection, or looking at the material sample. The material's appearance in terms of color, texture, and other physical characteristics provides quick, insightful information concerning the cause and extent of many flaws and defects. Visual inspection has its advantages and disadvantages, as do all the nondestructive methods. However, visual inspection should not be overlooked as an insightful inspection method, with or without the use of additional equipment. Some of the equipment used in visual inspection is presented here.

One application of visual inspection is in determining surface roughness. Many industrial operations require a specified surface roughness for appearance, lower friction in contact surfaces, and so on. One method of inspection involves scratching a fingernail over the surface and comparing the result to a reference card. Major categories of surface roughness can be identified using this method. Surface roughness can also be determined by visual inspection under magnification. However, the results vary with the observer. Microscopic inspection comparing two images can provide adequate data over small areas. In addition, surface roughness averaging machines are available that average the surface roughness over small areas. These machines run a stylus over the areas in question and provide feedback on peaks, valleys, and average conditions.

Borescopes are often used to inspect the inside of tubes and other enclosed structures where size or environmental factors may prohibit direct visual inspection, such as the inside of a jet engine, nuclear structure, or artery. They are sophisticated optical instruments designed to provide a visual image for remote viewing. They are manufactured in various sizes and configurations, including rigid, flexible, and micro types. The size of the borescope determines the minimum bore size into which it will fit. Figure 22–17 is a sample of the various borescopes commonly available.

(a)

Measurement scale

Eyepiece

Orbital scan control

Diopter

Reticle focus

Image focus

Integral light guide cable

(b)

(c)

Figure 22–17

A. Typical inspection borescopes. B. Olympus measuring borescope. C. Olympus IF8D4X1-20 fiberscope. D. Olympus borescope video printer system. E. Color video borescopic inspection system.

Part A. Courtesy of Instrument Technology Inc. Parts B.–E. Courtesy of Olympus America Inc., Industrial Fiberoptics Division.

Video Printer and typical internal inspection video system.

(d)

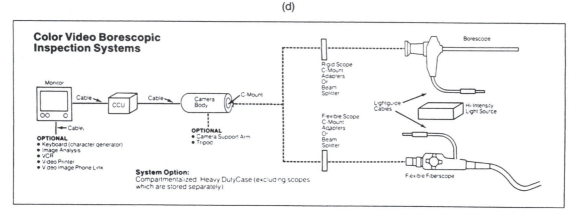

Figure 22–17, continued   (e)

The borescope is typically rather small in size. The image that it picks up must be magnified to be viewed properly by the inspector. The objective lens is located at the end of the borescope and acts like a camera lens. It picks up the image and transfers it to the relay lenses. Relay lenses use computer-controlled equipment to pass the image along the length of the borescope to the eyepiece or interface for viewing. The final magnification is equal to the product of the individual magnifications. Magnifications are specified as 2× at 1 in., 5× at 25 mm, and so on. The distance is given because the accuracy of the image and focus depends on the distance of the object from the objective lens.

Measurements are taken from specially made eyepieces equipped with graduated reticules or from the display. A high-resolution video camera can be connected to a borescope to provide recording or hard-copy documentation of the object. These systems are ideal for long-time inspection periods, providing viewing for more than one person, documentation of the inspection, and other characteristics associated

(a)

Figure 22–18
A. Fiberscopic inspection of a turbine. B. Operator feeding long Videoimagescope probe into piping. Shown is TV monitor view of interior. C. Schematic of Olympus Videoimagescope system.

Courtesy of Olympus America Inc., Industrial Fiberoptics Division.

with video systems. Modern systems can also use fiber-optic fiberscopes to provide image transfer. Figure 22–18 illustrates one configuration of this type of equipment and some of the many probes available for visual inspection.

In addition to borescopes, the laboratory-standard microscope is often used in inspection. Microscopes provide a wide range of magnification, although the size of the object under inspection is often limited by the platen size of the microscope. Stereo microscopes and macroscopes are used in many applications for determining surface features and flaws and to measure object features. Probably the most familiar of the applications of these are in criminal investigations, where forensic specialists use comparison microscopes, which project dual images, to run ballistics tests on bullets to determine if rifling marks align on two specimens, to examine fibers for a match, and to make similar comparisons. Comparison microscopes are simply two pieces of equipment combined to form one machine. These machines are fitted with interchangeable eyepieces, which are equipped with graduated reticules to allow the measure of object features. One special piece of equipment is the optical comparator, especially the interferometer. An interferometer is used to check for flatness

(b)

Figure 22–18, continued

and parallelism of an object. Figure 22–19 illustrates various microscopes and comparators used in visual inspection.

With special equipment, such as closed-circuit television and video recording equipment, measurements from microscopes, comparators, interferometers, and other such equipment can be displayed and recorded to provide hard-copy evidence of measurements taken.

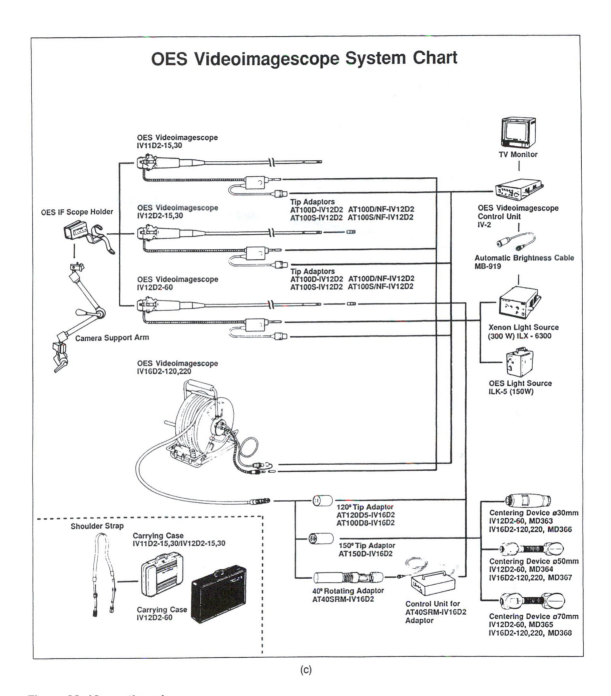

OES Videoimagescope System Chart

(c)

Figure 22–18, continued

(a)   (b)   (c)

Figure 22–19
Microscopes and comparators.
Courtesy of Meiji.

Holographic interferometry uses a laser to give monochromatic (one-color), coherent light to provide data concerning an object. Its use does not depend on the size, preparation, or positioning of the object. It is a reliable, simple-to-use, and relatively automatic form of visual inspection.

The beam of light coming from the laser is split in two, an object beam and a reference beam. The object beam is used to illuminate the object, whereas the reference beam is projected directly onto the recording film. These beams are adjusted so that they are in phase and are the same distance from the film. The beams can be directed independently through the use of mirrors. The intensity of the beams and exposure time depend on the characteristics of the beams and the film speed. The object is then photographed, and the film is exposed. The film is developed, and a hard copy is made of the object. This holographic image of the object contains all of the visual information concerning the test object. This image can then be superimposed on other objects to determine whether a match is made. Any differences between the original and subsequent objects can be seen in real time. Holographic cameras can be used in place of the film. This process is illustrated in Figure 22–20.

Holograms are used to inspect deformations in structural materials, vibrations in rotating objects and structural members, mechanical and thermal strains in various materials, and other similar materials. Using sophisticated video equipment, holography can be used to provide real-time feedback on part inspection and "machine vision."

(a)

(b)

Figure 22–20
A. Shearography system. B. Digital holography system. C. Unbond in the cowling of an
Airbus 330 detected by SC-4000 shearography system.

Courtesy of Laser Technology, Inc.

(c)

Figure 22–20, continued

22.6 PHOTOELASTICITY

A common problem in the design and analysis of structures and machine elements is the magnitude and distribution of stresses throughout the specimen. Photoelastic analysis allows technicians to study areas of interest to determine the areas where stresses are concentrated and the maximum stresses that occur at these areas. Equipped with these data, factors can be studied to determine the reasons why stresses are concentrated in the areas of interest. Reasons include the application of external forces, fasteners and fastening systems used, structural design conditions, and the shape of the structure, among other factors.

Photoelastic analysis is a means of experimentally determining the concentration and distribution of stresses within the elastic range of the specimen. Typically, a model of the structure is made using a suitable transparent material, such as a plas-

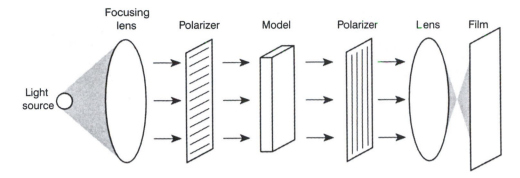

Figure 22–21
Photoelastic analysis apparatus.

tic. Polarized light is then passed through the model while appropriate stresses are applied. The light is refracted within the model, and a pattern of colored light and dark bands, called *fringes*, appears in the model. The characteristics of the pattern of fringes determine the distribution of stresses in the model. These observations, along with supporting data concerning applied stress, shape of the model, and so forth, help determine the stress relationships throughout the model and, therefore, the full-scale structure.

The apparatus used for making a photoelastic analysis is shown in Figure 22–21. Simply, a beam of polarized light is directed into the model. A light source is filtered using a polarizing lens or prism. The filtered light is then made parallel through a

Figure 22–22
Photoelastic analysis.

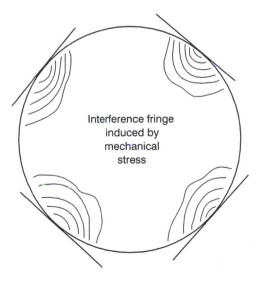

series of lenses or a lens system. After passing through the model, the light passes through an exit polarizing unit, which is set perpendicular to the original polarizing unit. Another lens system is then used to project the light pattern from the exit lens onto a viewing screen or monitor. The reason for setting the original and exit polarizing lenses perpendicular is simple: If there is no model between the lenses, no light will pass through to the screen or monitor. Figure 22–22 is a photograph of a photoelastic analysis.

If the model is subjected to uniform stress throughout, the image will be all dark, all light, or all of one color, depending on the magnitude of the difference in principal stresses and the characteristics of the light source used. Larger differences in the principal stresses produce larger differences in the fringes produced (phase differences in the light), and the model refracts the light source to produce different-colored fringes.

22.7 LIQUID PENETRANTS

Liquid penetrant tests are among the oldest, simplest, and least expensive tests to conduct. They are still widely used to detect surface flaws in nonmetals and nonferromagnetic materials. In some circumstances, such as when locating small surface defects, liquid penetrant tests are more reliable than other methods, such as radiography.

Liquid penetrants used for NDT must be able to penetrate fine flaws and yet still be able to remain in larger openings. They should flow easily onto and wet the surface but be easy to remove after testing. They should not cause any permanent change in the specimen, a primary objective of NDT. Liquid penetrant testing cannot be used on highly porous materials. In the case of large, subsurface defects with small surface openings, the penetrant reveals the surface opening only. Most of the liquid seeps into the larger subsurface opening. An indication that this is happening is excessive "bleeding" after the developer is applied. Most liquid penetrant tests are performed with a two-part system: a dye penetrant and a developer. The term *bleeding* refers to the absorption of the dye by the developer.

Generally, a part is prepared prior to the test. Preparation may include washing and degreasing, brushing or sanding, sandblasting, etching, or other suitable cleaning and preparation procedure. This procedure should remove any material or feature that would give a false reading. After the part is prepared, the dye penetrant (a liquid penetrant material containing a visible and fluorescent dye) is applied to the test area by brushing, spraying, or dipping. The part is then allowed adequate time to dry. This time may be specified for the system. While the penetrant is drying, it is absorbed into surface irregularities by capillary action. After drying, the excess dye penetrant is removed, leaving the penetrant in the flaws. A colored developer is then applied to the surface to draw some of the penetrant out of the flaw. This provides a wide band of color, which may be observed visually or with the aid of ultraviolet, or "black," light. The developer is brightly colored and contrasts sharply with the penetrant. Figure 22–23 shows a typical two-part liquid penetrant system.

Figure 22–23

Liquid penetrant system.

Courtesy of Met-L-Chek Company.

22.8 SUMMARY

Any procedure that does not render the sample unusable for its intended purpose is considered a nondestructive test. Nondestructive testing (NDT) is often used to determine if a product meets specifications. NDT can be used to find hidden or obvious flaws in products in a number of ways. Visual inspection is a simple procedure for just looking at the part for surface defects. Visual inspection should be one of the first procedures used, because of its simple, inexpensive, widespread, and fast application. Borescopes, microscopes, and other equipment are used to magnify the image to help in detecting flaws and defects in parts. In addition, techniques such as magnetic particle inspection can be used to find small surface flaws that cannot be detected by the unaided eye. A flaw will interrupt the magnetic flow and cause a concentration of magnetic flux around the flaw.

Liquid penetrants are also used to provide a way of detecting small flaws or cracks in parts that cannot be seen with the naked eye. The liquid penetrant test is often a two-part system, where a base coat is applied, followed by a developer. A strong fluorescent light source is then directed on the part. The flaw or crack will absorb the penetrant, and the light source will bring out the flaw.

Additional techniques for finding deeper flaws include X-ray and radiographic techniques, where the part is subjected to X rays or gamma rays and a photographic plate, film, or real-time detector is used to produce a picture of the test results. This provides an analysis of the internal features of the part. Flaws will deflect the rays and absorb the rays so that areas appear in contrast with the detector's background. Eddy current testing and similar electronic techniques use the conductivity of the part in finding flaws. An electric current is induced or applied to the part. Flaws interrupt the flow of current and can be detected by sensitive test equipment. Eddy current testing can also be used to provide data concerning a part's conductivity, permeability, heat treatment, and other related properties.

Photoelastic analysis provides data concerning the concentration and distribution of stresses in a transparent model of a full-size structure. Polarized light is directed through the model, and areas of stress show up as bands of light called fringes. By studying these fringes and correlating the data with other measurable factors concerning the model and application of stress, design factors and suitable judgments can be made concerning the full-size structure. A recent NDT method, holography, is now being used more widely. A laser is used to project an interference pattern on the surface of a part. When the part is subjected to a load, the interference pattern changes. If the part contains flaws, the part will be slightly more deformed where the flaw occurs. Holography can be used on a wide range of metals and nonmetals. Many other NDT techniques are used to provide inspection data concerning parts or samples. These tests are based on the many properties of the sample, including physical, chemical, mechanical, and other properties. Most tests require adequate training in the properties of materials and use of sophisticated test equipment.

QUESTIONS AND PROBLEMS

1. What differences do you perceive between destructive and nondestructive testing?
2. What is radiography? How is it used in NDT?
3. Describe the process of computerized axial tomography. How is it used in NDT?
4. Describe how the principles of magnetism are used in NDT.
5. Describe the eddy current testing procedure. How is it used?
6. What is the Hall effect? How is it used in testing?
7. Describe the ultrasonic testing procedure. How is it used in NDT?
8. Describe the procedure for acoustic emission testing. How is it used in NDT?
9. Describe the process of holography. How is it used in NDT?
10. Describe the process of photoelasticity and how it relates to NDT.
11. Describe the liquid penetrant process in NDT.
12. Which of the NDT methods would best be used to detect flaws in each material?
    a. Cast-iron castings
    b. Surface of a steel rod
    c. Inside a solid nonferrous casting
    d. Pipe weld joint
    e. Solid plastic casting
13. Of the different NDT techniques discussed, which would you consider for field testing and which are used for laboratory testing?
14. Which of these NDT techniques could be used for testing polymers?
15. Which of these NDT techniques could be used for testing nonferrous metals?
16. List five applications where nondestructive testing is preferred or required.
17. Describe a testing application requiring each of the following: (1) X rays, (2) acoustic waves, (3) magnetism, (4) liquid penetrant, and (5) visual inspection.
18. Develop and describe your own nondestructive test procedure for testing soda pop. The test should evaluate the fluid level, the pressure inside the can, and presence of foreign objects. Describe the test setup, grippers, equipment used, data collection, and expected results.

APPENDIXES

APPENDIX A
Sources of Further Information

A.1 WEB LINKS—INTRODUCTION

These sites are provided for further information, research, study, and clarification of the information and topics provided. An effort was made to choose sustaining links to free sites or those that provide public access. We cannot take responsibility for server access, the validity of information provided, or the continued status of these links.

Having stated that, please enjoy "surfing" these sites or searching for further information using one of a number of search engines available. Key-term searches using the headings provided should be a good start. Good luck and enjoy.

Link	Description
Ferrous Metals	
http://www.aise.org/	Association of Iron and Steel Engineers—AISE
http://www.asm-intl.org/	ASM International
http://www.avestasheffield.com/products/steelgrades.htm	Avesta Sheffield—Steel grades and properties
http://www.bethsteel.com/	Bethlehem Steel Corporation Steel
http://www.douglassteel.com/links.htm	Links from Douglas Steel Corp.
http://www.matweb.com/	MatWeb—Online reference to materials data
http://www.metalworld.com/metalworld/	MetalWorld—Information on buying, selling, trading various metals and products and organizations
http://www.principalmetals.com/properties/step1.asp	Principal Metals—Property information for selected ferrous and nonferrous metals
http://www.ssina.com/	Specialty Steel Industry of North America—Stainless Steel Information Center
http://www.steel.org/	American Iron and Steel Institute—AISI
http://www.ussteel.com/	US Steel Corporation—Division of USX Corporation

Nonferrous Metals

http://es.epa.gov/oeca/metd/nfm.html	Environmental Protection Agency—Information on the nonferrous metals industry
http://info.cas.org/ONLINE/DBSS/aaasdss.html	Contains nominal composition and composition limits, typical (average) mechanical and physical properties, and minimum tensile properties of commercial wrought aluminum alloys
http://nickeloid.com/	American Nickeloid Company—Chrome, nickel, brass, and copper
http://properties.copper.org/standard-designations/homepage.htm	Copper, brass, and bronze ASTM standard designations for wrought and cast copper and copper alloy
http://www.admiralmetals.com/	Admiral Metals—Free information on nonferrous metals but registration required
http://www.alcoa.com/	Aluminum Company of America—Integrated aluminum company producing and selling aluminum, semifabricated and finished aluminum products, and alumina
http://www.alliedmetalcompany.com/	Allied Metal Company—One of the largest smelters of aluminum and alloyers of zinc
http://www.aluminum.org/	The Aluminum Association Incorporated—Facts, figures, and recycling information
http://www.architecturalmetals.com/catalog.html	Nationwide Architectural Metals—Catalog of polished metal products
http://www.brz.com/specsheet.html	Bronze Bearings, Inc.
http://www.copper.org/	The Copper Page
http://www.copper.org/standard/euronumb.htm	European numbering system for copper and copper alloys
http://www.ericksonmetals.com/	Aluminum products from Erickson Metals
http://www.kaiseral.com/	Kaiser Aluminum—Aspects of its industry: mining and refining of bauxite, production of primary aluminum, and manufacture of fabricated aluminum products
http://www.mining-technology.com/	Mining industry information and resources
http://www.titanium.com/tech_sheets/index.html	Titanium Data and Reference Manual from Titanium Industries, Inc.

Plastics

http://www.asresin.com/	Allied Signal Plastics—Searchable site
http://www.bfgoodrich.com/default.asp	BF Goodrich
http://www.dow.com/homepage/index.html	Dow Chemical Company
http://www.dupont.com/enggpolymers/americas/index.html	DuPont Engineering Polymers

http://www.endura.com/	Endura Plastics—Thermoplastics information and selection
http://www.ge.com/plastics/	GE Plastics—Information on selection and properties of engineering thermoplastics
http://www.goodyear.com/	Goodyear Tire and Rubber Company
http://www.plasticsresource.com/	Plastics resource
http://www.plasticsusa.com/ polylist.html	Plastics USA—Comparative properties of various plastic materials
http://www.polymers.com/ dotcom/home.html	Polymers Dotcom—Information on properties, selection, uses, and equipment related to plastics
http://www.sdplastics.com/ plastics.html	San Diego Plastics—Introduction to plastic materials, properties, and testing

Glass

http://gom.mse.iastate.edu/	Glass and Optical Materials Lab—Iowa State University
http://www.agc.co.jp/	Asahi Glass Company, Limited—Japan
http://www.anchorhocking.com/ anchorhocking.co/	Anchor Hocking, Inc.—Glass products
http://www.bgf.com/	BGF Industries—Glass and composite fabrics
http://www.guardian.com/	Guardian Industries Corporation—Glass manufacturers
http://www.lrsm.upenn.edu/ ~frenchrh/opes.html	Optical properties and electronic structure of materials
http://www.olemiss.edu/depts/ compmatl/index.html	Composite Materials Research Group—University of Mississippi
http://www.owenscorning.com/	Owens Corning—Glass
http://www.pilkington.com/	Pilkington Libby-Owens-Ford—Glass manufacturers
http://www.ppg.com/default.asp	PPG Industries, Inc.
http://www.securityglass.com/	American Glass Products

Ceramics

http://mineral.galleries.com/ minerals/	The Minerals Gallery
http://minerals.usgs.gov/minerals/	USGS minerals information
http://minerals.usgs.gov/minerals/ pubs/mcs/	USGS mineral commodity summaries
http://web.wt.net/~daba/Mineral/ determin.html	Physical and optical properties of minerals
http://www.art-inc.com/	Advanced Refractory Technology, Inc.
http://www.ceramics.com/	Ceramic and Industrial Materials—Links to sites and suppliers
http://www.coorsceramics.com/ coorsceramics/home2.htm	Coors Ceramic Company—Products and services
http://www.incerco.com/ incerco.html	Incerco Ceramics—Ceramic materials and properties tables
http://www.nirim.go.jp/nirim/ english/	National Institute for Research in Inorganic Materials—Japan
http://www.porcelainenamel.com/	Porcelain Enamel Institute

http://www.sanken.osaka-u.ac.jp/labs/scm/about_resE.html	Niihara Laboratory, Osaka University, Japan—Ceramics research
http://www.science.ubc.ca/~geol202/petrology/rock.html	Petrology basics
http://www.stratamet.com/default.html	Stramet, Inc.—Ceramics and advanced materials
http://www.usgs.gov/themes/mineral.html	U.S. Geological Survey—Mineral information

Wood and Wood Products

http://www.bayer.com/bayer/index_e.htm	Bayer Corporation
http://www.bc.com	Boise-Cascade Corporation
http://www.calforests.org/home.html	California Forest Products Commission
http://www.forestweb.com/	Forest Web—Forest products information
http://www.foxgal.com/links/suppliers/supplierslinks.shtml	Foxworth-Galbraith Lumber Company—Links to major lumber suppliers
http://www.fs.fed.us/	USDA Forest Service
http://www.gp.com/	Georgia-Pacific Corporation
http://www.gp.com/build/products/engineered.html	Georgia-Pacific engineered board products
http://www.hammermillpaper.com/	Hammermill Papers—International Paper subsidiary
http://www.ipst.edu/	Institute of Paper Science and Technology, Atlanta, GA—One of the world's largest collections of scientific and technical information on pulp and paper
http://www.lumber.com/	American International Forest Products
http://www.mead.com/	Mead Corporation—Paper
http://www.menasha.com/	Menasha Corporation—Paper products
http://www.pbmdf.com/	World of Engineered Wood—Sponsored by the Composite Panel Association and the Composite Wood Council (formerly the PB-MDF Institute)
http://www.wii.com/	Willamette Industries, Inc.—Forestry products

Composites

http://research.et.byu.edu/acme/	Advanced Composites Manufacturing and Engineering Center
http://www.acg-us.com/products/products.htm	Advanced Composites Group, Inc.—Prepregs and properties
http://www.airtechintl.com/	AirTech Advanced Materials Group—Composites
http://www.bellcomb.com/	Bellcomb Technologies, Inc.—Manufacturers of composite, structural, lightweight panel systems
http://www.bgf.com/	BGF Industries—Glass and composite fabrics
http://www.grafil.com/	Grafil, Inc.—Standard modulus carbon fibers and properties
http://www.hexcel.com/	Hexcel Corporation—Glossary and properties of advanced composites

http://www.jdlincoln.com/	J. D. Lincoln, Inc.—Film and paste adhesives for prepregs
http://www.olemiss.edu/depts/compmatl/index.html	Composite Materials Research Group—University of Mississippi
http://www.owenscorning.com/owens/composites/lineup/	Owens-Corning—Properties of glass fibers for composites
http://www.ppg.com/fgs_main/excerpts.htm	PPG Industries—Fiberglass reinforcements for thermoplastic and thermoset composites
http://www.prepregs.com/PRODNDX.htm	Adhesive Prepregs for Composite Manufacturers—Product index and properties
http://www.stesalit.com/english/PROD/prod1.htm	Stesalit, Inc.—Fiber reinforced composite properties and applications
http://www.systems.textron.com/	Textron Systems—Advanced composite materials and properties
http://www.thiokol.com/TCR.htm	Thiokol Propulsion—TCR™ composite resin systems
http://www.toray.com/carbon/index.html	Toray, Inc.—Carbon fiber and prepregs
http://www.tpm2.com/html/products.html	TPM2, Inc.—Develops, produces, and sells thermoplastic advanced composite products fabricated by a patented pultrusion process
http://www.tricelcorp.com/	Tricel Honeycomb, Corporation—Sandwich panel construction and properties

Cement, Concrete, and Asphalt

http://www.asphalt.akzonbel.com/	Akzo Nobel Asphalt Applications, Inc.—Asphalt and bitumen products and testing
http://www.asphaltpages.com/	Asphalt Pages—Directory of asphalt-related information
http://www.firestonebpco.com/	Firestone Building Products Company—Roofing, waterproofing applications, and so on
http://www.kaisercement.com/	Kaiser Cement Corporation
http://www.lafargecorp.com/index.html	Lafarge Corporation—Cement and concrete products
http://www.natural-stone.com/	Natural Stone—Stone industry information
http://www.portcement.org/	Portland Cement Association
http://www.quikrete.com/	Quikrete Companies

Adhesives and Coatings

http://www.alliedcorrosion.com/	Allied Corrosion Industries, Inc.—Corrosion protection and information
http://www.cems.umn.edu/crc/	Corrosion Research Center—University of Minnesota
http://www.clihouston.com/	InterCorr International, Inc.—Corrosion protection and testing services
http://www.corrosion.com/	Protective coatings and linings resources
http://www.cp.umist.ac.uk/	Corrosion Information Server—ICC/UMIST—From the International Corrosion Council and the Corrosion and Protection Center, UMIST

http://www.dulux.com/trade/ index.html	Dulux Paint Assistant—Provides information, suggestions, and a calculator related to paints
http://www.galvanizeit.com/	American Tinning and Galvanizing Company
http://www.gluguru.com/	Adhesives for Industry Technology Center— GluGuru
http://www.paint.org/	National Paint and Coatings Association
http://www.ppg.com/	PPG Industries corporate homepage
http://www.ppgaf.com/school.htm	Pittsburgh Paint and Glass (PPG) Paint School
http://www.sherwin-williams.com/ default.asp	The Sherwin Williams Paint Company

Fuels and Lubricants

http://www.arco.com/ PRODUCTS/	Atlantic Richfield Company—Petroleum products
http://www.basf.com/	BASF Corporation
http://www.bnsf.com/business/ coal/html/coal_mines.html	Burlington-Northern Santa Fe Railway—Coal and coal mining information
http://www.bpamoco.com/	British Petroleum—Amoco
http://www.chevron.com/	Chevron Corporation
http://www.citgo.com/	Citgo, Inc.
http://www.coastalcorp.com/	The Coastal Corporation
http://www.conoco.com/	Conoco, Inc.
http://www.exxon.com/exxoncorp/ index.html	Exxon Corporation
http://www.hghouston.com/ refining.html	Hendrix Group—Petroleum refining processes
http://www.kendallmotoroil.com/	Kendall Motor Oil
http://www.maplic.com/	Marathon Ashland Petroleum LLC
http://www.mda.state.mo.us/ elc.htm	Missouri State Department of Agriculture Petroleum Quality Inspection
http://www.mobil.com/	Mobil Corporation
http://www.pennzoil.com	Pennzoil Company
http://www.phillips66.com	Phillips Petroleum—Fuels and lubricants
http://www.quakerstate.com/	Quaker State Motor Oil
http://www.shellus.com/	Shell Oil Company
http://www.southlandoilco.com/	Southland Oil Company
http://www.sunocoinc.com/	Sunoco, Inc.
http://www.texaco.com/	Texaco, Inc.
http://www.tosco.com/	Tosco Corporation—Refines and markets fuels and lubricants
http://www.valvoline.com/	Valvoline

Societies and Organizations

http://205.197.9134/	American Forest and Paper Association
http://asc.ime.ucla.edu/	American Society for Composites
http://home.doe.gov/	U.S. Department of Energy
http://web.ansi.org/default.htm	American National Standards Institute
http://wwpa.org/	Western Wood Products Association
http://www.4spe.org/	Society of Plastics Engineers

http://www.abms.com/index.html	American Bureau of Metal Statistics—Subscription service that provides detailed statistics on the metal industries. Sample reports available.
http://www.acers.org/	American Ceramic Society
http://www.aci-int.org/	American Concrete Institute
http://www.aec.org/cyberg/ aluminumoverview.htm	Aluminum Extruders Council—Overview of aluminum production, use, and recycling
http://www.afandpa.org/	American Forest and Paper Association
http://www.afsinc.org/	American Foundrymen's Society
http://www.aimeny.org/	The American Institute of Mining, Metallurgical, and Petroleum Engineers (AIME)
http://www.aioa.org/	American Iron Ore Association
http://www.aisc.org/home.asp OR http://www.aiscweb.com/home.asp	American Institute of Steel Construction, Inc.
http://www.aluminum.org/	The Aluminum Association, Inc.
http://www.amm.com/index.htm	Cahner's AMM Online is the international news service of American Metal Market, the daily newspaper of the metals and recycling industries.
http://www.ansi.org/	American National Standards Institute
http://www.apawood.org/	APA—The Engineered Wood Association
http://www.api.org/	American Petroleum Institute
http://www.apiencompass.org/	API EnCompass
http://www.asce.org/	American Society of Civil Engineers
http://www.asee.org/	American Society of Engineering Education
http://www.asme.org/	American Society of Mechanical Engineering
http://www.asm-intl.org/	ASM International—The Materials Information Society
http://www.asnt.org/	American Society for Nondestructive Testing
http://www.asphaltinstitute.org/	The Asphalt Institute
http://www.astm.org/	American Society for Testing and Materials
http://www.awc.org/	American Wood Council
http://www.awpi.org/	American Wood Preserver's Institute
http://www.aws.org/	American Welding Institute
http://www.bia.org/	Brick Institute of America
http://www.castmetals.com/	American Foundrymen's Society—Cast Metals Institute
http://www.cems.umn.edu/ ~mangipud/main/main.htm	The Adhesion Society
http://www.cfa-hq.org/	Composite Fabricators Association
http://www.copper.org/	International Copper Association
http://www.enr.com/	Engineering News Record—News of the construction industry and searchable database of news articles, directories, firms, and products
http://www.fmametalfab.org/	Fabricators and Manufacturers Association, International
http://www.forestprod.org/	Forest Products Society
http://www.fs.fed.us/	USDA Forest Service
http://www.gri.org/	Gas Research Institute

http://www.hardwood.org/	Hardwood Manufacturers Association
http://www.hardwoodhandbook.org/	Southeastern Lumber Manufacturers Association—Hardwood Uses Handbook
http://www.hotmix.org/	National Asphalt Pavement Association
http://www.hpva.org/	Hardwood Plywood and Veneer Association
http://www.ifant.net/	International Foundation for the Advancement of Nondestructive Testing
http://www.imiweb.org/	International Masonry Institute
http://www.iso.ch/	International Organization for Standardization
http://www.issource.org/	Iron and Steel Society
http://www.leadinfo.com/	Lead Industries Association
http://www.metalworld.com/	MetalWorld—International Metals Exchange—Directories, information, and related to all types of metals: ferrous, nonferrous, precious, and a variety of finished shapes
http://www.mfgtech.org/	Association for Manufacturing Technology
http://www.mmm.com/	3M Corporation
http://www.mrs.org/	Materials Research Society
http://www.nace.org/nacehome.html	NACE International—The Corrosion Society
http://www.nait.org/	National Association of Industrial Technology
http://www.nauticom.net/www/grind/	Abrasive Engineering Society
http://www.ndtma.org/	Nondestructive Testing Management Association
http://www.nist.gov/	National Institute of Standards and Technology
http://www.npc.org/	National Petroleum Council
http://www.ntma.org/	National Tooling and Machining Association
http://www.opec.org/	OPEC Online—Organization of the Petroleum Exporting Countries
http://www.pavement.com/	American Concrete Pavement Association
http://www.plasticsresource.com/	American Plastics Council
http://www.portcement.org/	Portland Cement Association
http://www.sae.org/index.htm	Society of Automotive Engineers
http://www.silversmithing.com/	Society of American Silversmiths
http://www.sme.org/	Society of Manufacturing Engineers
http://www.smenet.org/	Society for Mining, Metallurgy, and Exploration
http://www.socplas.org/	The Society of the Plastics Industry
http://www.spe.org/	Society of Petroleum Engineers
http://www.sspc.org/	The Society for Protective Coatings
http://www.steel.org/	American Iron and Steel Institute
http://www.steelforge.com/	Specialty Steel and Forge
http://www.swe.org/	Society of Women Engineers
http://www.tappi.org/	Technical Association of the Pulp and Paper Industries
http://www.thomasregister.com/	Thomas Register of Industrial Manufacturers and Suppliers—Searchable index of manufacturers and products
http://www.tms.org/	The Minerals, Metals, and Materials Society
http://www.usgs.gov/	U.S. Geological Survey

http://www.zinc.org/	American Zinc Association
http://www2.coatingstech.org/	Federation of Societies for Coating Technology

Other Links

http://about.com/	About.com—Latest news on materials
http://aurorametals.com/	Aurora Metals, Inc.—Casting facility for precision metal casting
http://bh1.com/fusion/metals.htm	Fusion, Inc.—Metals industry information and links. Searchable.
http://biznet.maximizer.com/ lindberg-stl/	Lindberg Heat Treating—St. Louis, MO
http://info.cas.org/	Chemical Abstracts Service—Database of material properties and testing data
http://lgic.glass-info.com/	North American Laminated Glass Information Center
http://primusweb.com/forest/	IRIS—The Internet's Forest Products Resource
http://www.advmat.com/links.html	The Composites Corner—Turner Moss Publishing Company
http://www.amm.com/	American Metal Market—International news service of American Metal Market, a daily newspaper of the metals and recycling industries. Subscription required for full service.
http://www.artrans.com/rmsg/ fuel/fuel.htm	Rocky Mountain Survival Group—Fuel information
http://www.aws.org/	American Welding Society
http://www.bir.org/uk.index.htm	Bureau of Industrial Recycling—Information on recycling of various materials including ferrous and nonferrous metals, paper, textiles, plastics, rubber, and glass
http://www.chatillon.com/	Chatillon Materials Testing Systems
http://www.discovery.com/	Discovery Channel Online
http://www.distdyn.com/	Distribution Dynamics Labs, Inc.—Test resources and testing equipment
http://www.enduratec.com/ stinfo.htm	Endura TEC Systems Corporation—Materials testing
http://www.europages.com/ nmc04-en.html	Europages—The European business directory. Searchable.
http://www.fdinc.com/Default.htm	Fatigue Dynamics, Inc.—Fatigue testing
http://www.forging.org/	Forging Industry Association—Facts about the forging industry and technology
http://www.gcc.net/commerce/ industry/industry.htm	Global Chamber of Commerce—Industry and Service
http://www.ge.com/	General Electric—General materials selection. Enter "datasheets" in search box for product information.
http://www.geneq.com/catalog/ mat_en.htm	Geneq, Inc.—Materials testing equipment
http://www.glassonline.com/	Glass Online—Searchable links to glass industries, dictionary of terms, and news of the glass industry

http://www.griffgrips.com/	Griff Grips for most modern testing machines and systems
http://www.harvardndt.com/	Harvard Nondestructive Testing, Inc.
http://www.heattreatingonline.com/	Heat Treating Online—Information, articles, and links to heat treatment information
http://www.hocking.com/Theory_applications.htm	Hocking eddy current testing information
http://www.hounsfield.com/	Hounsfield Test Equipment, LTD
http://www.ih.navy.mil/testeval.htm	Navy Sea Systems Command—Indian Head Division—Testing and evaluation of metal parts
http://www.instron.com/	Instron Corporation—Material testing systems
http://www.instron.com/apps/glossary/A.htm	Instron glossary of materials testing terms
http://www.kodak.com/	Eastman-Kodak Company—Films and paper
http://www.matls.com/	MatWEb—Free online material database of properties
http://www.metal-mart.com/	Metalmark Incorporated—Includes dictionary of metals terminology
http://www.ms.ornl.gov/	Oak Ridge National Laboratory—Metals and Ceramics Division
http://www.mts.com/	MTS Systems Corporation—Materials testing systems
http://www.pfg.co.za/PFGlow/welcome.htm	PFG Building Glass
http://www.polymer-search.com/	Polymer Search on the Internet—Elastomers and polymers search engine
http://www.principalmetals.com/	Principal Metals Online—Properties, data, weight calculator, and conversions
http://www.suppliersonline.com/properties/step1.asp	Materials Suppliers Online—Free online properties database
http://www.thermalceramics.com/	Thermal Ceramics, Inc.—Materials for refractory, thermal, and acoustical insulation applications
http://www.thomasregional.com/newtrd/nc_home.html	Thomas Regional directory of suppliers, services, and products. Searchable.
http://www.tiniusolsen.com/	Tinius Olsen Testing Machine Company, Inc.
http://www.woodresource.com/	Searchable site for products and companies related to the wood industry
http://www.woodweb.com/	Wood Web—Information about the forestry industry and processing

Other Links—Educational Institutions

http://arvind.coe.drexel.edu/	Department of Materials Engineering—Drexel University—Philadelphia
http://mrcemis.ms.nwu.edu/	Materials Research Center—Northwestern University
http://www.alfred.edu/research/html/research_centers_programs.html	Research Centers for Glass and Ceramics—Alfred University—New York
http://www.brunel.ac.uk/depts/mt/	Department of Materials Engineering—Brunel University—United Kingdom

http://www.ccm.udel.edu/	Center for Composite Materials—University of Delaware
http://www.cit.cmu.edu/	Carnegie Institute of Technology—Carnegie Mellon University—Pittsburgh
http://www.cnde.iastate.edu/	Center for Nondestructive Evaluation—Iowa State University
http://www.dmse.mit.edu/	Massachusetts Institute of Technology—Department of Materials Science and Engineering
http://www.ims.uconn.edu/	Institute of Materials Science—University of Connecticut
http://www.mate.calpoly.edu/	Materials Engineering Dept.—California Polytechnic State University
http://www.mse.berkeley.edu/	Materials Science and Minerals Engineering— UC Berkeley
http://www.twi.co.uk/	The Welding Institute—United Kingdom

A.2 SELECTED REFERENCES FOR FURTHER INFORMATION

Allen, D. K. 1969. *Metallurgy: Theory and practice*. Homewood, IL: American Technical Publishers.

American Iron and Steel Institute. 1976. *Steel products manual*. Washington, DC: American Iron and Steel Institute.

Askeland, D. R. 1989. *The science and engineering of materials*. Boston: PWS.

Bethlehem Steel Corporation. 1972. *Modern steels and their properties*. Bethlehem, PA: Bethlehem Steel Corporation.

Bikales, N. M. 1971. *Mechanical properties of polymers*. New York: Wiley-Interscience.

Bikerman, J. J. 1961. *The science of adhesive joints*. New York: Academic Press.

Billmeyer, F. W. 1971. *Textbook of polymer science*, 2nd ed. New York: John Wiley.

Bowen, H. K., W. D. Kingery, and D. R. Uhlman. 1976. *Introduction to ceramics*, 2nd ed. New York: John Wiley.

Bradstreet, S. W., and L. W. Davis. 1970. *Metal and ceramic matrix composites*. Boston: Cahners.

Brooks, C. R. 1979. *Heat treatment of ferrous alloys*. New York: Hemisphere.

Budinski, K. G. 1989. *Engineering materials: Properties and selection*, 3rd ed. Upper Saddle River, NJ: Prentice Hall.

Chandler, M. 1968. *Ceramics in the modern world*. Garden City, NY: Doubleday & Co.

Clapp, W. H., and D. S. Clark. 1938. *Engineering materials and processes*. Scranton, PA: International Textbook Co.

Committee on Engineering Materials. 1958. *Engineering materials*. New York: Pitman.

Cottrell, A. H. 1967. *An introduction to metallurgy*. New York: St. Martin's Press.

Cottrell, A. H. 1964. *The mechanical properties of matter*. New York: John Wiley.

Cowan, H. J., and P. R. Smith. 1988. *The science and technology of building materials*. New York: Van Nostrand Reinhold.

Davis, H. E., G. F. W. Hauck, and G. E. Troxell. 1982. *The testing of engineering materials*, 4th ed. New York: McGraw-Hill.

Davis, H. E., G. E. Troxell, and C. T. Wiskocil. 1964. *The testing and inspection of engineering materials*, 3rd ed. New York: McGraw-Hill.

DeLollis, N. J. 1980. *Adhesives, adherends, adhesion*. New York: Robert E. Krieger.

Dieter, G. E. 1961. *Mechanical metallurgy*. New York: McGraw-Hill.

Enos, G. M., and W. E. Fontaine. 1966. *Elements of heat treatment*. New York: John Wiley.

Fellers, W. O. 1990. *Materials science, testing, and properties for technicians*. Upper Saddle River, NJ: Prentice Hall.

Flinn, R. A., and P. K. Trojan. 1975. *Engineering materials and their applications*. Boston: Houghton-Mifflin.

Francis, W. 1965. *Fuels and fuel technology*. New York: Pergamon Press.

Gensamer, M. 1941. *Strength of materials under combined stresses*. Cleveland: American Society for Metals.

Guy, A. G. 1976. *Essentials of materials science*. New York: McGraw-Hill.

Hall, C. M. 1981. *Polymer materials*. New York: John Wiley.

Halmshaw, R. 1987. *Non-destructive testing*. London: Edward Arnold.

Harper, C. A. 1975. *Handbook of plastics and elastomers*. New York: McGraw-Hill.

Jackson, N., and R. K. Dhir, eds. 1988. *Structural engineering materials*. New York: Hemisphere Publishing.

Jastrzebski, Z. D. 1987. *The nature and properties of engineering materials*, 3rd ed. New York: John Wiley.

Kazanas, H. C., R. S. Klein, and J. R. Lindbeck. 1974. *Technology of industrial materials*. Peoria, IL: Bennett.

Keyser, C. A. 1968. *Materials science in engineering*. Columbus, OH: Merrill.

Kingery, W. D. 1958. *Ceramic fabrication processes*. New York: John Wiley.

Kohn, M., and M. Starfield. 1952. *Materials and processes*. New York: Macmillan.

Krauss, G. 1980. *Principles of heat treatment of steels*. Metals Park, OH: ASM.

Leslie, W. C. 1981. *The physical metallurgy of steels*. New York: Hemisphere.

Lubin, G., ed. 1982. *Handbook of composites*. New York: Van Nostrand Reinhold.

McLean, D. 1962. *Mechanical properties of metals*. New York: John Wiley.

Nickelsen, L. E. 1974. *Mechanical properties of polymers and composites*. New York: Decker.

Pollack, H. W. 1988. *Materials science and metallurgy*, 4th ed. Upper Saddle River, NJ: Prentice Hall.

Ray, M. S. 1987. *The technology and applications of engineering materials*. Upper Saddle River, NJ: Prentice Hall.

Reed-Hill, R. E. 1964. *Physical metallurgy principles*. Princeton, NJ: Van Nostrand Reinhold.

Roberson, E., and R. Herbert. 1963. *Fuel: The conquest of man's environment*. New York: Harper & Row.

Rowe, G. W. 1965. *An introduction to the principles of metal-working*. New York: St. Martin's Press.

Schlenker, B. R. 1969. *Introduction to materials science*. New York: John Wiley.

Schultz, J. 1974. *Polymer materials science*. Upper Saddle River, NJ: Prentice Hall.

Smith, C. O. 1986. *The science of engineering materials*, 3rd ed. Upper Saddle River, NJ: Prentice Hall.

Smith, W. F. 1986. *Principles of materials science and engineering*. New York: McGraw-Hill.

Somiya, S. 1984. *Advanced technical ceramics*. New York: Academic Press.

Tegart, W. J. 1966. *Elements of mechanical metallurgy*. New York: Macmillan.

Tietz, T. E., and J. W. Wilson. 1965. *Behavior and properties of refractory materials*. Stanford, CA: Stanford University Press.

Tobolsky, A. V., ed. 1965. *Structure and properties of polymers*. New York: Wiley-Interscience.

Tooley, F. V. 1974. *Handbook of glass*. New York: Books for Industry.

Tottle, C. R. 1966. *The science of engineering materials*. New York: Chemical Publishing Co.

Ulrich, H. 1982. *Introduction to industrial polymers*. New York: Macmillan.

United States Steel Corporation. 1951. *The making, shaping, and treating of steels*. Pittsburgh: United States Steel Corporation.

Van Vlack, L. H. 1964. *Physical ceramics for engineers*. Reading, PA: Addison-Wesley.

Weismantel, G. S. 1981. *Paint handbook*. New York: McGraw-Hill.

Williams, D. J. 1971. *Polymer science and engineering*. Upper Saddle River, NJ: Prentice Hall.

Wilson, F. 1984. *Building materials evaluation handbook*. New York: Van Nostrand Reinhold.

Young, R. J. 1981. *Introduction to polymers*. London: Chapman & Hall.

A.3 OTHER MATERIALS FOR GENERAL REFERENCE AND FURTHER INFORMATION

Aluminum Standards and Data, Aluminum Association

Journal of Applied Polymer Science

Journal of Materials Science

Metals Handbook, ASM, Metals Park, OH

Modern Plastics Encyclopedia

APPENDIX B
Measurement and Properties of Materials

METRIC SYSTEM

Traditionally, length was measured in yards, feet, and inches; weight, mass, and force were measured in pounds and ounces; electrical energy was measured in joules; heat energy was measured in British thermal units (Btu); work was measured in horsepower (hp), and temperature was measured in degrees Fahrenheit.

In the old metric system, length was measured in meters; weight, mass, and force were measured in grams; electric energy was measured in joules; heat energy was measured in calories; work was measured in joules, and temperature was measured in degrees Celsius.

In SI (Système International d'Unités), length is measured in meters; mass is measured in kilograms; electrical energy, heat energy, and work are measured in joules; temperature is measured in degrees Celsius, and force is measured in newtons.

Because SI units may be unfamiliar to many people, the following abbreviation conversions are given to aid in understanding the examples and calculations in the text.

Common Abbreviations

Btu	British thermal units
°C	degrees Celsius
°F	degrees Fahrenheit
K	degrees Kelvin
ft	foot
g	gram
gal	gallon
GPa	gigapascals
Hz	hertz
in.	inch
J	Joules
kPa	kilopascals
L	liter
lb	pound

m	meter
mi	mile
N	newtons
oz	ounce
Pa	pascals
W	watts

COMMONLY USED SI QUANTITIES, UNITS, AND SYMBOLS

Quantity	Unit	Symbol
Acceleration	meter per second squared	m/s^2
Area	square meter	m^2
Density	kilogram per cubic meter	kg/m^3
Electric current	ampere	A
Force	newton	N
Frequency	hertz	$Hz\ (s^{-1})$
Length	meter	m
Mass	kilogram	kg
Pressure	pascal	Pa
Specific volume	cubic meter per kilogram	m^3/kg
Time	second	s
Velocity	meter per second	m/s
Volume	cubic meter	m^3

COMMON SI PREFIXES

Factor	Prefix	Symbol
10^{-18}	atto	a
10^{-15}	femto	f
10^{-12}	pico	p
10^{-9}	nano	n
10^{-6}	micro	μ
10^{-3}	milli	m
10^{-2}	centi	c
10^{-1}	deci	d
10	deka	da
10^2	hecto	h
10^3	kilo	k
10^6	mega	M
10^9	giga	G
10^{12}	tera	T
10^{15}	peta	P
10^{18}	exa	E

CONVERSION FACTORS

Temperature

$C = \frac{5}{9}(F-32)$, where $C = °C$ and $F = °F$
$F = (9/5)C + 32$
$K = C + 273.15$, where $K = $ degrees Kelvin

Length

1 mi = 1.609 km
1 km = 0.62137 mi
1 m = 3.281 ft
1 m = 39.37 in.
1 ft = 0.3048 m
1 in. = 25.4 mm

Area

$1 \text{ m}^2 = 10.764 \text{ ft}^2$
$1 \text{ ft}^2 = 0.0929 \text{ m}^2$

Volume

$1 \text{ m}^3 = 35.315 \text{ ft}^3$
$1 \text{ ft}^3 = 0.02832 \text{ m}^3$
$1 \text{ L} = 0.001 \text{ m}^3$
$= 0.264 \text{ gal}$

Mass

1 kg = 2.205 lb
1 lb = 0.454 kg

Density

$1 \text{ kg/m}^3 = 0.0642 \text{ lb/ft}^3$
$1 \text{ lb/ft}^3 = 16.018 \text{ kg/m}^3$

Speed and Velocity

1 m/s = 196.85 ft/min
= 2.237 mi/h
= 3.6 km/h
1 ft/min = 0.00508 m/s
1 mi/h = 0.447 m/s

Force, Pressure, and Stress

$1 \text{ Pa} = 1 \text{ N/m}^2$
1 N = 0.225 lb
1 lb = 4.448 N
$1 \text{ kPa} = 0.145 \text{ lb/in.}^2$
$1 \text{ lb/in.}^2 = 6.895 \text{ kPa}$
$1 \text{ MPa} = 145 \text{ lb/in.}^2$
$1000 \text{ lb/in.}^2 = 6.895 \text{ MPa}$

Standard atmospheric pressure is equal to 101.3 kPa or 14.7 lb/in.^{2} or 760 mm of mercury or 30 in. of mercury.

PROPERTIES OF SELECTED MATERIALS

Given here are some of the more important properties of selected materials. When reviewing these properties, look for the relationships that exist among the values. Further information on these properties and how they relate to applications can be found in the specific chapters concerning the materials.

Table B–1

| | | | Properties of Nonferrous Metals and Alloys | | |
| | | | Comparison of Selected Nonferrous Metals and Alloys | | |

Material	Specific Gravity	Melting Point °F (°C)	Tensile Strength lb/in² × 10³ (MPa)	Yield Strength lb/in² × 10³ (MPa)	Modulus of Elasticity lb/in² × 10⁶ (GPa)
Aluminum	2.70	1,200 (654)	12 (83)	4 (28)	10 (69)
Beryllium	1.84	2,370 (1,300)	45 (312)	32 (221)	45 (310)
Chromium	7.21	3,434 (1,890)	—	—	36 (248)
Copper (hard)	8.96	1,981 (1,083)	44–55 (303–379)	45 (310)	17 (117)
Copper (soft)			30–38 (207–262)	10 (69)	15 (103)
Gold	19.32	1,945 (1,063)	19 (131)	—	11 (76)
Lead	11.30	621 (326)	2.6 (18)	1.3 (9)	2 (14)
Magnesium	1.74	1,204 (651)	37 (255)	22–29 (152–200)	6.5 (45)
Nickel	8.89	2,630 (1,439)	65 (449)	23 (156)	30 (200)
Platinum	21.45	3,225 (1,773)	19 (131)	—	21.3 (147)
Silver	10.49	1,762 (961)	24 (165)	8 (55)	11 (76)
Tin	7.29	449 (232)	3.1 (21)	—	6 (41)
Titanium	4.54	3,038 (1,670)	93 (623)	83 (569)	16 (107)
Tungsten	19.3	6,161 (3,405)	120 (827)	—	46 (317)
Zinc	7.1	786 (419)	24 (166)	—	15 (103)
Zirconium	6.45	3,355 (1,846)	35 (241)	15 (103)	14 (96)

Table B–1, continued

Comparison of Selected Ferrous Metals and Alloys

Material	Specific Gravity	Melting Point °F (°C)	Tensile Strength lb/in² × 10³ (MPa)	Yield Strength lb/in² × 10³ (MPa)	Modulus of Elasticity lb/in² × 10⁶ (GPa)
Iron	7.87	2,797 (1,536)	42 (290)	19 (131)	29 (200)
Low-carbon steel (hot-rolled)	7.80	2,750–2,775 (1,510–1,524)	50–70 (345–483)	29–45 (200–310)	30 (203)
Low-carbon steel (cold-worked)			56–80 (386–552)	33–69 (228–476)	
Medium-carbon steel (hot-rolled)	7.80	2,700–2,750 (1,482–1,510)	80–98 (552–676)	50–60 (345–414)	30 (203)
Medium-carbon steel (cold-worked)			85–103 (586–710)	72–90 (496–620)	
Medium-carbon steel (heat-treated)			98–120 (676–827)	70–90 (483–620)	
High-carbon steel (annealed)	7.80	2,600–2,675 (1,427–1,468)	97–120 (670–827)	52–66 (359–455)	30 (203)
High-carbon steel (hot-rolled)			109–142 (752–980)	65–84 (448–580)	
High-carbon steel (heat-treated)			150–192 (1,034–1,324)	105–142 (724–980)	
Gray cast iron	7.12	2,050–2,150 (1,121–1,177)	29–45 (200–310)	—	13–16 (90–110)
Malleable cast iron-ferritic	7.12	2,050–2,150 (1,121–1,177)	53 (365)	35 (241)	25 (172)
Ferritic stainless steel (annealed)	7.70	2,600–2,750 (1,427–1,510)	75 (517)	40–45 (276–310)	29 (200)
Ferritic stainless steel (cold-worked)			75–90 (517–620)	45–80 (310–552)	
Martensitic stainless steel (annealed)	7.70	2,700–2,790 (1,482–1,532)	65–75 (448–517)	35–45 (241–310)	29 (200)
Martensitic stainless steel (heat-treated)			90–190 (620–1,310)	50–150 (345–1,035)	
Austenitic stainless steel (annealed)	8.00	2,500–2,552 (1,371–1,400)	80–95 (552–655)	35–55 (241–379)	28 (193)
Austenitic stainless steel (cold-worked)			100–150 (690–1,035)	50–125 (345–862)	
Maraging steel (oil-quenched)	7.78	—	275 (1,896)	268 (1,848)	26 (183)

Properties of Selected Glasses and Glass Products

Material	Specific Gravity	Softening Point °F (°C)	Tensile Strength lb/in² × 10³ (MPa)	Modulus of Elasticity lb/in² × 10⁶ (GPa)	Index of Refraction
Aluminosilicate	2.35	2,000 (1,102)	4–10 (28–69)	18 (124)	1.51
Borosilicate	2.23	1,508 (820)		8.8 (61)	1.47
Fiberglass		1,545 (841)	500 (3,450)	10.5 (72)	
Fused silica	2.18	3,000 (1,662)	4–10 (28–69)	12 (83)	1.46
Lead alkaline	2.85	1,165 (634)		9.0 (62)	1.54
Soda lime	2.47	1,300 (700)		9.8 (68)	1.51

Properties of Selected Polymers

Polymer Type	Density lb/ft³ (g/cm³)	Tensile Strength lb/in² (MPa)	Modulus of Elasticity lb/in² × 10³ (MPa)	Maximum % Elongation
Thermoplastics				
Polyethylene	56.8–60.2 (0.91–0.97)	1,000–5,500 (6.9–37.9)	14–160 (0.97–1.1)	15–700
Polyvinyl chloride (PVC)	68.7–87.4 (1.1–1.4)	1,500–9,000 (10.3–62.0)	200–600 (1.4–4.1)	2–400
Polypropylene	56.2 (0.9)	4,300–5,500 (29.6–37.0)	1,400–1,700 (9.6–11.7)	210
Polystyrene	64.9–67.4 (1.04–1.08)	5,000–10,000 (34.5–69.0)	400–600,000 (2.8–4,137)	1–3
Polymethylmethacrylate (acrylic-Plexiglas)	73.7–74.9 (1.18–1.20)	7,000–11,000 (48.3–75.8)	350–500 (2.4–3.4)	2–10
Polytetrafluoroethylene (PTFE-teflon)	131–144 (2.1–2.3)	2,000–4,500 (13.8–31.0)	32–65 (0.22–0.45)	200–400
Polyamide (nylon)	70.5–71.8 (1.13–1.15)	9,000–9,500 (62.0–65.5)	—	60–300
Polyaramid fibers (Kevlar)	90.5 (1.45)	410,000 (2,827)	20,000 (138)	—

Table B–1, continued

Properties of Selected Polymers *(Continued)*

Polymer Type	Density lb/ft³ (g/cm³)	Tensile Strength lb/in² (MPa)	Modulus of Elasticity lb/in² × 10³ (MPa)	Maximum % Elongation
Polycarbonate	74.9 (1.2)	8,000–9,500 (55.2–65.5)	290–325 (2.0–2.2)	20–100
Polyester (PET)	84.9 (1.36)	8,000–10,500 (55.2–72.4)	400–600 (2.76–4.13)	50–300
Thermosetting				
Epoxy	69.0 (1.1)	4,000 (27.6)	300 (2.1)	2–6
Melamine-formaldehyde	91.8–111.1 (1.47–1.78)	5,500–13,000 (37.9–89.6)	1,300–1,950 (8.96–13.4)	0.6–0.9
Phenol-formaldehyde	84.9–89.3 (1.36–1.43)	6,000–9,000 (41.4–62.0)	900–1,300 (6.2–8.96)	0.5–1.0
Urea-formaldehyde	91.8–94.9 (1.47–1.52)	5,500–13,000 (37.9–89.6)	1,300–1,400 (8.96–9.65)	0.6–0.8
Elastomers				
Acrylonitrile-butadiene-styrene	66.2 (1.06)	7,000 (48.3)	—	1–2
Butadiene	58.7 (0.94)	3,500 (24.1)	—	3,000
Butadiene-styrene	62.4 (1.0)	600–3,000 (4.1–20.7)	—	600–2,000
Chloroprene	78.0 (1.25)	3,500 (24.1)	—	800
Isoprene	58.5 (0.93)	3,000 (20.7)	—	800
Silicone	93.6–174.8 (1.5–2.8)	3,000–4,000 (20.7–27.6)	—	0–700

Properties of High-strength Fibers

Fiber Type	Density lb/ft³ (g/cm³)	Tensile Strength lb/in² (MPa)	Modulus of Elasticity lb/in² × 10³ (MPa)
Graphite—Low Density	110 (1.76)	410,000 (2.83)	30–37 (207–255)
Graphite—High Density	121 (1.94)	300,000 (2.07)	50–75 (345–517)
Kevlar	90.5 (1.45)	410,000 (2.83)	20 (138)
Glass	159 (2.54)	500,000 (3.45)	10–11 (69–76)

Properties of Selected Woods

Wood Type	Specific Gravity	Modulus of Elasticity $lb/in^2 \times 10^6$ (GPa)	Compressive Strength lb/in^2 (MPa)
Hardwoods			
Maple	0.676	1.83 (12.6)	7,800 (53.8)
Oak	0.710	1.77 (12.2)	7,440 (51.3)
Poplar	0.427	1.50 (10.3)	5,540 (38.2)
Walnut	0.562	1.68 (11.6)	7,580 (52.3)
Softwoods			
Cedar	0.340	1.10 (7.6)	5,000 (34.5)
Douglas fir	0.512	1.90 (13.1)	7,440 (51.3)
Hemlock	0.440	1.10 (7.6)	6,230 (43.0)
Pine	0.375	1.27 (8.8)	4,800 (33.0)
Spruce	0.432	1.42 (9.8)	5,500 (37.9)

Properties of Selected Technical Ceramics

Note: Technical ceramics are also known as advanced, engineering, or structural ceramics and include carbides, borides, and nitrides as well as some oxides.

Ceramic Type	Density lb/ft^3 (g/cm³)	Tensile Strength $lb/in.^2 \times 10^3$ (MPa)	Flexural Strength $lb/in.^2 \times 10^3$ (MPa)	Compressive Strength $lb/in.^2 \times 10^3$ (GPa)	Modulus of Elasticity $lb/in^2 \times 10^6$ (GPa)
Al_2O_3	248.5 (3.98)	30 (207)	80 (552)	400 (2.76)	56 (386)
BeO	—	14 (97)	—	185 (1.28)	50 (345)
B_4C	—	22 (152)	—	420 (2.90)	42 (290)
MgO	—	20 (138)	—	120 (0.83)	40 (276)
SiC (sintered)	193.5 (3.1)	25 (172)	80 (552)	560 (3.86)	60 (414)
Si_3N_4 (reaction bonded)	156.1 (2.5)	20 (138)	35 (241)	150 (1.03)	30 (207)
Si_3N_4 (hot-pressed)	200 (3.2)	80 (552)	80 (552)	500 (3.45)	45 (310)

Properties of Selected Technical Ceramics *(Continued)*

Ceramic Type	Density lb/ft³ (g/cm³)	Tensile Strength lb/in.² × 10³ (MPa)	Flexural Strength lb/in.² × 10³ (MPa)	Compressive Strength lb/in.² × 10³ (GPa)	Modulus of Elasticity lb/in² × 10⁶ (GPa)
TiC	—	65 (448)	—	300 (2.07)	50 (345)
WC	—	130 (896)	—	650 (4.48)	80 (552)
ZrO$_2$	362 (5.8)	65 (448)	100 (690)	270 (1.86)	30 (207)

Comparison of Ceramic to Metal Hardness

Material	Hardness (kg/mm²)
Lead	3
Carbon	15
Aluminum	20
Copper	40
Low-carbon steel	80
Magnesium oxide	700
Tool steel	750
Silicon dioxide	820
Chrome plating	900
Titanium dioxide	1,100
Zirconium oxide	1,150
Tantalum carbide	1,750
Tungsten carbide	2,000
Aluminum oxide	2,100
Silicon nitride	2,200
Titanium carbide	2,500
Boron carbide	2,720
Silicon carbide	2,800
Cubic boron nitride	5,000
Diamond	8,000

Properties of Selected Composites

Composite Type	Density lb/ft³ (g/cm³)	Tensile Strength lb/in.² × 10³ (MPa)	Flexural Strength lb/in.² × 10³ (MPa)	Compressive Strength lb/in.² × 10³ (MPa)	Modulus of Elasticity lb/in² × 10⁶ (GPa)
Fiberglass Composites					
BMC (bulk molding compound) polyester mat with 15–25% glass fiber content	0.055 (0.0020)	10 (69)	15 (103)	20 (138)	1.6 (11)
SMC (sheet molding compound) polyester mat with 30–40% glass fiber content	0.062 (0.0022)	20–25 (138–172)	20–25 (138–172)	25–28 (172–193)	1.7–1.9 (11.7–13.1)
Filament wound epoxy	0.07–0.08 (0.0027)	150–200 (1,034–1,379)	90–200 (621–1,379)	45–80 (310–552)	4.5–7.5 (31–52)
High-Performance Composites					
Graphite/epoxy	0.056–0.058 (0.0021)	113–250 (779–1,724)	90–270 (621–1,862)	90–230 (621–1,586)	20–47 (138–324)
Graphite/polyimide	0.056 (0.002)	160–200 (1,103–1,379)	220–230 (1,517–1,586)	—	17 (117)
Aramid/epoxy	0.049 (0.0018)	75–180 (517–1,241)	50–90 (345–621)	12–40 (83–276)	4.5–12 (31–83)

Appendix 3
Refractory and Industrial Materials Melting Point Chart

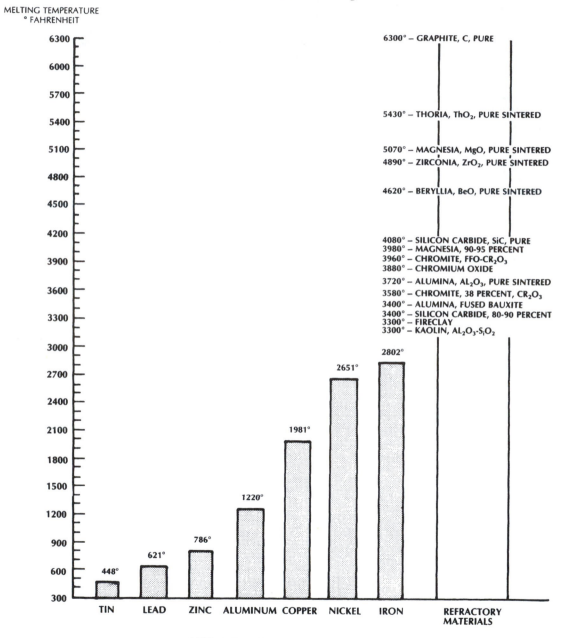

MELTING TEMPERATURE
° FAHRENHEIT

6300° – GRAPHITE, C, PURE

5430° – THORIA, ThO₂, PURE SINTERED

5070° – MAGNESIA, MgO, PURE SINTERED
4890° – ZIRCONIA, ZrO₂, PURE SINTERED

4620° – BERYLLIA, BeO, PURE SINTERED

4080° – SILICON CARBIDE, SiC, PURE
3980° – MAGNESIA, 90-95 PERCENT
3960° – CHROMITE, FFO-CR₂O₃
3880° – CHROMIUM OXIDE
3720° – ALUMINA, AL₂O₃, PURE SINTERED
3580° – CHROMITE, 38 PERCENT, CR₂O₃
3400° – ALUMINA, FUSED BAUXITE
3400° – SILICON CARBIDE, 80-90 PERCENT
3300° – FIRECLAY
3300° – KAOLIN, AL₂O₃-S₁O₂

2802°
2651°
1981°
1220°
786°
621°
448°

TIN LEAD ZINC ALUMINUM COPPER NICKEL IRON REFRACTORY MATERIALS

KEY INDUSTRIAL MATERIALS

Figure B–1

Courtesy of The Refractories Institute. From *Refractories*, copyright 1987 by The Refractories Institute, Appendix 3, page 39.

468

APPENDIX C
Laboratory Exercises

TABLE OF CONTENTS

C.1 INTRODUCTION

These laboratory exercises are designed to help you develop the knowledge and skills required to design and conduct testing and inspection procedures on materials. They are meant to supplement the textbook and other sources, not replace them. The primary focus of these laboratory exercises is the study of the theory and applications of the fields of materials science, materials technology, and materials testing. This study includes the determination of a material's chemical, physical, mechanical, and dimensional properties. Specifically, you will be asked to identify, describe, and address the structure and properties of metals, polymers, ceramics, and composite materials, among others.

Specimens will be tested individually; some will be tested as a group. Instruction has been provided for describing the purpose of the test, describing the procedure(s) used during the test, describing the apparatus required, recording the data, and interpreting the findings based on that data. Whenever possible, sample test results and procedures are given as examples.

C.1.a Safety

IMPORTANT: Operator safety and observer safety before, during, and after testing should be your foremost concern. It is your responsibility to make sure that all guards are in place, all shields are secure, all safety procedures have been followed, everyone is wearing proper safety equipment (including safety glasses, gloves, and safe, appropriate clothing), and everyone is aware that you will be testing. Use common sense. During testing, some materials will shatter; some will not. When shattering occurs, pieces fly in all directions, possibly causing injury. MAKE SURE YOU FOLLOW ALL SAFETY PROCEDURES AND HAVE CHECKED THE EQUIPMENT. Follow *all* test procedures and make sure you know how to stop a test or test machine if things aren't going as expected. The time to stop a test is before someone gets hurt. There isn't any penalty for stopping to make sure everything is all right. Do it right the first time; you may not have a second chance. DON'T BE IN A HURRY. USE YOUR HEAD.

C.1.b Metals and Mechanical Properties

Among the primary concerns when testing materials are the mechanical properties of the material. Mechanical properties of metals include tensile strength, yield strength, ductility, toughness, and hardness. These properties are the most commonly investigated and are among the most important properties involved in design.

Most data concerning a material's strength are collected through tensile-type tests. The compression test may also yield an important value for design. The tensile test is of primary concern in many applications.

Due to the crystalline structure of metals, they exhibit elastic properties. Most metals are also ductile, resulting from slippage within the crystal structure. Ductile metals can, therefore, exhibit a larger deformation in their plastic regions.

Figure C–1

Stress-strain curves for iron, mild steel, cast iron, and brass.

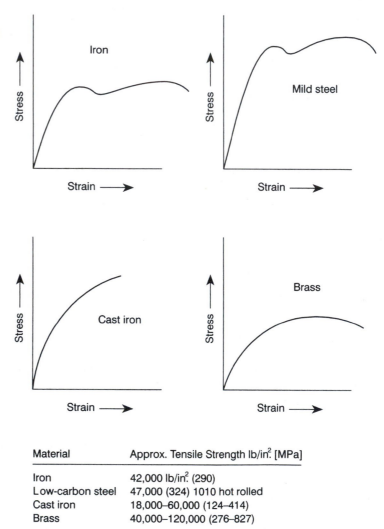

Material	Approx. Tensile Strength lb/in^2 [MPa]
Iron	42,000 lb/in^2 (290)
Low-carbon steel	47,000 (324) 1010 hot rolled
Cast iron	18,000–60,000 (124–414)
Brass	40,000–120,000 (276–827)

The strengths exhibited by different metals (based on alloying elements, alloy percentages, flaws/imperfections, and so on) depend on the direction in which the strengths or properties are measured. Due to uneven internal and external stresses, metals and metal specimens may exhibit significantly different properties, depending on the grain orientation in the measured axis and the manufacturing technique used on the material. For example, the test results of a casting with a random grain arrangement along equal axes should be the same regardless of the direction or axis along which the test was done. Further, a forged product may exhibit different properties than the same product in a cast form or one that has been machined. This fact is reinforced by the curves in Figure C–1.

The stress-strain curves in Figure C–1 represent the correlation between applied stress and the resultant strain that these materials exhibit under loading. The linear portion of the curve represents elastic deformation. The slope within this linear portion of the curve is the modulus of elasticity. The greater the slope of this linear portion, the stiffer the material is said to be. Moduli of elasticity for metals typically range from 1.5 million to 60 million lb/in.², or from 10,350 to 410,000 MPa.

There are several points on the stress-strain curve that are important in the fields of testing and design. These include the yield strength and the yield point. The yield strength of a material is usually determined based on some arbitrary strain value such as 0.1% or 0.2% (the offset). The offset method for determining the yield point includes drawing a line parallel to the elastic region of the material at the desired offset. The stress applied, coinciding with the point at which the offset line intersects the stress-strain curve (yield point), is used to calculate the yield strength for that material at the designated offset. This process is illustrated in Figure C–2.

Referring to Figure C–1, the curve for cast iron shows little or no linearity between stress and strain. Therefore, it exhibits little or no elastic action. Cast iron is, therefore, considered a brittle material. The mechanical behavior for cast iron is a function of the material's graphite phase. Percent elongation for cast iron is typically around 0.5%. The tensile strength for cast iron can be calculated by dividing the load at fracture by the original cross-sectional area of the specimen.

Figure C–2
Offset method for determining yield point.

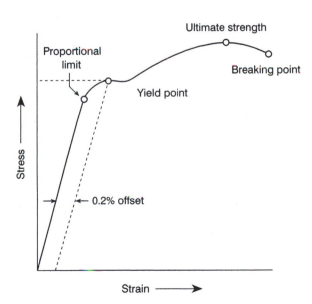

Table C–1
Yield strengths of different metals and alloys.

Metal	Yield Strength 10^3 lb/in.^{2} (MPa)
Very High Strength	
Cobalt and its alloys	26–290 (180–2,000)
Nickel and its alloys	10–230 (70–1,585)
Stainless steels (martensitic)	60–275 (415–1,895)
Very high strength steels	170–270 (1,170–1,860)
High Strength	
Bronze	14–114 (95–785)
Carbon steels	58–188 (400–1,295)
Cast alloy steels	112–170 (770–1,170)
Cast nodular iron	40–150 (275–1,035)
Cast stainless steels	31–165 (215–1,140)
Molybdenum	82–210 (565–1,450)
Tantalum and its alloys	48–168 (330–1,090)
Titanium and its alloys	27–191 (185–1,315)
Tungsten	220 (1,515)
Medium Strength	
Aluminum alloys (2000)	10–66 (70–455)
Aluminum alloys (3000)	6–36 (40–250)
Aluminum alloys (4000)	46–50 (315–345)
Aluminum alloys (5000)	6–59 (40–405)
Aluminum alloys (6000)	7–55 (50–380)
Aluminum alloys (7000)	14–91 (95–625)
Brasses	10–92 (70–625)
Carburized carbon steels	46–77 (315–530)
Cast carbon steels	38–55 (260–380)
Cast ductile iron (austenitic)	28–65 (195–450)
Copper	10–72 (70–495)
Copper-nickel-zinc	18–90 (125–620)
Cuprous nickels	13–85 (90–585)
Magnesium alloys	13–44 (90–305)
Silver	8–44 (55–303)
Stainless steels (ferritic)	45–80 (310–550)
Uranium	35–50 (240–345)
Low Strength	
Aluminum alloys (1000)	4–24 (30–165)
Cast magnesium alloys	12–30 (85–205)
Gold	30 (205)
Lead	2–8 (13–55)
Platinum	2–27 (15–185)
Tin	1.5–6 (6–40)

Metals can be divided into four categories: low strength, medium strength, high strength, and very high strength, as shown in Table C–1. The strength of most metals in their purest states is low. Typically, the strength of the material will rise with alloy content with the purposeful addition of specified impurities. Low-strength materials are often the most economical and practical based on their machinability and low cost. They may also be chosen based on properties other than strength. These other properties include formability, weight, and corrosion resistance, which may be more important than strength in the particular application. For instance, we could make a hang glider out of cast iron or aluminum. Both materials would work. However, the cast-iron glider would weigh much more than the aluminum glider and be more brittle. In this application, weight and ductility would be of greater importance than strength, although strength would be a factor to consider. Table C–1 lists some metals and alloys with their yield strengths.

Most materials are alloyed to increase their strength beyond that of their pure form. Materials above the low-strength level are almost all alloys. These alloys are added in various numbers and quantities. Each addition of an alloy or greater quantity of existing alloy elements increases the cost of the metal and generally causes a decrease in some other property. For example, increasing the carbon content of steel increases the strength but reduces the ductility and weldability of the steel. Other alloys can be used that do not cause as severe a change in properties.

C.2 TENSILE TESTING OF METALS

C.2.a Introduction

In the commercial tensile test of metals, the properties usually determined are yield strength, tensile strength, ductility, and type of fracture. In more complete tests, determinations of stress-strain relations, modulus of elasticity, and other mechanical properties are included.

The tensile test is one of the most common procedures used to measure a material's behavior under pulling forces. The data from a tensile test can often be used directly in design or to evaluate a design. Typical test materials include steel, plastics, cables, rope, wire, adhesives, cord, string, and any other material for which data on tensile strength are required.

Specimens for tensile testing come in a variety of sizes and shapes. The most common types are

1. Flat specimens (1/2 in. maximum thickness)
2. Smooth and round specimens (3/8–7/8 in. diameter)
3. Threaded-end specimens (same as previous type)
4. Bolts (7/16 in. diameter maximum, typically)

Flat specimens require the flat tensile jaws with the serrated grippers. Smooth and round specimens require the tensile jaws with the V-center grippers. Threaded specimens require threaded grippers. Bolts require test blocks, which are donut-shaped. These specimens are typically tested on a typical universal testing machine, which has a minimum tensile limit of 20,000 lb/in.^{2} (140 MPa).

The primary objective of the tensile test is to determine the maximum load a specimen can resist before taking a permanent set or before rupture occurs. Ductile materials will neck down through the plastic range before rupture occurs. Brittle materials do not neck down significantly; instead, they fail sharply and abruptly at the maximum load, because brittle materials do not exhibit much plasticity.

During the test, data should be recorded concerning applied load in pounds (kg) and deformation in inches (mm). The number of data points and the interval between these points is arbitrary. However, you should plan to record at least 10 data points for a reasonably accurate representation of the material's behavior under tensile loading. More than 10 points may not be practical, and if fewer than 10 data points are taken, major effects during testing may be missed between intervals.

Tensile tests involving ductile materials provide information regarding the elastic and plastic behavior of the material as well as the ultimate tensile strength of the material. While within a material's elastic range, if the applied load is released, the material will snap back into its original condition. The elastic range for ductile metals is relatively short and deals with the attractive forces between the material's atoms. Once the load exceeds the point where the material can return to its original condition, it takes a permanent set and is said to exhibit *plastic behavior.* The point at which the material's behavior crosses from elastic to plastic behavior (yet still exhibits elastic behavior) is called the *elastic limit.*

Once loading goes beyond the elastic limit, the material enters its plastic range. The material deforms plastically and will not return to its original condition if the load is relaxed. Plastic behavior is very useful. Because ductile metals exhibit plasticity, they can be shaped by stamping, punching, or pressing or drawn through dies to become wire. The point at which a material enters the plastic range is called the yield point. Once a material has yielded, testing should continue until fracture.

C.2.b Required Equipment

Universal testing machine with appropriate grippers
Test specimen
Gage length–indicating device
Scale, rule, or appropriate measuring device
Micrometer or calipers
Safety glasses and shields
Data sheet

C.2.c Procedure

1. Prior to applying load to a specimen, measure its dimensions. Measure the cross sectional area of the specimen. If elongation measurements are to be made, you need to scribe, or lay out, the gage length. On ductile specimens of ordinary size, do this with a center punch, but on thin sheets or brittle materials, make fine scratches. Care should be taken to make the marks lightly so that they do not influence the test. Typically, marks are made 2 in. apart. Refer to the appropriate ASTM standard for the material under test.
2. Before operating a testing machine for the first time, you should familiarize yourself with the machine, its controls, its speeds, the action of the weighing mechanism, and the value of the graduations on the load indicator. Before testing a specimen, check for zero-load indication, and adjust the machine if necessary.
3. After placing a specimen in the testing machine, check for proper alignment of the grippers. Make sure all guards and shields are in place and that all safety features work properly. Place the specimen so that it is convenient to take gage length measurements.
4. When using the extensometer, determine the value of the divisions on the indicator and the multiplication ratio before placing the extensometer on the specimen. Place it centrally on the specimen and align it properly. Place a small load on the specimen before setting the extensometer to zero.
5. The speed of testing should not be greater than that at which load and other readings can be taken with the desired degree of accuracy. A common range for load application rate is from 0.01 to 0.05 in. (0.003 to 0.015 mm) of crosshead travel per minute. Refer to the appropriate ASTM standard for exact loading rates.
6. When using the extensometer, either apply the load in increments and read the load and deformation at the end of each increment, or apply the load continuously at a slow rate and observe the load and deformation simultaneously. The latter is preferable.
7. Remove the extensometer after reaching the proportional limit.
8. Continue applying load until the test specimen fails. After the test specimen fails, remove it from the testing machine; if elongation is needed, fit the broken ends of the specimen together and measure the distance between gage points with a scale or dividers to the nearest 0.01 in. (0.003 mm). The diameter of the smallest section is calipered, preferably with a set of vernier or dial calipers for determining the reduction in area.
9. Make a record of your observations regarding applied stress and strain and graph a stress-strain diagram.
10. Determine the elongation—the increase in the gage length—expressed as a percentage of the original gage length. Report both the original gage length and the percent of increase. If breakage occurred beyond the gage points, specifications often call for a retest.
11. Determine the reduction in area by calculating the difference between the area of the smallest cross section (at the break) and the original cross-sectional area, expressed as a percentage of the original cross-sectional area.

12. Classify the fracture in regard to form, texture, and color. Types of fracture in symmetrical form are cup-cone, flat, and irregular, or ragged. Asymmetrical fractures are partial cup-cone, flat, and irregular, or ragged. Various descriptions of texture are silky; fine grain, coarse grain, or granular; fibrous or splintery; crystalline; and glassy or dull.

C.2.d Specimens

Figure C–3 gives typical tensile test specimens from a variety of situations and materials.

C.2.e Data

Figure C–4 shows an example of what data should be recorded and what a test data sheet might look like. This is only one example, but it should be used as a model to develop your own test data sheet to fit the particular test and test material.

C.2.f Results

In the results section, the data obtained and the stress-strain curve plotted based on the data obtained are used to identify and calculate the yield point, yield strength, ultimate tensile

(a) Smooth-end round tensile test specimen

(b) Threaded-end tensile test specimen

(c) Flat tensile test specimen

Figure C–3
Tensile test specimens.

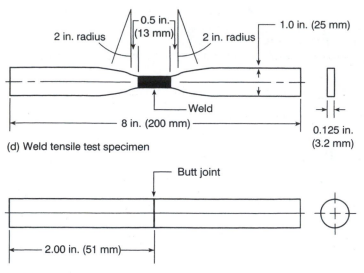

(d) Weld tensile test specimen

(e) Brazed or silver-soldered tensile test specimen

(f) Fixture tensile test of adhesives

Figure C–3, continued

strength, percent elongation, percent reduction in area, modulus of elasticity, and rupture strength for the material.

Sample data obtained are shown in Figure C–5.

The stress-strain curve for this test is shown in Figure C–6.

Calculations based on the data and stress-strain curve are given in Figure C–7.

Also included in the Results section are your observations about the test. Were there any problems? Were there any limiting factors (time, material, and so on)? Was the test valid, reliable, and significant? If you had it to do over again, what would you change? Questions similar to this should be asked and answered to provide better insight into the test.

Material:	1020 cold-drawn steel	Date:		

Original diameter: 0.505 in.   Gage length: 2.0000 in.

Reading	Load (lb)	Stress (lb/in.²)	Δ length (in)	Strain (in./in.)
1	1,000	5,100	0.0004	0.0002
2	2,000	10,200	0.0008	0.0004
3	3,000	15,300	0.0010	0.0006
4	4,000	20,400	0.0014	0.0008
5	5,000	25,500	0.0018	0.0009
6	6,000	30,600	0.0022	0.0011
7	7,000	35,700	0.0025	0.0012
8	8,000	40,800	0.0029	0.0014
9	9,000	45,900	0.0033	0.0017
10	9,500	48,000	0.0035	0.0018
11	10,000	50,500	0.0037	0.0019
12	10,500	53,000	0.0042	0.0021
13	11,000	56,000	0.0100	0.005
14	12,000	61,100	0.0160	0.008
15	13,000	66,300	0.0260	0.013
16	14,000	72,000	0.0620	0.031
17	11,800	60,000	0.4000	0.250
Fracture				

Yield strength at 0.2% offset: 52,750 psi
Proportional limit: 45,500 psi
Ultimate strength: 68,000 psi
Rupture strength: 60,000 psi

Figure C–4
Example of tensile test data.

You should also make recommendations and draw conclusions concerning the material's fitness for use. Here the possibilities are endless. You could study the effects of heat treating, annealing, tempering, different shapes, or alloy contents; test conditions, such as temperature and humidity, could be varied; and different materials (ferrous and nonferrous metals, plastics, composites, wood, and so on) could be tried and compared. The limits of the test are set by the tester's creativity and imagination.

A typical test reporting procedure might include the following:

Title.   Tensile Testing of Metals: 4140 Steel

Introduction.   The following is given as an example of how to write a test report. Use it as an example or model in preparing your own report.

Material:	1020 cold-drawn steel	**Date:**		
Original diameter:	0.505 in.		**Gage length:**	2.0000 in.

Reading	Load (lb)	Stress (lb/in.²)	Δ length (in)	Strain (in./in.)
1	1,000	5,100	0.0004	0.0002
2	2,000	10,200	0.0008	0.0004
3	3,000	15,300	0.0010	0.0005
4	4,000	20,400	0.0014	0.0007
5	5,000	25,500	0.0017	0.0009
6	6,000	30,600	0.0021	0.0010
7	7,000	35,700	0.0024	0.0012
8	8,000	40,800	0.0028	0.0014
9	9,000	45,900	0.0033	0.0016
10	9,500	47,500	0.0035	0.0017
11	10,000	50,100	0.0037	0.0018
12	10,500	52,500	0.0080	0.0040
13	11,000	56,000	0.0110	0.0055
14	12,000	61,100	0.0200	0.0100
15	13,000	66,300	0.0560	0.0280
16	14,000	70,000	0.1900	0.0950
17	11,400	57,000	0.4530	0.2270
Fracture				

Final diameter: 0.401 in.
Final length: 2.3000 in.

Figure C–5
Sample test data.

Object. To determine the strength and several elastic and nonelastic properties of a specimen; to observe the behavior of the material under a tensile load; and to study the fracture developed at rupture.

Equipment.
  Universal testing machine
  Extensometer
  Micrometer or vernier caliper
  Dividers
  Safety glasses and shields

Procedure.
1. The original cross-sectional area of the specimen was measured and recorded.
2. A 2-in. gage length was marked on the specimen.
3. The proper grips were selected and mounted on the testing machine.

Figure C–6
Sample test graph.

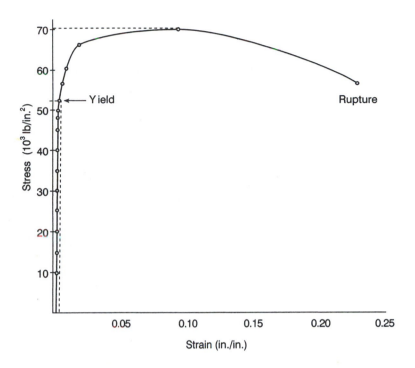

Figure C–7
Calculations based on test data.

From the test graph:

Yield strength	$= 53{,}000 \text{ lb/in}^2$
Ultimate tensile strength	$= 70{,}500 \text{ lb/in}^2$
Rupture strength	$= 60{,}000 \text{ lb/in}^2$
Proportional limit	$= 50{,}000 \text{ lb/in}^2$

Calculations:

$$\% \text{ reduction in area} = \frac{A_o - A_f}{A_o} \times 100 = 37\%$$

$$\% \text{ elongation} = \frac{l_f - l_o}{l_o} \times 100 = 15\%$$

$$E = \frac{\sigma}{\varepsilon} = \frac{50{,}100}{0.0018} = 2.8 \times 10^7 \text{ lb/in}^2$$

(modulus of elasticity)

4. The crosshead of the testing machine was then adjusted to the appropriate length for mounting the specimen.
5. The load pointer was set to zero.
6. The specimen was then properly mounted in the grips, making sure that it was seated properly. The crosshead was moved until all slack in the grips and specimen was eliminated.
7. An extensometer was mounted on the testing machine to measure the crosshead movement. Care was taken so that damage would not occur to the extensometer.
8. Tensile loading was applied to the specimen in a slow but steady manner, approximately 5000 lb (2270 kg) per minute.
9. Once the yield point of the material was observed and recorded, the extensometer was removed, and the displacement was then recorded using the dividers.
10. Loading continued until the specimen ruptured. The ultimate strength and the rupture strength were recorded.
11. Measurements of the final gage length and the final cross-sectional area were taken and recorded.

Data. Data were recorded concerning the applied load and extension at appropriate intervals. Construct a plot of the stress-strain curve for this material and compare to those given in Chapter 14. From the test data and the stress-strain curve, the following features were determined:

1. Elastic limit
2. Proportional limit
3. Yield point
4. Yield strength at 0.2% offset
5. Ultimate tensile strength
6. Modulus of elasticity
7. Percent elongation
8. Percent reduction in area
9. Nature and type of fracture

Results. Questions that arose during the test concerning the procedure and test results included the following:

1. Did the test specimen used conform to ASTM standards in shape and dimensions?
2. Was the test carried out in accordance with ASTM standards?
3. Would the material be satisfactory and in conformity with ASTM specification A7? In what way(s) did it fail to meet the standard?
4. Why was it necessary to state the gage length when reporting the percentage of elongation?
5. Discuss the variation in the percentage elongation with the size and shape of the bar.
6. How are the properties of steel, such as yield point, tensile strength, ductility, modulus of elasticity, and work to rupture, affected by changes in carbon content?
7. How could the work required to rupture this specimen be determined?
8. What errors might have been introduced if the axis of the extensometer and that of the specimen did not coincide?
9. Which wedge grips were used? Why?
10. Distinguish between the proportional limit and the elastic limit for this material. Which is the more important indicator of a material's mechanical behavior?

11. Distinguish between yield point and yield strength on a stress-strain curve. Which gives the more accurate indication of a material's fitness for a particular tensile application?
12. What are the advantages of a stress-strain diagram over a load-elongation diagram for showing the results of a test?

Answers to these and other questions and problems that occurred during the test or occurred to the person(s) conducting the test would appear in the Results section. It would be helpful to include a graph of the data and results in the Data section for analysis in the Results section.

C.3 TENSILE TESTING OF WIRE AND CABLE

C.3.a Introduction

The purpose of this test is to determine the strength and efficiency of a cable and the strength and ductility of the wire from which it is made. Specifically, the items to be found are the following:

1. Details of the wire and cable, including its diameter, number of strands, number and diameter of the wires in the strands, type of center (hemp or steel), lay (regular or long, right or left), and pitch of wires and strands
2. Tensile strength and percentage elongation of each wire in one strand
3. Tensile strength and efficiency of cable

C.3.b Required Equipment

Universal testing machine
Extensometer
Micrometer or vernier caliper
Dividers
Safety glasses and shields
Data sheet

C.3.c Procedure

1. Obtain a piece of cable between 0.25 and 0.5 in. (65 and 125 mm) in diameter and approximately 12 to 18 in. (30 to 45 cm) long.
2. Measure the pitch of the wires and observe the lay. Measure the diameter of the cable, which is defined as the diameter of the circle that will just enclose it.
3. Secure the heads of the testing machine and mount the specimen in the machine. Apply load at a slow, constant rate.
4. Record the load when the first wire breaks, the maximum load, and the character and location of the fracture.
5. Continue to pull the specimen apart and then measure the nominal size of the center. Note the kind of material from which it was made.
6. Next, take one of the strands from the cable at least 12 in. (30 cm) long and mount it in the machine.
7. Apply a load that will stress the wire to no more than 25% of its ultimate strength. Measure the average diameter of the wire.
8. Increase the load to bring the stress to 50% of the breaking strength. Attach the strainometer and set the dial to zero.
9. Apply the load continuously at a speed of no more than 1 in. (2.5 cm) per minute. Record the maximum load and the elongation observed at the point of rupture. The break must occur within the gage length, no closer than 1/4 in. (65 mm) from either jaw, or a retest must be made.
10. Test each wire in the strand following the procedure in steps 6 through 9.

C.3.d Data

1. Tabulate the wire test results, showing the diameter, maximum load, tensile strength, and percentage of elongation for each wire. Compute the average tensile strength and percentage of elongation of the wires tested.
2. Compute the efficiency of the cable and include all properties (diameter, maximum load, tensile strength, and percentage of elongation).

C.3.e Results

1. Were the strength and elongation characteristics of the wires uniform?
2. In what respects, if any, did the rope and wire tests differ from ASTM standards?
3. Why is the efficiency of a cable always less than 100%?
4. Name the principal factors that tend to shorten the life of a cable in service.
5. What are the possible causes of single-strand breaks in a wire cable tensile test?

C.4 TENSILE TESTING OF CORDAGE

C.4.a Introduction

The purpose of this test is to study and observe the tensile strength of cord and cordlike materials, such as rope, twine, and threads. In addition, the purpose is to learn one procedure for performing acceptable tensile tests on cordage and to interpret the results of the tests.

The breaking strength of cordage is used to determine a safe working load, as is common with other materials. However, cord materials vary widely in their load-bearing capacity, and tests must be made to verify that a particular sample has the required strength for an intended application. In addition, both natural and manufactured cord materials show deterioration with age, particularly when exposed to the elements. It is, therefore, necessary to determine the breaking strength of cordage to ensure its adequacy for service.

Also, knots have an effect on cordage strength. Knots tend to lower the breaking strength. If we consider the breaking strength of cordage to be 100%, the strength with a knot is as follows:

Eye splice	90%
Short splice	85%
Half hitch	65%
Clove hitch	60%
Bowline	60%
Sheet bend	55%
Square knot	45%

C.4.b Required Equipment

Universal testing machine
Extensometer
Cordage grips or suitable alternative
Safety glasses and shields
Data sheet

C.4.c Procedure

1. Record the appropriate data concerning diameter, material, knots, and design (single strand, multiple strands, type of center, and so on) for the cordage to be tested.
2. Select the proper grips and mount the grips on the testing machine.
3. Set the machine to the appropriate range and zero the machine.
4. Mount the specimen in the machine. If a round bar is to be used for the cordage, the cordage must be attached with a clove hitch.
5. Mount the extensometer.
6. Apply the load in a slow, steady manner. Note any fraying or slippage while applying the load.
7. Continue loading until failure. Record the breaking load.

C.4.d Data

1. Plot the load-versus-deflection curve.
2. Calculate the true breaking load by adjusting the recorded value by any factors, such as knots, in the test procedure.

3. Calculate the working load for the specimen and compare the result with the expected value. The working load is typically rated at 20% of the breaking load.

C.4.e Results

1. Were the strength and elongation characteristics of the cords uniform?
2. In what respects, if any, did the cordage tests differ from ASTM standards?
3. Name the principal factors that tend to cause a short life of a cord in service.
4. What are the possible causes of single-strand breaks in a tensile test of a multiple-strand cord?

C.5 TENSILE TESTING OF WELDED SPECIMENS

C.5.a Introduction

The purpose of this test is to determine the appropriateness and functionality of the weld procedure used. This test is not used to qualify welders; rather, it is used to qualify the procedure used to join two materials. The weld should not fail if the weld metal has a higher tensile strength than the base metal. If the weld fails, check to ensure that proper penetration occurred and that there weren't any slag inclusions, weld porosity, or other unexpected factors that affected the test results.

C.5.b Required Equipment

    Universal testing machine
    Test specimen
    Data sheet
    Safety glasses and shields

C.5.c Procedure

1. Obtain two pieces of $3/8 \times 1 \times 4$ in. ($10 \times 25 \times 100$ mm) hot-rolled steel. Bevel each piece $60°$ along one 1-in. side. Weld the two pieces together along the beveled sides. Chip and run a bead down the back side. After the piece has cooled, grind the bead flush with the surface of the test pieces.
2. Measure and record the dimensions of the test piece. Measure and record the cross-sectional area of the test piece. Place appropriate gage marks on the specimen.
3. Set up the testing machine. Select the proper grippers. Set the machine to zero. Place the test specimen in the machine, making sure that the specimen is properly aligned in the machine.
4. Apply a small load to the specimen to seat it in the grippers. Rezero the machine.
5. Prepare your data sheet to record data. The data will be load readings only.
6. Load the piece until failure, and record the maximum load applied.
7. Remove the piece from the machine.

C.5.d Data

1. Plot the load-versus-deflection curve.
2. Calculate the breaking strength of the specimen used in this test procedure.

C.5.e Results

1. Calculate the ultimate tensile strength for the specimen. Examine the failed specimen closely to determine where it failed. Pay particular attention to the point of failure relative to the weld.
2. Calculate the deformation and percent elongation.
3. Determine the percent reduction in area.
4. Further testing can be made using different materials, different filler materials, different weld techniques, and other factors.
5. What would be the significance of a fracture through the welded area?

C.6 TENSILE TESTING OF BRAZED AND SOLDERED SPECIMENS

C.6.a Introduction

Brazing and soldering are similar processes, differing in the temperatures and filler materials used. Both techniques are used to attach two base metals together without dissolving the surface of the base metal. This can be accomplished using a variety of processes, including an oxyacetylene torch, a carbon-arc apparatus, an air-MAPP gas torch, electric induction, or any other similar method used to provide heat for melting the filler metal.

Typical brazing alloys for ferrous metals include copper and zinc. Silver solders range from a few percent up to almost pure silver. Typically, for ferrous metals, a 45% silver solder is used. Tensile strengths for these materials range from 50,000 to 100,000 lb/in.^{2} (345 to 690 MPa).

The butt joint is used for this test. Specimens are two 0.5-in. diameter↔2-in. (13-mm diameter↔50-mm) lengths of cold-drawn rod. The ends to be joined are machined smooth and square. Spacing between the base metals ranges from 0.002 to 0.005 in. (0.05 to 0.13 mm). This distance is critical to the strength of the joint and varies according to the filler material used.

C.6.b Required Equipment

Universal testing machine
Test specimen
Data sheet
Safety glasses and shields

C.6.c Procedure

1. Prepare an appropriate test specimen as shown in Figure C–8. Measure and record the dimensions of the test piece. Measure and record the cross-sectional area of the test piece.
2. Set up the testing machine. Select the proper grippers. Set the machine to zero. Place the test specimen in the machine, making sure that the specimen is properly aligned in the machine.
3. Apply a small load to the specimen to seat it in the grippers. Rezero the machine.
4. Prepare your data sheet to record data. The data will include load readings only.
5. Load the piece slowly until failure and record the maximum load applied.
6. Remove the test piece from the machine. Inspect it, looking for reasons for failure.

C.6.d Data

1. Plot the load-versus-deflection curve.
2. Calculate the breaking strength of the specimen used in this test procedure.

Figure C–8
Brazed or soldered tensile specimen.

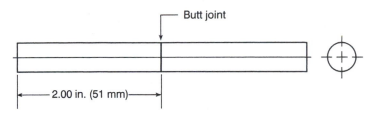

C.6.e Results

1. Calculate the ultimate tensile strength for the specimen. Examine the failed specimen closely to determine where it failed. Pay particular attention to the point of failure relative to the joint.
2. Determine the type of failure that occurred.
3. Estimate the percentage of the total cross-sectional area that failed due to tin, or coating with filler metal.
4. Estimate the percentage of the total area that failed due to the filler metal pulling away from the base metal.
5. Estimate the percentage of the total area that failed due to the filler metal breaking, leaving both base metals tinned.
6. Additional tests can be made using different filler metals and different spacing between the joint faces.
7. Tubing can also be tested where the ends are either flattened or plugged.

C.7 TENSILE TESTING OF ADHESIVES

C.7.a Introduction

The purpose of this test is to determine the tensile strength, or holding ability, of different types of adhesives. Many adhesives have remarkable strength in shear applications, whereas others offer good tensile resistance. Two types of adhesives are of particular interest: epoxies and cyanoacrylates (super glues). A special fixture is required for this test (Figure C–9).

Adhesion has two parts: one mechanical, the other chemical. The mechanical component of adhesion takes place when an adhesive flows into the surface topography of the base material. It fills the hollows and voids in the material to be joined and then "freezes," or solidifies, there. This action makes a mechanical lock between the adhesive and the base material.

The chemical component involves three molecular forces that act between the adhesive and the base material. Adhesion requires that the two base materials be in reasonably intimate contact. Internal forces within the base material tend to attract each other (Force 1). The function of the adhesive is to wet both surfaces and eliminate all air and voids between the two surfaces to be joined. This happens when (Force 2) forces between the adhesive molecules and the base material's molecules are greater than (Force 3) the forces between the adhesive molecules. Ideally, all of the forces should be equal (within the adhesive, between the adhesive and the base material, and within the base material).

C.7.b Required Equipment

Universal testing machine
Testing fixture
Test specimen
Data sheet
Safety glasses and shields

C.7.c Procedure

1. Measure and record the dimensions of the test piece. Measure and record the cross-sectional area of the test piece.

Figure C–9
Adhesive testing fixture.

2. Set up the testing machine. Select the proper grippers. Set the machine to zero. Place the test specimen in the machine, making sure that the specimen is properly aligned in the machine.
3. Apply a small load to the specimen to seat it in the grippers. Rezero the machine.
4. Prepare your data sheet to record data. The data will include only load readings.
5. Load the piece slowly until failure and record the maximum load applied.
6. Remove the test piece from the machine. Inspect it, looking for reasons for failure.

C.7.d Data

1. Plot the load-versus-deflection curve.
2. Calculate the breaking strength of the specimen used in this test procedure.

C.7.e Results

1. Calculate the ultimate tensile strength for the specimen. Examine the failed specimen closely to determine where it failed. Pay particular attention to the point of failure relative to the joint.
2. Determine the type of failure that occurred.
3. Estimate the percentage of the total cross-sectional area that was not coated with the adhesive.
4. Estimate the percentage of the total area that failed due to the adhesive pulling away from the base metal.
5. Estimate the percentage of the total area that failed due to the adhesive breaking, leaving both base metals tinned.
6. Additional tests can be made using different adhesives and different spacing between the joint faces.
7. Further testing can be done, altering the temperature at which the test is conducted, applying adhesive to just one surface instead of both surfaces, applying different epoxy mixes and proportions, or changing the surface conditions of the joint, etchants, and so on.

C.8 COMPRESSION TESTING

C.8.a Introduction

The compression test of a material, in theory, is the opposite of the tensile test with respect to direction and sense of the applied stress. The following are some limitations to the compression test:

1. It is difficult to apply a truly concentric or axial load.
2. This type of loading is relatively unstable as contrasted with tensile testing (due to buckling, column action, and so on).
3. Friction between the heads of the testing machine or bearing plates and the ends of the specimen due to lateral expansion of the specimen may affect the results of the test.
4. A relatively larger cross-sectional area of the compression test specimen is needed to obtain a proper degree of stability of the piece. This larger piece results in the need for a larger-capacity machine; otherwise, specimens are so small and, therefore, so short that it is difficult to obtain strain measurements of suitable precision.
5. Specimens are kept relatively short to reduce the column action of structural members and, thus, to test only the simple compression characteristics of the specimen.

C.8.b Procedure

1. Obtain a material specimen approximately 1.5 times as long as the diameter. The selection of the ratio between length and diameter is a compromise between several undesirable conditions. As the length of the specimen is increased, there is a tendency toward bending of the piece, with subsequent nonuniform distribution of stress. A height-diameter ratio of 10 is suggested as a practical upper limit. As the length of the specimen is decreased, the effect of the frictional restraint at the ends becomes relatively important. Also, for lengths less than about 1.5 times the diameter, the diagonal planes along which failure would take place in a longer specimen intersect the base, with the result that the apparent strength is increased.
2. Dimensions vary according to the material being tested. Wood is tested perpendicular to the grain, and a $2 \times 2 \times 8$-in. specimen is used. Concrete compression samples are typically 6 in. in diameter by 12 in. (15 cm in. diameter by 30 cm) in height, but a 3-in. diameter $\times$ 6-in. (8-cm diameter $\times$ 15-cm) specimen can be used.
3. In commercial tests, the only property ordinarily determined is the compressive strength. In brittle materials where fracture occurs, the ultimate strength can be determined. For materials where there is no fracture or other phenomenon to mark ultimate strength, arbitrary limits or deformation are taken as criteria of strength.
4. In making stress-strain determinations, three markings 120° apart are used and averaged.
5. Observations include identification, dimensions, critical loads, compressometer readings (if available), and type of failure; sketches are made in a form appropriate for the test and extent of the data taken.
6. The speed of testing is important in compression testing. The loading rate for a particular test can be found in the ASTM standards. However, a good general speed is approximately 2000 lb/in.² (13 MPa) per min.

C.8.c Data

Data recorded during a compression test usually involve load and deflection. This information can then be used to calculate other figures of merit, and a stress-strain curve can be plotted. Figure C–10 shows typical data taken during compression testing.

Date:

Material: 0.375 × 0.75 in. round aluminum specimen

Original area: 0.4418 in.²

Reading	Load (lb)	Stress (lb/in.²)	Δ length (in.)	Strain (in./in.)
1	500	1,132	0.0015	0.0020
2	1,000	2,264	0.0025	0.0033
3	1,500	3,395	0.0035	0.0047
4	2,000	4,527	0.0050	0.0067
5	2,500	5,659	0.0090	0.0120
6	3,000	6,791	0.0150	0.0200
7	3,500	7,922	0.0270	0.0360
8	4,000	9,054	0.0620	0.0827
9	4,250	9,620	0.0720	0.0960
Rupture	4,300	9,733		

Figure C–10
Sample compression data.

C.8.d Results

Discuss the loading rate and its effects on the test results. Explain the effects that nonparallel load-bearing surfaces would have on results. Look for evidence of slippage and defects, and describe the type and features of the failure. Figure C–11 shows typical compression testing results.

Figure C–11
Sample compression testing results.

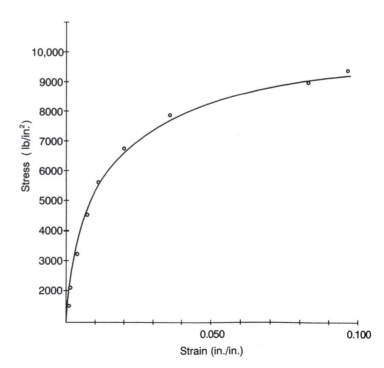

C.9 COMPRESSION TESTING OF WOOD PARALLEL TO THE GRAIN

C.9.a Introduction

The purpose of this test is to study the action of wood under compressive loading parallel to the grain. Because wood is an organic material composed primarily of long hollow cells, or fibers, that are aligned vertically along the trunk and along the limbs of the parent tree, wood is easily pulled apart or crushed by forces perpendicular to the fibers' direction of growth. However, compressive forces encounter a higher resistance.

C.9.b Required Equipment

Universal testing machine
Compressometer
Micrometer or vernier caliper
Dividers
Test specimen
Drying oven
Weighing scales
Data sheet

C.9.c Procedure

1. Obtain a $2 \times 2 \times 8$-in. ($50 \times 50 \times 200$-mm) piece of clear wood. If the ends are not flat and at right angles to the axis of the specimen, reject the specimen and obtain others. Note any defects in the specimens.
2. Measure the cross section and length of the specimen to the nearest 0.01 in. (0.25 mm) and weigh to the nearest ounce (gram).
3. Determine the average number of annual growth rings per inch, the percentage of summerwood, and the percentage of sapwood.
4. Determine the gage length and multiplication ratio of the compressometer. Determine the strain corresponding to the least reading of the dial.
5. Mount the compressometer on the specimen and mount the specimen on the universal testing machine.
6. Apply the load continuously at a speed of about 0.020 in. (0.5 mm) per min. until failure. Read the compressometer dial and the load at intervals of about 2000 lb (900 kg). Draw a sketch revealing, in perspective, the grain and manner of failure.
7. Cut a moisture sample of about 1 in. in length from the specimen after failure. Remove all splinters and weigh to the nearest 0.1 g. Place in a drying oven controlled at 103°C. After the moisture specimen has dried to a constant weight, which may take 2 to 4 days, weigh it again.

C.9.d Data

1. Construct a stress-strain diagram and mark the proportional limit, ultimate strength, and the yield strength on the curve.

2. Calculate the moisture content as a percentage of the oven-dry weight. Calculate the following:
   a. Approximate specific gravity, wet and dry
   b. Annual growth rings per inch
   c. Percentage of summerwood
   d. Percentage of sapwood
   e. Modulus of elasticity
   f. Modulus of resilience

C.9.e Results

1. Discuss what effects, if any, the following would have on the test results:
   a. Moisture content
   b. Approximate specific gravity, wet and dry
   c. Annual growth rings per inch
   d. Percentage of summerwood
   e. Percentage of sapwood
2. What is the effect of moisture on the strength of wood in compression?
3. What is the relation between specific gravity and strength?
4. What effect does time have on the loading of the wood?
5. What is the effect of the percentage of summerwood on strength?

C.10 COMPRESSION TESTING OF CONCRETE

C.10.a Introduction

The purpose of this test is to study the behavior of concrete under compressive loading and to determine the following physical and mechanical properties:

1. Proportional limit
2. Yield strength at 0.01% offset
3. Compressive strength
4. Initial tangent modulus of elasticity
5. Secant moduli of elasticity at stresses of 500, 1000, 1500, and 2000 lb/in.^{2}
6. Weight per cubic foot

C.10.b Required Equipment

Universal testing machine
Hardened steel plates
Tamping rod with a bullet end
Compressometer or dial indicator
Concrete compression specimen mold [6 × 12-in. (15 × 30-cm) or 3 × 6-in. (8 × 15-cm) molds]
Concrete mix
Weighing scale
Safety glasses and shields
Data sheet

C.10.c Procedure

1. Prepare enough mixed concrete to fill the required molds.
    a. Fabricate the molds.
    b. Tape the seams together.
    c. Mix a batch of 1:2:3 portland cement, sand, gravel mix by weight.
    d. Add water until a stiff, dry mixture is gained.
2. Fill the molds approximately one-third full.
3. Rod the concrete with 25 strokes using the tamping rod. The strokes should be evenly distributed throughout the mold.
4. Fill the molds up to two-thirds full.
5. Repeat step 3.
6. Completely fill the mold and heap concrete above the top of the mold.
7. Repeat step 3.
8. Screed the top of the mold to remove the excess concrete.
9. Repeat steps 2 through 8 for all molds.
10. Set the mold aside for 24 hours and allow the concrete to harden. Remove the concrete from the mold.
11. Allow the specimens to cure for a total of 28 days. If desired, curing may be done under water to ensure the proper hydration of the concrete.

Use the following procedure during testing:

1. Measure and record the cross-sectional area of the specimen.
2. Install the hardened plates on the testing machine. (These plates protect the crossheads of the machine.)
3. Place the test specimen between the hardened plates with the specimen oriented so that the load is applied equally throughout the specimen.
4. Adjust the position of the crossheads for minimum clearance between the upper hardened plate and the test fixture and set the pointer to zero.
5. Mount the compressometer on the specimen or mount the dial indicator on the fixed column of the testing machine to record the crosshead deflection.
6. Apply the load to the specimen in a slow but steady manner. Record the force and deflection.
7. Observe carefully and record the yield strength of the material. Remove the compressometer at the first sign of fracture.
8. Continue loading the specimen until it breaks. Record maximum load.

C.10.d Data

1. Plot the load-versus-deflection curve.
2. Calculate and record the following:
   a. Compressive strength at failure
   b. Compressive yield point
   c. Modulus of elasticity
   d. Ultimate compressive strength

C.10.e Results

1. Describe briefly, in your own words, the significance of the test.
2. What factors affect the development of strength of concrete?
3. How is the compressive strength affected by moisture content at time of test?
4. How is the modulus of elasticity affected by age and by moisture content at test time?
5. Why is the compression test the one most frequently made for concrete?
6. What other tests would be applicable to concrete?
7. What are the effects of hydration?
8. Several tests might be conducted on specimens with different curing periods: hydrated and nonhydrated specimens or different materials ratios (different from the 1:2:3 standard). The slump test can be used to determine the consistency of the batch. The consistency of the batch can be varied, and the effects of varying the consistency can be studied.

C.11 COMPRESSION TESTING OF BRITTLE METALS

C.11.a Introduction

The purpose of this test is to determine the compressive strength of a brittle material. There are a variety of brittle materials that could be studied. Two in particular are of interest because of their availability and characteristics under test conditions: cast aluminum and cast iron.

C.11.b Required Equipment

Universal testing machine
Compression test blocks
Micrometer or calipers
0.375 × 1.125-in. (10 × 30-mm) specimen
Safety glasses and shields
Data sheet

C.11.c Procedure

1. Measure and record the cross-sectional area of the specimen.
2. Install the hardened plates on the testing machine. (These plates protect the crossheads of the machine.)
3. Place the test specimen between the hardened plates, with the specimen oriented so that the load is applied equally throughout the specimen.
4. Adjust the position of the crossheads for minimum clearance between the upper hardened plate and the test fixture, and set the pointer to zero.
5. Mount the compressometer on the specimen or mount the dial indicator on the fixed column of the testing machine to record the crosshead deflection.
6. Apply the load to the specimen in a slow but steady manner. Record the force and deflection.
7. Observe carefully and record the yield strength of the material. Remove the compressometer at the first sign of fracture.
8. Continue loading the specimen until it breaks. Record the maximum load.

C.11.d Data

1. Plot the load-versus-deflection curve.
2. Describe the type and nature of the material's failure.
3. Calculate and record the following:
   a. Compressive strength at failure
   b. Compressive yield point
   c. Modulus of elasticity
   d. Ultimate compressive strength

C.11.e Results

1. What significance do the results of this test have?
2. What other tests would you conduct on brittle materials to determine compression-related properties?
3. What factors should you consider when designing a compression test for brittle metals?

C.12 COMPRESSION TESTING OF DUCTILE METALS

C.12.a Introduction

The purpose of this test is to determine the compressive strength of a ductile material. Some examples of ductile metals are hot- and cold-drawn steel, brass, copper, and aluminum. These materials exhibit an elastic and a plastic range. In compressive testing, ductile materials exhibit a "bulging" or "barreling" through the cross section after yielding.

One method for determining the yield point of a ductile material under a compressive load is to apply a compressive load to the specimen and then relax it, checking for a permanent set. A second method involves carefully watching the compressometer to determine where further strain occurs without a coincident increase in load or stress.

C.12.b Required Equipment

Universal testing machine
Compression test blocks
Micrometer or calipers
0.375×1.125-in. (10×30-mm) specimen
Safety glasses and shields
Data sheet

C.12.c Procedure

1. Measure and record the cross-sectional area of the specimen.
2. Install the hardened plates on the testing machine. (These plates protect the crossheads of the machine.)
3. Place the test specimen between the hardened plates, with the specimen oriented so that the load is applied equally throughout the specimen.
4. Adjust the position of the crossheads for minimum clearance between the upper hardened plate and the test fixture, and set the pointer to zero.
5. Mount the compressometer on the specimen or mount the dial indicator on the fixed column of the testing machine to record the crosshead deflection.
6. Apply the load to the specimen in a slow but steady manner. Record the force and deflection.
7. Observe carefully and record the yield strength of the material.
8. Continue loading the specimen into the plastic range and observe the barreling that takes place in the specimen.

C.12.d Data

1. Plot the load-versus-deflection curve.
2. Describe the type and nature of the material's failure (yielding).
3. Calculate and record the following:
   a. Compressive strength at the yield point
   b. Modulus of elasticity
   c. Ultimate compressive strength

C.12.e Results

1. What significance do the results of this test have?
2. What other tests would you conduct on ductile materials to determine compression-related properties?
3. What factors should you consider when designing a compression test for ductile metals?

C.13 SHEAR TESTING

C.13.a Introduction

A shearing stress is one that acts parallel to a plane, as distinguished from tensile or compressive stresses, which act normal to a plane. If a specimen is subjected to a tensile or compressive stress acting in only one direction, the shear stresses at 45° are one-half the magnitude of the applied direct stress. In general, the maximum shear stresses are equal to one-half the difference between the maximum and minimum principal stresses and act on planes inclined at 45° with these stresses. The strain that accompanies shear arises from the effort of thin parallel slices of a body to slide, one over the other. The types of shear tests in common use are the direct shear test and the torsion test.

A shear test is conducted by placing a suitable test specimen in the testing machine so that bending stresses are minimized across the plane along which the shearing load is applied (direct shear test). Although this method suffices for determining the shear strength of rivets, crankpins, wooden blocks, and so on, owing to bending or the friction between parts or both, it gives only an approximation of the correct shearing strength.

The punching shear test is also a form of direct shear test; its use is restricted to tests of flat stock, principally metals. When a metal plate is punched, the punched area is removed by a slicing motion within a narrow ring of material adjacent to the cutting edge of the punch. The results of punching shear tests are unsatisfactory as measures of shear strength and should be considered as giving simply a representation of overall load to cause punching.

For a more accurate representation, the torsion test is made using either solid or hollow specimens of circular section. In such a test, the specimen can be of such length that a strainometer can be attached to assist in determinations of proportional limit and yield strength in shear, shearing resilience, and stiffness, stiffness being the angle of twist and the applied torque. The ductility of the material is determined from the amount of twist to rupture, the toughness is represented by the amount of twist and the strength, and uniformity is indicated by the spacing, distribution, and appearance of the lines of twist. For accurate determination of the elastic strength, a tubular specimen should be used. Anyone who has ever twisted a bolt off trying to get "one more turn" has demonstrated the torsion shear test to failure.

C.13.b Procedure

C.13.b.1 Direct Shear Test. A bar is usually sheared in some device that clamps a portion of the specimen while the remaining portion is subjected to load by means of suitable dies. Specimens for ductile materials are cut from standard rods [0.25 in., 0.313 in., 0.375 in., and so on (6.4 mm, 8 mm, 9.5 mm, and so on)] with a minimum length of 1.5 in.

In direct shear testing, the testing device should hold the specimen firmly and preserve good alignment, and the load should be applied evenly at right angles to the axis of the piece. The speed for applying the load should not exceed 0.05 in. (1.3 mm) per minute.

In the direct shear test, the one critical value that can be observed is the maximum load applied. The area can be calculated, and the load/area can be determined. The shape and texture of the fractured surface should be reported.

C.13.b.2 Torsion Test. The criteria for selecting a torsion test specimen are as follows:

1. The specimen should be of such size as to permit the desired strain measurements to be made with suitable accuracy.

2. It should be large enough to eliminate the effect of stresses due to gripping the ends from that portion of the specimen on which measurements are made.

Actual specimen sizes are commonly chosen to suit both the size and type of testing machine available as well as the product to be tested.

C.13.c Data

Figure C–12 shows sample shear testing data for torsional and direct shear testing.

(a) Round ductile shear test specimen

(b) **Direct Shear Testing of Metallic and Plastic Materials**

Material:
Original dimensions:
Maximum load (lb):

Shear strength (psi):

Torsional Shear Testing of Metallic Solid and Tube Materials

Material:
Original dimensions (ID and OD):

Reading	**Applied Torque (in.-lb)**	**Detrusion (radians)**

Figure C–12
Sample direct and torsional shear testing data.

Direct Shear Test of 1 in.–diameter Aluminum pin

Material: Annealed aluminum round rod   Date:_____
Original dimensions: 1 × 6 in.
Maximum load (lb):   9,000
Shear strength (lb/in.²):   11,459

Torsional shear test of 0.5 in.–diameter solid bar

Material: Solid bar   Date:_____
Length: 10 in.
Original OD: 0.5 in.

Reading	Torque (in.-lb)	Detrusion angle (°)	Stress lb/in.²	Strain (°)
1	100	2	4,075	0.0009
2	200	4	8,150	0.0017
3	300	6	12,200	0.0026
4	400	8	16,300	0.0035
5	500	10	20,400	0.0044
6	600	12	24,400	0.0052
7	625	15	25,450	0.0065
8	650	18	26,500	0.0080
9	675	25	27,500	0.0108
10	685	Rupture	27,800	

A torsion stress-strain curve can be constructed from these data.

Figure C–13
Sample shear testing data.

C.13.d Results

Figure C–13 is a sample of the results typically obtained from shear testing.

C.14 DIRECT SHEAR TESTING OF METALS

C.14.a Introduction

The purpose of this test is to study and observe the direct shear test, to develop skills in performing direct shear tests, and to interpret the results of direct shear tests.

C.14.b Required Equipment

Universal testing machine
Double-shear test fixture
Test specimen
Hardened plates
Micrometer or calipers
Dial indicator
Data sheet

C.14.c Procedure

1. Measure and record the original cross-sectional area of the specimen.
2. Install the hardened plates on the testing machine.
3. Insert the test specimen through all three holes of the double-shear test fixture.
4. Place the test fixture with the specimen on the lower hardened plate on the testing machine.
5. Mark the position of the punch in regard to the holder.
6. Adjust the position of the crossheads for minimum clearance between the upper hardened plate and the test fixture, and set the load pointer to zero.
7. Mount the dial indicator on the fixed column of the testing machine in such a position as to record crosshead deflection.
8. Apply the load to the specimen in a slow but steady manner. Record the force and deflection.
9. Carefully observe and record the yield strength of the material. Continue reading the displacement.
10. Continue loading the specimen until it breaks. Record the maximum load.
11. Continue applying the load to bring the punch into the holder so that the sheared slug passes into the relieved area in the lower part of the holder.

C.14.d Data

1. Plot the load-versus-deflection curve.
2. Calculate and plot the stress-versus-strain curve.
3. Calculate and record the following:
    a. Proportional limit in shear
    b. Yield point in shear
    c. Yield strength in shear
    d. Shear strength
    e. Shear modulus of elasticity

C.14.e Results

1. What factors affect a material's shear strength?
2. How would you use shear test results in materials design?
3. Design a different shear test than the one used here. Be practical and economical.
4. Illustrate the type and characteristics of the material's failure.
5. A variety of metals can be tested. The effects of different manufacturing techniques on shear strength can be studied. The effects of annealing, tempering, hardening, and so on, can be tested. Other materials, such as plastics, polymers, and elastomers, can be studied.

C.15 DIRECT SHEAR TESTING OF ADHESIVES (WOOD)

C.15.a Introduction

The purpose of this test is to study and observe the direct shear test, to develop skills in performing direct shear tests, and to interpret the results of direct shear tests of adhesives on wood.

C.15.b Required Equipment

Universal testing machine
Adhesives (several different types)
Hardened plates
Micrometer or calipers
Dial indicator
Test specimens (2): 0.75 × 1.5 × 3 in. (19 × 38 mm × 76 mm); (see Figure C–14)
Data sheet

C.15.c Procedure

1. Install the hardened plates on the testing machine.
2. Place the test fixture (if any) with the specimen on the lower hardened plate on the testing machine.
3. Adjust the position of the crossheads for minimum clearance between the upper hardened plate and the test fixture, and set the load pointer to zero.
4. Apply the load to the specimen in a slow but steady manner. Record the force and deflection.
5. Continue loading the specimen until it breaks. Record the maximum load.

Figure C–14
Shear test specimen for adhesives.

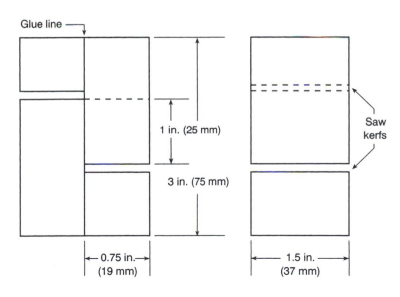

Adhesive test specimen
(wood)

C.15.d Data

1. Plot the load-versus-deflection curve.
2. Calculate the shear strength by dividing the maximum load by the shear area.
3. Carefully examine the failure and determine the type of failure that occurred, especially the percentage of failure due to destruction of the wood fibers and the percentage of failure present in the adhesive.

C.15.e Results

1. Discuss how moisture content would affect the test results.
2. How could the shear strength of wood joints be increased?
3. A variety of glues and woods can be tested. The effects of different gluing techniques on shear strength can be studied. Other materials, such as plastics, polymers, and elastomers, can be studied.

C.16 DIRECT SHEAR TESTING OF ADHESIVES (METALS)

C.16.a Introduction

The purpose of this test is to study and observe the direct shear test, to develop skills in performing direct shear tests, and to interpret the results of direct shear tests of adhesives on metals.

A large variety of adhesives are used to join metals. Epoxies are thermosetting, cross-linking polymers that usually come in two tubes. The contents of the tubes are mixed together in equal quantities and allowed to cure. Cyanoacrylates, or super glues, cure rapidly through polymerization, which is stimulated by traces of moisture found on the material's surface.

Typical surface preparation entails vapor or solvent degreasing, immediately followed by acid etching and baking the material dry. If these processes are not available, the metal surfaces can be prepared by scrubbing them with a non-chlorine-based commercial cleaner. Surface preparation is very important to the test results.

C.16.b Required Equipment

Universal testing machine
Adhesives (several different types)
Tensile grippers
Micrometer or calipers
Extensometer
Test specimens (2): 4 × 1.5 in. (102 × 38 mm); 18 to 24 gage
(see Figure C–15)
Data sheet

C.16.c Procedure

1. Prepare the specimens as shown, following the adhesive manufacturer's directions for adhesive application.
2. Place the test specimen and the test grippers in the testing machine.
3. Apply a small load to the specimen so that it is properly seated in the grippers and crosshead. Rezero the machine.

Figure C–15
Adhesive test specimen.

Adhesive test specimen
(metal)

4. Apply the load to the specimen in a slow but steady manner. Record the force and deflection.
5. Continue loading the specimen until it breaks. Record the maximum load.

C.16.d Data

1. Plot the load-versus-deflection curve.
2. Calculate the shear strength by dividing the maximum load by the shear area.
3. Carefully examine the failure and determine the type of failure that occurred, especially the percentage of failure due to destruction of the base metals and the percentage of failure present in the adhesive.

C.16.e Results

A variety of glues and metals can be tested. The effects of different gluing techniques on shear strength can be studied, as can other materials, such as plastics, polymers, and elastomers.

C.17 FLEXURE, OR BEND, TESTING

C.17.a Introduction

If forces act on a piece of material in such a way that they tend to induce compressive stresses over one part of a cross section of the piece and tensile stresses over the remaining part, the piece is said to be *bending*. Bending may be accompanied by direct stress, transverse shear, or torsional shear. Bending action in beams is often referred to as *flexure*. The term flexure applies to bending tests of beams subject to transverse loading. The deflection of a beam is the displacement of a point on the neutral surface of a beam from its original position under the action of applied loads. Deflection is a measure of the overall stiffness of a given beam and can be considered a function of the stiffness of the material and the proportions of the piece.

C.17.b Procedure

1. The test specimen should be of such shape that it permits use of a definite and known length of span.
2. The areas of contact with the material under test should be such that unduly high stress concentrations do not occur.
3. There should be provision for longitudinal adjustment of the position of the supports so that longitudinal restraint will not be developed as loading progresses.
4. There should be provision for some lateral rotational adjustment to accommodate beams having a slight twist from end to end, so that torsional stresses will not be induced.
5. The arrangement of parts should be stable under load. Many flexure tests are conducted in universal testing machines, with the supports placed on the platen or an extension of it and with the loading block fastened to or placed under the moving head.
6. A load-versus-deflection diagram should be plotted.
7. The speed of applying the load is important and should not produce failure too rapidly. Test specifications for specific materials and specimen characteristics should be referenced.
8. Flexure testing provides a crude estimation of the ductility of the material tested.

Figure C–16
Sample flexure testing data.

Flexure Testing of Metallic Materials

Date:
Material:
Original dimensions:
Maximum load (lb):
Maximum deflection (in.)
Test span (in.):

Reading	Load (lb)	Stress (lb/in.²)	Deflection (in.)

Stress is plotted against deflection rather than strain.

Figure C–17
Sample flexure testing results.

Flexure Testing Results

Date: _____
Material: Aluminum bar
Original dimensions: ¼ × ½ × 9 in.
Max. load (lb): 90
Max. deflection (in.): 0.020
Test span (in.): 6

Reading	Load (lb)	Stress (lb/in.²)	Deflection (in.)
1	25	2,000	0.003
2	35	2,800	0.005
3	45	3,600	0.008
4	55	4,400	0.0100
5	65	5,200	0.0150
6	90	(7,200) Rupture	

C.17.c Data

Sample flexure test data are given in Figure C–16.

C.17.d Results

Sample flexure testing results are given in Figure C–17.

C.18 FLEXURE TESTING OF METALS

C.18.a Introduction

The purpose of this test is to study and observe the flexure testing of metallic materials and to develop skills in performing acceptable flexure tests and interpreting results of a flexure test.

C.18.b Required Equipment

Universal testing machine
Transverse loading fixture
Specimen: cast aluminum $1 \times 2 \times 8$ in. $(25 \times 50 \times 200 \text{ cm})$
Safety glasses and shields
Deflectometer or dial indicator
Data sheet

C.18.c Procedure

1. Measure and record the diameter of the bar at the midsection and record the maximum, minimum, and mean values.
2. Fix the specimen span at an appropriate length.
3. Place the specimen in position with its minimum diameter vertical.
4. Place the deflectometer in place.
5. Apply load at a uniform rate.
6. Record loads and deflections.
7. Record the maximum load, the maximum deflection, and the manner of failure.
8. Record the vertical and horizontal diameters at the fracture.

C.18.d Data

1. Plot the load-versus-deflection curve.
2. Calculate the following:
   a. Beam stiffness
   b. Ultimate bending strength
   c. Fracture strength
   d. Ductility
   e. Modulus of elasticity

C.18.e Results

1. What is the effect of variations in span length on the modulus of rupture and the modulus of elasticity of the flexure bar?
2. Is the modulus of rupture a true representation of the ultimate fiber stress? Explain.
3. Discuss the shape of the load-deflection diagram for different metals. What effect does this have on the computed modulus of elasticity?
4. Does the strength of the flexure-test specimen represent the actual strength of the metal? Explain.
5. List the advantages of the flexure test over tension or compression tests for metals.
6. Different metals, spans, and loading rates can be used to judge the effects of these variables.

C.19 FLEXURE TESTING OF CONCRETE

C.19.a Introduction

The purpose of this test is to determine the mechanical properties of concrete, reinforced and nonreinforced, when subjected to transverse loads. Concrete exhibits a high compressive strength but very low tensile strength. Therefore, a cross section that is subjected to tensile loads will fail very quickly. To adjust for this, reinforcement materials are cast into the concrete to compensate for this inherent weakness. Because tensile forces are at work on the bottom half of the flexural specimen, the reinforcement should be cast in the bottom half of the specimen. The same material can be used as was made in the compression test. Review the compression test if you have a question concerning the batch or process.

C.19.b Required Equipment

Universal testing machine
Concrete mixture
Concrete flexural specimen mold
Tamping rod
Reinforcement rods
Safety glasses and shields
Deflectometer or dial indicator
Data sheet

C.19.c Procedure

1. Prepare enough concrete for all of the required specimens.
2. Fill the mold one-third full.
3. Rod the concrete 30 strokes.
4. Fill mold two-thirds full.
5. Repeat step 3.
6. Completely fill the mold and heap concrete on top of the mold.
7. Repeat step 3.
8. Screed the top of the mold to remove excess concrete.
9. Repeat steps 2 through 8 for all molds.
10. Set the molds aside for 24 hours and allow the concrete to harden. Remove the specimens from molds.
11. Allow the concrete specimens to cure for a total of 28 days. If desired, the concrete may be cured under water to ensure proper hydration.

C.19.c.1 Testing Procedure.
1. Measure and record the cross-sectional area of the specimen.
2. Load the transverse loading fixture into the test machine. Set the appropriate span length.
3. Center the concrete beam onto the test fixture in the appropriate manner. Secure the beam, but do not restrain longitudinal motion of the beam.
4. Load the concrete beam slightly to seat the specimen and fixture. Rezero the machine.
5. Slowly load the specimen until failure. Reinforced specimens should be loaded until cracks develop.

C.19.d Data

1. Plot the load-versus-deflection curve.
2. Calculate and record the following:
   a. Bending stress at failure
   b. Modulus of elasticity
   c. Modulus of rupture

C.19.e Results

1. Tests can be conducted using more than one reinforcing rod and by varying the size of the reinforcing rods.
2. The slump test can be used to determine the consistency of the batch. The consistency can then be varied, and the effects of such variation can be studied.
3. The effects of curing time and hydration can also be analyzed.

C.20 FLEXURE TESTING OF WOOD

C.20.a Introduction

The purpose of this test is to determine the mechanical properties of wood subjected to bending, to observe the behavior of the material under load, and to study its failure. Also included are the following concepts:

1. Proportional limit stress in outer fiber
1. Modulus of rupture
2. Modulus of elasticity
3. Average work to proportional limit, inch-pounds per cubic inch
4. Maximum shearing stress
5. Average total work to ultimate load, inch-pounds per cubic inch
6. Type of failure

C.20.b Required Equipment

Universal testing machine
Transverse testing fixture
Wood specimen: $1.5 \times 3.5 \times 24$ in. ($38 \times 89 \times 610$ mm)
Safety glasses and shields
Data sheet

C.20.c Procedure

1. Measure and weigh each specimen, count the number of annual growth rings per inch, and estimate the percentage of summerwood and of sapwood.
2. Mark the center and endpoints of a 24-in. span.
3. Attach the deflectometer, and adjust both it and the testing machine to zero.
4. Apply the load continuously through a standard wooden block at midspan at a rate of 0.10 in. (2.5 mm) per minute or 400 lb (182 kg) per minute, taking simultaneous load and deflection readings. Try to obtain 20 readings before failure.
5. Describe the failure.
6. Cut a moisture specimen 1 in. (25 mm) long and weigh it to the nearest 0.1 g. Place in an oven of 103°C. After the moisture specimen has dried to constant weight, about 2 to 4 days, reweigh it.

C.20.d Data

1. Plot a load-versus-deflection diagram.
2. Determine the moisture content, approximate specific gravity (wet and dry), and all other properties listed previously.
3. Calculate the flexural strength (modulus of rupture) of the specimen.

C.20.e Results

1. Are the values obtained from tests on smaller beams applicable to beams of larger size? Explain.
2. Is the modulus of elasticity in bending the same as in compression or tension parallel to the grain? Explain. Would you expect it to be higher or lower? Why or why not?
3. Discuss the relationship between the modulus of rupture of a small wooden beam and the ultimate compressive strength of a short block loaded parallel to the grain.
4. How does the speed of load application affect the observed strength?
5. What other methods can be used to determine the moisture content of wood other than the one used here? Explain.
6. Additional tests can be made by adjusting the span and dimensions of the test specimen. Also, holes can be drilled in the wood to simulate joists, rafters, and so on. The number and sizes of the holes can be varied, and the effects of such variation can be studied.

C.21 HARDNESS TESTING

C.21.a Introduction

The general concept of hardness as a quality of matter having to do with solidity and firmness of outline is easily comprehended, but no single measure of hardness, universally applicable to all materials, has yet been devised. A number of different arbitrary definitions of hardness form the theoretical base from which the various hardness testing procedures now performed are derived:

1. Resistance to permanent indentation under static or dynamic loads (indentation hardness)
2. Energy absorption under impact loads (rebound hardness)
3. Resistance to scratching (scratch hardness)
4. Resistance to abrasion (wear hardness)
5. Resistance to cutting or drilling (machinability)

The principle behind the hardness test involves the idea that hardness is measured by the resistance to indentation and serves as the basis for a variety of instruments. The indenter, either a ball or a plain or truncated cone or pyramid, is usually made of hard steel or diamond and is ordinarily used under a static load. Either the load that produces a given depth of indentation or the indentation produced under a given load can be measured. In the rebound test, as in the scleroscope, a dynamic or impact load is dropped onto the surface of the test specimen. The amount of rebound determines the hardness of the specimen. This is a rebound-type test.

Probably the most commonly used hardness tests for metals in this country are the Brinell and Rockwell tests. However, the growing use of harder steels and hardened steel surfaces has brought a number of other tests into common use. These other tests include the Shore scleroscope and the Vickers, Monotron, Rockwell superficial, and Herbert machines. Also, the need for determining the hardness of very thin materials, very small parts, and the hardness gradients over very small distances has led to the development of so-called microhardness testers, such as the Knoop indenter.

C.21.b Test Descriptions

C.21.b.1 Brinell Test. The Brinell test involves pressing a hardened steel ball into a test specimen. It is customary to use a 10-mm ball under a 3000-kg load for hard metals, a 1500-kg load for metals of intermediate hardness, and a 500-kg or lower load for soft materials. Various types of machines for making the Brinell test are available. They may differ according to (1) the method of applying load (e.g., oil pressure, gear-driven screw, or weights with lever); (2) the method of operation (e.g., hand or motive power); (3) the method of measuring the load (e.g., piston with weights, bourdon gage, dynamometer, or weights with lever, and (4) the size [e.g., large or small (portable)].

The Brinell test can be made in a universal testing machine by using an adapter for holding the ball. Tests of sheet metal can be made using a hand-held device much like a pair of pliers having a 3/64-in. ball and a 22-lb spring.

To conduct a Brinell test, the specimen is placed on the anvil and raised to contact with the ball. Load is applied by pumping oil into the main cylinder, which forces the main piston down and presses the ball into the specimen. The plunger has a ground fit, so frictional losses are negligible. A bourdon gage is used to give a rough indication of the load. When the desired load is applied, the balance weight on top of the machine is lifted by action of the small piston; this action ensures that an overload is not applied to the ball.

In the standard test, the diameter of the ball indentation is measured by use of a micrometer microscope, which has a transparent engraved scale in the field of view. The Brinell test is a good measure of hardness, but it has its limitations. One limitation is that it is not well adapted for very hard materials. The ball deforms too much. It also isn't very appropriate for very thin pieces, where the indentation may be much greater than the thickness of the material. It is not suited for case-hardened materials, where the indentation may have a greater depth than the thickness of the case, so that the yielding of the soft core invalidates the results. In the standard test, the full load is applied for a minimum of 30 seconds for ferrous metals and 60 seconds for softer metals.

The Brinell hardness number is nominally the pressure per unit area, in kilograms per square millimeter, of the indentation that remains after the load is removed. It is obtained by dividing the applied load by the area of the surface indentation, which is assumed to be spherical.

Hardness numbers for common materials are as follows:

Ordinary steels	100 to 500
Medium-carbon steel	130 to 160
Very hard special steels	800 or 900

Note: The Brinell hardness test is not recommended for materials beyond 630 BHN.

C.21.b.2 Rockwell Test. The Rockwell test is similar to the Brinell test in that the hardness number found is a function of the degree of indentation of the test piece by action of an indenter under a given static load. Various loads and indenters are used, depending on the conditions of the test. It differs from the Brinell test, because the indenters and the loads are smaller, and the indentation made by the load is smaller and shallower. It is applicable to testing of materials beyond the scope of the Brinell test, and it is faster because it gives arbitrary direct readings (ASTM E 18).

The Rockwell test is conducted in a specially designed machine that applies load through a system of weights and levers. The indenter or penetrator may be either a hardened steel ball or a diamond cone with a somewhat rounded point. The hardness number indicated on the dial is an arbitrary value that is inversely related to the depth of indentation.

In the operation of the machine, a minor load of 10 kg is first applied, which causes an initial indentation that sets the indenter on the material and holds it in position. The dial is then placed onto the "set" mark on the scale, and the major load is applied. The major load typically ranges from 60 to 100 kg when the steel ball is used, although other loads may be necessary, and is 150 kg when the diamond cone is used. The ball indenter is normally 1/16 in. in diameter, but 1/8-, 1/4-, and 1/2-in. diameters are available for softer materials. After the major load is applied and removed, the hardness reading is taken from the dial while the minor load is in position.

There is no Rockwell hardness value consisting of a number alone, because the value depends on the indenter and load employed in the test. Therefore, a letter designation is also required for the test value to have any meaning. The dial on the machine has two sets of figures, one red and the other black, which differ by 30 hardness numbers. The dial was designed to accommodate the B and C scales, which were the first ones standardized and are the most commonly used. The red scale is used for readings obtained through the use of ball indenters, regardless of size of ball or magnitude of major load. The black figures are used only for the diamond cone.

The B scale is for testing materials of medium hardness; the working range for the B scale is 0 to 100. If the ball indenter is used for values over 100, there is a chance it will be flattened, thus giving a false reading.

The C scale is the one most commonly used for materials with hardness greater than B100. The hardest steels have values of about C70. The useful range of the C scale is from C20 upward. Variations in the grinding of the diamond cone make readings below C20 unreliable for smaller indentations. Choose a scale to employ the smallest ball that can be properly used, because sensitivity is lost as the indenter size increases.

Rockwell scales are divided into 100 divisions, and each division or point of hardness equates to 0.002 mm in indentation; thus, the difference in the readings B53 and B56 is 3↔ 0.002 mm, or 0.006 mm. Because the scales are reversed, higher numbers indicate harder materials.

The test piece should be flat and free of scale, oxide, pits, and foreign material on top and bottom. The thickness of the piece should be such that the piece will not bulge or deform on the anvil side when undergoing the test. Such deformation will give a false reading. On round stock, a flat place can be filed or ground into the material. All tests should be conducted on a single thickness of material.

C.21.b.3 Rockwell Superficial Hardness Test. The Rockwell superficial hardness test is designed to test the surface hardness, where only shallow indentations are possible or desired. The superficial tester operates on the same principle as the regular Rockwell hardness tester, although it employs lighter minor and major loads and has a more sensitive depth-measuring system. The superficial tester employs a minor load of 3 kg and major loads of 15, 30, or 45 kg. One point of hardness on the superficial machine corresponds to a difference in indentation depth of 0.001 mm.

The superficial tester uses the N and T scales. The W, X, and Y scales are used for very soft materials. These machines have only one set of dial graduations; thus, the readings must be interpreted according to the indenter and major load used.

C.21.b.4 Vickers Hardness Test. The machine used for the Vickers hardness test is similar to the Brinell in that the hardness number is based on the ratio of load in kilograms to the surface area of the indentation in square millimeters. The indenter used is a square-based diamond pyramid, in which the angle between the opposite faces is 136°. The load used may vary from 5 to 120 kg in 5-kg increments.

In conducting the test, a specimen is placed on the anvil and raised by a screw until it is close to the point of the indenter. By tripping a starting lever, a 20:1-ratio loading beam is unlocked, and the load is slowly applied to the indenter and then released. A foot lever is used to reset the machine. After the anvil is lowered, a microscope is swung over the specimen, and the diagonal of the square indentation measured to 0.001 mm. The machine may also employ 1- and 2-mm ball indenters.

One advantage claimed by some Vickers machine operators is in the measurement of the indentation: A much more accurate measure can be made of the diagonal of a square than can be made on a circle, where the measurement is made between two tangents to the circle. It is a fairly rapid method and can be used on metal as thin as 0.006 in. It is claimed to be accurate for hardness as high as 1300 (about 850 Brinell).

C.21.b.5 Durometers. Durometers come in various types, depending on the test material. All durometers are the same, differing only in the sharpness of the indenter point and the magnitude of the load applied to the indenter by a calibrated spring. The durometer hardness is also a measure of indentation depth; it varies from 100 at zero indentation to 0 at an indentation of 0.100 in. The load acting on the indenter varies inversely with the depth of penetration, being a maximum load at zero penetration and minimum load at full indentation (0.100 in.). Test specimens should be at least 1/4-in. thick with no tests made within 1/2 in. of the edge of the material. Results made on one type of durometer cannot be correlated to other types of durometers.

C.21.b.6 Hardness Test of Wood.   The only standardized hardness test for wood is an indentation type (ASTM D 143). The hardness is determined by measuring the load required to embed a 0.444-in. steel ball to one-half its diameter into the wood. The wood hardness value is for comparative purposes only. The approximate range for wood hardness values is from 400 lb for poplar to 4000 lb for persimmon. The hardness for Douglas fir is about 900 lb.

C.21.b.7 Dynamic Hardness Tests.   Most dynamic hardness testers are indentation tests and depend on the calculation of energy absorbed by the specimen during the test. Therefore, they are questionable, using a fixed test procedure and yielding arbitrary results.

Among the first dynamic hardness tests were those of Rodman, who experimented with a pyramidal punch in 1861. Later experiments used a hammer with a spherical end; they were made to verify and substantiate Rodman's results. These tests proved that the work of the falling hammer is proportional to the volume of the indentation. The hardness is, therefore, the work required to produce a unit volume of indentation. This method is very useful in determining the hardness of specimens at high temperatures, because the hammer is not in contact with the specimen for any considerable length of time and is, therefore, not affected appreciably by the heat.

A number of foreign machines use a dynamic load. Typically, the hardness number is calculated by dividing the net energy of the blow by the volume of the indentation. One of the more important is the Pellin hardness tester, in which the indentation is produced by a falling rod of known mass, which has at its lower end a steel ball 2.5 mm in diameter. Another dynamic tester is the Whitworth autopunch, which is a hand-held Brinell machine actuated by release of a spring in the handle that supplies a standard striking energy to the ball indenter in the bottom of the punch. The diameter of the indentation is measured in a manner similar to the Brinell test. The Waldo hardness tester utilizes a conical-point steel indenter weighing 0.1 lb and is dropped from a height of 12 in. Hardness is determined based on the diameter of the indentation. The duroskop hardness tester depends on the rebound of a pendulum hammer. Finally, the Avery hardness tester is a modification of the Izod impact machine for dynamic hardness testing.

At present, the Shore scleroscope is probably the most widely used dynamic hardness tester. The hardness test conducted on a scleroscope is often referred to as *rebound hardness.* Scleroscope hardness is a number representing the height of rebound of a small pointed hammer that falls within a glass tube from a height of 10 in. against the surface of the specimen. The standard hammer is approximately 1/4 in. in diameter, is 3/4 in. long, and weighs 1/12 oz, with a diamond tip rounded to a 0.01-in. radius.

The scale is graduated into 140 divisions, a rebound of 100 being equivalent to the hardness of martensitic high-carbon steel. For this material, the area of contact between the hammer and specimen is only about 0.0004 in.^{2}, and the stress developed is in excess of 400,000 lb/in^2. These hardness numbers are arbitrary and comparable only when obtained from similar material.

C.21.b.8 Scratch Hardness. An arbitrary scale of hardness is the ability of one material to scratch another material with a lower hardness number. The first material can also be scratched by material of a higher hardness number. One well-known scale is the Mohs' scale used by mineralogists:

1	Talc
2	Gypsum
3	Calcite
4	Fluorite
5	Apatite
6	Feldspar
7	Quartz
8	Topaz
9	Corundum
10	Diamond

The ability of these materials to scratch a substance is used to determine the relative hardness of the materials. There is no set relationship between the classes, though. In other words, gypsum is not twice as hard as talc or half as hard as calcite. The scale should not be used to form conclusions based on a relationship that might exist between the classes.

A sclerometer is a device that attempts to quantify either the pressure required to make a given scratch or the size of scratch produced by a stylus drawn across a test surface under a fixed load. Sclerometer tests are hard to standardize and interpret.

The file test is widely used as a qualitative, or inspection, test for hardened steel. The inspector or tester runs a file over the test specimen. The file will pass over materials of the proper hardness and dig into soft materials. Thus, acceptance or rejection can be determined by the file test.

C.22 ROCKWELL HARDNESS TEST

C.22.a Introduction

The purpose of this test is to determine the hardness of a material through the measurement of the material's resistance to penetration by an indenter of known parameters.

C.22.b Required Equipment

Rockwell hardness tester
Anvil
Penetrator or indenter
Standard test specimen to calibrate
Safety glasses and shields
Data sheet

C.22.c Procedure

1. *Selecting the penetrator:* The 1/16-in. ball chuck is used for testing brass, bronze, cast iron, and soft steel. The ball is held in place by a screw cap, which leaves about one-third of the ball exposed for penetration into the specimen. If the ball should be damaged, the cap can be unscrewed, the damaged ball removed, and a new ball inserted. The cap should then be replaced and screwed finger-tight to hold the ball.
   Note: For softer materials, the 1/8-in., 1/4-in., and 1/2-in. ball penetrators are available.
2. The spheroconical diamond penetrator (brale) must be used when testing hardened tool steel, case-hardened steel, and heat-treated spring steel.
   Note: The operator should avoid striking the brale against the anvil, as the brale may break. Striking the steel ball against the anvil may flatten the ball.
3. *Selecting the weight:* A 60-kg weight pan, a 40-kg weight marked with a red circle, and a 50-kg weight marked with a black circle permit tests of 60, 100, and 150 kg. The weights are made so that they can only be stacked in this order.
   Note: The 100-kg load is used with the 1/16-in. ball chuck for testing softer materials. The 150-kg load is usually used with the brale diamond penetrator for testing harder materials.
4. *Anvils:* The plane anvil should be used for testing flat-bottom pieces. The spot anvil with the small elevated flat should be used for small pieces or those having a bottom that is not truly flat. In testing pieces that are not truly flat, place them on the anvil with the more concave side up so that they make better contact with the anvil at the test point. Use the anvil specified for round pieces. The shallow V anvil is used for round pieces under 1/4-in. diameter.
5. *Caution:* A round piece should *never* be supported on its cylindrical surface by a flat anvil because of the danger of its rolling out when the major load is applied. This may break the penetrator and will definitely invalidate the test. An exception to this rule may be when the test piece is supported or held in position by other means to prevent rolling.
6. When changing anvils, run the elevating screw to a low position and remove the anvil carefully to avoid striking the penetrator.
7. Before starting the test, make sure the crank handle is pulled forward, counterclockwise, as far as it will go to raise the major load. This lifts the power arm and weights.
8. Select the proper penetrator and insert in the plunger rod.

9. Place the proper anvil on the elevating screw.
10. Place the test specimen in position on the anvil.
11. Raise the specimen into contact with the penetrator by turning the capstan handwheel clockwise. Continue the motion until the small pointer on the dial is near the set dot. Continue until the large pointer on the dial is in the vertical position.
12. Turn the bezel of the dial gage until the "set" mark is directly behind the large pointer.
13. Release the weights by tipping the crank handle clockwise toward the back of the test machine, thus releasing the major load.
14. After a short period of time, the large pointer will come to rest; then return the crank handle to the starting position. This removes the major load but leaves the minor load still applied.
15. Read the scale letter and the Rockwell number from the dial gage.
16. Remove the minor load by turning the capstan handwheel counterclockwise to lower the elevating screw and specimen so that they clear the penetrator.
17. Remove the specimen.

C.22.d Data

Data taken are the indenter used, the major load applied, and the hardness number.

C.22.e Results

1. Create a table of Rockwell hardness numbers for a variety of materials in a variety of ranges using the different scales.
2. Compare and contrast the scope and limitations of the Brinell, Vickers, Rockwell, and Tukon-Knoop hardness test procedures.

C.23 BRINELL HARDNESS TEST

C.23.a Introduction

The purpose of this test is to determine the Brinell hardness number for a variety of materials.

C.23.b Required Equipment

Brinell hardness tester
Indenter
Test specimen
Safety glasses and shields
Data sheet

C.23.c Procedure

1. The specimen should be flat and clean.
2. Select the proper load, and apply it for 15 seconds for ferrous metals and 30 seconds for nonferrous metals.

Ball Diameter	Load	Recommended Range
10 mm	3000 kg	160–300 BHN
10 mm	1500 kg	80–300 BHN
10 mm	500 kg	25–100 BHN

3. Release the load after the proper time period has elapsed. Read the diameter of the impression to the nearest 0.01 mm using a micrometer microscope.
4. The Brinell hardness number can be found by dividing the applied load by the area of the indentation. Tables are provided for conveniently finding this number.

C.23.d Data

The Brinell hardness number is found either by calculating it or by reading it from a chart, based on the actual load applied and the indentation diameter.

C.23.e Results

1. Compare the results of different hardness tests. Which were more reliable? Significant? Accurate? Precise? Economical? Practical? Lab-based? Field-based?
2. Prepare a chart that lists the different tests and try to set up a correlation among them. In other words, make a chart showing how the different test results compared.
3. Test a variety of materials using different methods and try to form an opinion about which is best for the different applications.

C.24 HARDNESS TESTING FOR WOOD

C.24.a Introduction

The purpose of this test is to determine the hardness of a variety of different types of wood and wood products (ASTM D 143).

C.24.b Required Equipment

Universal testing machine
0.444 in. (11.3 mm) hardened steel ball penetrator
Safety glasses and shields
Data sheet

C.24.c Procedure

1. Prepare the testing machine and specimen.
2. Place the specimen in the machine.
3. Place a load on the specimen and continue loading until the indenter has penetrated halfway into the wood.
4. Read the load from the machine's dial.
5. Remove the load and the specimen from the machine.

C.24.d Data

Data taken involve reading the machine dial to determine the load required to drive the indenter halfway into the wood.

C.24.e Results

1. What is the significance of the hardness number found using this method?
2. Test a variety of wood types and create a table of your results.
3. Compare your results with others. Are they similar or different? Why?
4. Calculate the pounds per square inch required to drive the indenter into the wood.
5. Repeat the test using a deflectometer or dial indicator to measure the deflection produced during the test. Use this information to plot a load-versus-deflection curve. What can you deduce from the curve?

C.25 FILE HARDNESS TESTING

C.25.a Introduction

The purpose of this test is to determine the relative hardness of a material by passing a file over its surface.

C.25.b Required Equipment

File
Test specimen
Safety glasses and shields
Data sheet

C.25.c Procedure

1. Place the material specimen in a suitable holding fixture.
2. Run the file over it, using an even, constant motion.
3. Remove the test piece.

C.25.d Data

Information is taken regarding the visual characteristics of the file's effectiveness in scratching the material's surface and the depth of these scratches.

C.25.e Results

1. What factors affect the test results in this method?
2. Make several tests and try to design a standard chart for this procedure.
3. Test a variety of materials, primarily metals, using different files and filing techniques. Report your findings.
4. The following should be used as a guide:
   a. If the file removes material easily: approximately Rockwell C20.
   b. If the file resists cutting: approximately Rockwell C30.
   c. If the file cuts with difficulty: approximately Rockwell C40.
   d. If the file barely cuts the material: approximately Rockwell C50.
   e. If file glides over without cutting: approximately Rockwell C60.

C.26 IMPACT TESTING

C.26.a Introduction

As the velocity of a moving body is changed, a transfer of energy must occur; work is done on the parts receiving the blow. The mechanics of the impact involve not only the question of stresses induced but also a consideration of energy transfer and of energy absorption and dissipation.

The energy of the blow can be absorbed in a number of different ways:

1. Through elastic deformation of the members or parts of a system
2. Through plastic deformation in the parts
3. Through hysteresis effects in the parts
4. Through frictional action between parts
5. Through effects of inertia of moving parts

In the design of structures and machines, an attempt is made to provide for the absorption of as much energy as possible through elastic action, relying secondarily on some form of damping to dissipate it. In impact testing, the object is to use the energy of the blow to rupture the test specimen.

The property of a material that relates to the work required to cause rupture is called *toughness.* Toughness depends chiefly on the ductility and strength of a material. Impact testing is an adequate measure of a material's toughness.

In making an impact test, the load can be applied in flexure, tension, compression, or torsion. Flexure loading is the most common; tensile is less common; compression and torsional loadings are used only in special circumstances. The impact blow can be delivered through dropping weights, swinging a pendulum, or rotating a flywheel. Some tests involve rupturing the test specimen in a single blow; others involve repeated blows. In tests involving repeated blows, some are multiple blows of the same magnitude, and others, such as the increment-drop test, gradually increase the height of the weight drop until rupture is induced.

Perhaps the most common are the Izod and Charpy impact tests. Both employ a pendulum and are made on small notched specimens broken in flexure. In the Charpy test, the specimen is supported as a single beam, whereas in the Izod, the specimen is supported as a cantilever. In these tests, a large part of the energy absorbed is taken up in a region immediately adjacent to the notch. A brittle type of fracture is often induced.

For wood, the Hatt-Turner test is used. This test is a flexural-impact test of the increment-drop type. The height of drop at which failure occurs is taken as a measure of toughness. From the data obtained from the Hatt-Turner test, the modulus of elasticity, the proportional limit, and the average elastic resilience can be found.

C.26.b Test Descriptions

Items that require standardization in impact testing are the foundation, anvil, specimen supports, specimen, striking mass, and velocity of the striking mass. Principal features of a single-blow testing machine are as follows:

1. A moving mass whose kinetic energy is great enough to cause the rupture of the test specimen placed in its path
2. An anvil and a support on which the specimen is placed to receive the blow
3. A means for measuring the residual energy of the moving mass after the specimen has been ruptured

The kinetic energy is determined from, and controlled by, the mass of the pendulum and the height of free fall measured from the center of the mass. The pendulum should be supported to reduce or restrain the lateral play and friction that may be felt as it swings in an arc toward the test specimen. The release mechanism should be constructed to reduce any binding, acceleration, or vibratory effects.

The anvil should be heavy enough in relation to the energy of the blow so that an undue amount of energy is not lost by deformation or vibration. The specimen should be supported firmly and in the correct position throughout the test.

C.26.b.1 Charpy Test. A Charpy test machine usually has a capacity of 220 ft-lb for metals and 4 ft-lb for plastics (ASTM E 23). The pendulum consists of a relatively light, although rigid, rod or piece of channel, at the end of which is a heavy disk. This pendulum swings between two upright supports and has a rounded knife-blade edge at the end aligned so that it contacts the specimen over its full depth at the time of impact.

The standard test specimen is $10 \times 10 \times 55$ mm, notched on one side in the center. Some tests require keyhole notches; others require U-shaped notches. The specimen is supported between two anvils so that the knife strikes opposite the notch at the midswing point. The pendulum is lifted to horizontal and is held in place by a catch. It is then released and allowed to fall and rupture the specimen.

C.26.b.2 Izod Test. The common Izod test machine is made with a 120 ft-lb capacity. The test is similar to the Charpy test, although the specimen placement and features are different. In the Izod test, the specimen is $10 \times 10 \times 75$ mm, having a 45° notch cut 2 mm deep. The impact strength of the specimen is based on the angle of rise after rupture occurs.

C.26.b.3 Hatt-Turner Test. The Hatt-Turner test is used primarily for flexure-impact tests of wood in which the height of drop is increased by increments until failure occurs (ASTM D 143). A tup weighing 50 lb is held by an electromagnet, which is raised by a motor. The tup drops through vertical guides when the current is stopped by opening a relay.

The specimen is a clear piece of wood having nominal dimensions of $2 \times 2 \times 30$ in. The piece is supported over a 28-in. span so that the tup drops directly over the center of the span. The first drop is 1 in., and each succeeding drop is increased by 1-in. increments. If the piece has not ruptured after 10 in. is reached, 2-in. increments are maintained until either rupture or a 6-in. deflection occurs.

Figure C–18
Sample impact test data.

Impact Testing of Metallic Materials

Date: _____

Material: Pendulum weight (lb):
Original dimensions: Radius of swing (ft):
Absorbed energy (ft-lb):

                    or

Initial angle:
Final angle:
Impact strength:

Figure C–19
Sample impact test results.

Charpy Impact Test

Date: _____
Material: 1020 steel
Original dimensions: 10 × 10 × 55 mm
Pendulum weight: 60 lb
Radius of swing (ft): 2.63
Initial angle: 90°
Rise angle: 29°

Absorbed energy = 60 lb · 2.63 ft · (cos 29° − cos 90°)
= 138 ft-lb

C.26.c Data

Sample testing data for impact testing are given in Figure C–18.

C.26.d Results

Sample impact results are given in Figure C–19.

C.27 IZOD IMPACT TESTING

C.27.a Introduction

The purpose of this test is to study, observe, and experiment with impact testing of common materials, to develop skills in performing commercially acceptable impact tests, and to interpret performance data.

C.27.b Required Equipment

Izod impact testing machine
Test specimen
Safety glasses and shields
Data sheet

C.27.c Procedure

1. Measure and record the dimensions of the test specimen.
2. Place the friction pointer in the starting position.
3. Place the specimen accurately into position on the machine. The notch should be facing the direction from which the pendulum will be coming, and the center of the notch should be at the center of the support.
4. Break the specimen by lifting and releasing the suspended load.
5. Record the energy value required to break the specimen and the type of surface at the break.

C.27.d Data

1. Record the type and size of specimen used including type of notch.
2. Determine the energy absorbed by the specimen and an estimate of the energy loss due to friction.
3. Observe the shape of the fractured surface and its texture.
4. Find the number of specimens failing to break and the causes of nonfracture.

C.27.e Results

1. Discuss the significance and advantages of impact tests in comparison with static tests.
2. Why are impact specimens notched?
3. What physical property is determined by means of an impact test?
4. Repeat the test using the Charpy test, where the specimen is supported as a simple beam rather than a cantilever (ASTM E 23). Compare the results.

C.28 FATIGUE TESTING

C.28.a Introduction

Most structural assemblies are subject to variations in applied loads, causing fluctuations in the applied stresses in these parts. If these fluctuations are of sufficient magnitude, even though they may be considerably less than the static strength of the material, failure may occur when the stress is repeated enough times.

The stress required for fatigue failure should be designated by the degree of stress variation and type of stress. The stresses may be axial, shearing, torsional, or flexural. The stress at which a material fails by fatigue is called *fatigue strength*. The limiting stress, below which a load can be repeated indefinitely without failure, is called the *endurance limit*.

Fatigue tests are conducted over long periods of time, sometimes months or even years. They are, therefore, generally not used for quality control or inspection because of the time and effort required to collect the necessary data.

C.28.b Test Descriptions

Many different machines are employed for fatigue testing. Among these are the following:

1. Machines for axial stresses
2. Machines for flexural stresses
3. Machines for torsional shearing stresses
4. Universal machines for axial, flexural, or torsional shearing stresses or a combination of these

The general procedure for a fatigue test is to prepare several representative specimens. The first specimen is treated with a high amount of stress so that it fails rapidly. The second specimen is subjected to less stress until it fails. This procedure continues, decreasing stress and increasing repetitions, until the stress is below the endurance limit of the specimen, so it will not fail. A plot is then made showing the maximum stress applied versus the number of cycles.

C.28.c Data

Sample test data for fatigue testing are given in Figure C–20.

C.28.d Results

Sample fatigue testing results are given in Figure C–21.

Figure C–20
Sample fatigue test data.

Fatigue Testing of Metals

Date:
Material:
Original dimensions:
Applied stress (lb/in.^{2}):
Number of cycles to failure:
Frequency:
Load variation:

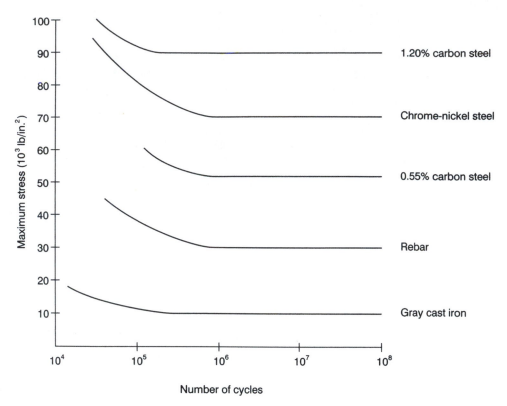

Figure C–21
Sample fatigue testing results.

C.29 CREEP TESTING

C.29.a Introduction

Creep tests at high temperatures seem to be the only satisfactory guide for materials in high-temperature service. Creep tests are inherently long-time tests and are, therefore, unsatisfactory for acceptance tests.

C.29.b Test Descriptions

In general, creep testing involves placing specimens at a constant elevated temperature under a fixed tensile stress and observing the strain produced. Such tests cover a long time period. Time periods for tests depend on the expected life of the material in service. Tests that extend beyond 10% of the life expectancy of the material in service are preferred. Specimens may be tested over a range of tensile stresses and temperatures.

The requirements for a creep test include the following:

1. A heat source
2. An extensometer
3. A tensile load

Round test specimens typically have diameters of 0.505, 0.320, 0.252, or 0.100 in. and gage lengths of approximately 4 times the diameter. The 0.505-in. specimen is preferred.

C.29.c Data

Sample test data for creep testing are given in Figure C–22.

C.29.d Results

Sample creep test results are given in Figure C–23.

Figure C–22
Sample creep test data.

> **Creep Testing Data**
>
> Date:
> Material:
> Original dimensions:
> Applied stress (lb/in.^{2}):
> Time to rupture (h):
> Creep rate (90/h):
> Temperature (°F):

Figure C–23
Sample creep test results.

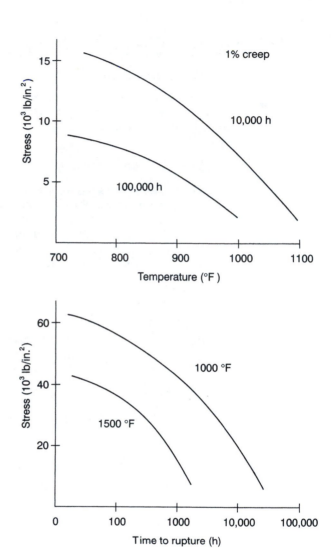

C.30 CREEP TEST FOR POLYMERS

C.30.a Introduction

The purpose of this test is to observe the creep phenomenon and to study the various aspects of time-dependent material properties.

C.30.b Required Equipment

Grippers for supporting specimen
Extensometer or dividers and scale
Sufficient dead weight
Clock
Data sheet

C.30.c Procedure

1. Record the appropriate data for the specimen to be tested.
2. Mark the gage length on the specimen in such a manner that the measuring length is easily distinguished.
3. Calculate the required load based upon the tensile area and the stress level for the creep test.
4. Collect the appropriate weight and verify the total, including the weight of the grip.
5. Mount the specimen in the upper grip and attach the upper grip in a secure location in such a manner so that the length measurement can be made conveniently.
6. Apply the load to the specimen in a slow but steady manner. Immediately measure and record the distance between gage marks. This is the elongation at zero time.
7. Continue recording deformation and time until the specimen fails or it is no longer practical to continue. Record the final elongation.

C.30.d Data

1. Plot the load-versus-deformation curve.
2. Determine the time to the onset of secondary and tertiary creep.
3. Calculate the rate of secondary creep.

C.30.e Results

1. What is the purpose and value of the creep test? How would you use the values gained through a creep test?
2. Discuss the effects of temperature on the test results. Draw graphs predicting the effects of temperature on various loads.
3. What other factors affect creep test results?

C.31 SLUMP TESTING FOR CONCRETE

C.31.a Introduction

The purpose of this test is to study and observe the slump testing of concrete and to develop an understanding of how the slump test can be used to predict the acceptability of a concrete mixture.

C.31.b Required Equipment

Concrete mix
Concrete slump mold
Tamping rod with a bullet end
Measuring stick with 1/4-in. divisions
Safety glasses and shields
Data sheet

C.31.c Procedure

1. Dampen the mold and place it on a flat, rigid surface. The mold must be held firmly in place while being filled.
2. Fill the mold approximately one-third full.
3. Rod the concrete with 25 strokes using the tamping rod. The strokes should be evenly distributed over the mold.
4. Fill the mold to approximately two-thirds full.

Slump cone.

5. Repeat step 3.
6. Completely fill the mold and heap concrete above the top of the mold.
7. Repeat step 3.
8. Screed the top of the mold to remove the excess concrete.
9. Carefully remove the mold.
10. Place the mold next to the concrete so that the mold is in its original configuration. Lay the tamping rod across the top of the mold in the direction of the concrete.
11. Measure and record the vertical distance between the top of the concrete and the bottom of the tamping rod.

C.31.d Data

Record the slump value associated with the concrete mix.

C.31.e Results

1. How would you use the slump test to ascertain the value of a concrete mixture? What is the merit of the slump test?
2. What precautions should be taken when conducting a slump test?
3. Suggest some appropriate slump values for different uses of concrete.

C.32 MEASUREMENT OF ENTRAINED AIR IN CONCRETE

C.32.a Introduction

The purpose of this test is to study and observe the entrained air test for concrete, to develop skills in performing acceptable entrained air calibration and testing, and to interpret the results of an entrained air test.

C.32.b Required Equipment

Concrete mix
Concrete air meter test apparatus (Figure C–24)
Tamping rod with bullet end
Safety glasses and shields
Data sheet

Figure C–24
Entrained air test apparatus.

C.32.c Procedure

1. Place the clean, dry, standardizing vessel open end down into the bottom of the air meter.
2. Fill the air meter with water.
3. Place the lid on the air meter and securely clamp the lid into position. Care should be taken to keep the contact surface of the lid and container free of dirt and other extraneous matter.
4. Add water through the funnel until the level of the water is slightly above the mark on the plastic tube. Valve T must be open while the water is being added to vent the air meter.
5. Open valve S to adjust the top of the water column in the tube until it reads exactly zero.
6. Close all valves except air inlet valve U and apply pressure with the air pump until the reading on the tube is 5%.
7. Release the pressure by opening valve T, and add as much water through the funnel as is necessary to restore the water column in the tube to the zero mark. Close all valves except valve U and apply pressure with the air pump until the standardizing vessel volume is again read on the water column in the tube. Lightly tap the air pressure gage and note its reading.
8. Repeat step 7 until several readings of identical pressures are obtained.
9. Empty the air meter of water and thoroughly dry the air-meter container and the standardized vessel.
10. Place the air meter on a flat, rigid surface.
11. Fill the container approximately one-third full.
12. Rod the concrete with 25 strokes using the tamping rod. The strokes should be evenly distributed throughout the container.
13. Fill the container approximately two-thirds full.
14. Repeat step 3.
15. Completely fill the container and heap concrete above the top of the container.
16. Repeat step 3.
17. Screed the top of the container to remove the excess concrete.
18. Clean the lip of the container to remove any residual concrete.
19. Place the lid on the air meter and securely clamp the lid into position. Care should be taken to keep the contact surfaces clean and free of contamination.
20. Add water through the funnel until the level of the water is slightly above the zero mark on the plastic tube. Valve T must be open while water is being added to vent the air meter.
21. Lightly tap the container and tilt and roll it while adding water to remove any trapped air.
22. Open valve S to adjust the top of the water column in the tube until it reads exactly zero.
23. Close all valves except air inlet valve U and apply pressure with the air pump until the reading on the pressure gage is the same as that determined by the calibration procedure.
24. Lightly tap the pressure gage and note its reading.
25. Read the plastic tube to determine the percent of entrained air.
26. Repeat steps 14 through 16 for verification.

C.32.d Data

Record the percentage of entrained air in the concrete mixture.

C.32.e Results

1. What is the value of the measurement of entrained air? How would you use it as a figure of merit for concrete?
2. Determine a range of entrained air for different applications.

C.33 TESTING OF LUBRICANTS

C.33.a Introduction

The purpose of this test is to study the viscosity of different lubricants and oils.

C.33.b Required Equipment

    Oils (10, 20, 30, 40, 50, 90 weights)
    Multigrade oils (10W-30, 10W-40, 20W-50, and so on)
    Heavy grease
    Hot plate or heating unit
    2-qt pan to hold liquid
    Large pan or jar to hold liquid
    Large funnel
    Stopwatch
    Safety glasses, shields, and gloves
    Data sheet

C.33.c Procedure

1. Select the oil to be tested.
2. Place the funnel over the pan or jar to catch the liquid that goes through the funnel.
3. When the oil is poured into the funnel, start the stopwatch. When the last of the oil runs through the funnel, stop the stopwatch.
4. If your results do not vary enough, place some appropriate restriction in the bottom of the funnel to slow down the oil.
5. Repeat the procedure for the oils you want to test.
6. Repeat the procedure after first heating the oil to a predetermined temperature.
7. WEAR SAFETY GLASSES, SHIELDS, AND GLOVES during the heating. Be careful not to overheat the oils or grease. When they overheat, they begin to smoke. Use adequate ventilation and reduce the heat immediately if the liquid begins to smoke.
8. Prepare a chart showing the various times in seconds required for the oil to run through the funnel.

C.33.d Data

Data taken are the temperatures at which the oil is placed in the funnel and the time (in seconds) it takes the oil to run through the funnel.

C.33.e Results

1. Compare the heated results to the unheated results. Explain the differences, if any.
2. Compare the multigrade results with the standard-grade results. What differences did you observe? Why?
3. Other than temperature and grade, what other variables could affect your results?

C.34 TEST GLOSSARY

Brinell hardness A measure of the indentation hardness of a material. It is calculated by dividing the applied load in kilograms by the calculated surface area under penetration (the area enclosed by the impression remaining after the load is removed). The greater the BHN, the harder the material is said to be. Variables for the Brinell hardness test include the magnitude of applied load, the indenter diameter, and the indenter material.

Ordinary loads are 500, 1500, and 3000 kg using a 10-mm indenter. The 500-kg load is used for BHNs of 25–100; 1500-kg loads are used in the 80–300 BHN range; and 3000-kg loads are used in the 160–600 BHN range. Up to 630 BHN, a carbide indenter may be used, whereas a hardened steel indenter should not be used above 450 BHN. Smaller loads may be used for softer materials.

The load is applied for a minimum of 15 seconds on ferrous metals and 30 seconds on most nonferrous metals and may be varied on softer materials (as long as it remains constant throughout the testing).

The Brinell hardness test, in comparison with the Rockwell and Vickers tests, is particularly applicable in situations where it is desirable to minimize the effects of surface conditions on the specimen or a deep penetration is necessary due to a lack of homogeneity in the specimen. The Brinell test is not well suited to very hard or thinner materials, which can be better tested using other methods. Also, the BHN cannot be read directly, as on other testers, such as the Rockwell.

Charpy impact test The Charpy impact test is used to determine the amount of energy absorption required in the fracture of a material specimen. To conduct the test, a rectangular or round specimen is supported as a beam and struck transversely by a hammer. A notch is cut in the specimen to aid in rupture. Notches can be V-, keyhole-, or U-shaped. Notches for nonferrous and ductile metals are deeper than for brittle and ferrous metals.

Specimens vary, depending on the material used, according to size and shape. However, all Charpy test specimens are struck in the center of the notch but on the opposite side, away from the notch.

Specimens may be subjected to greater or less than normal temperatures during testing.

Compression test The compression test is used to determine the behavior of a material subjected to a uniaxial compressive load. The specimen is squeezed or compressed at a constant rate until it fails. Information concerning applied load and deformation is recorded at specific intervals. Compressive stress and strain can be calculated. From these data, a stress-strain curve is plotted. From the stress-strain curve, the elastic limit, modulus of elasticity, proportional limit, yield point, yield strength, and often the compressive strength can be found. The yield strength and compressive strength can be determined without the use of a stress-strain curve.

The yield strength or yield point can be determined by observing the load at which deformation occurs without a continued increase in loading. For compressive strength, the maximum compressive load is recorded at some predetermined deformation or at rupture.

Typically, the compression test specimen is a solid cylinder, which varies in height and girth according to the material. The effects of temperature on compressive strength are also commonly studied.

Creep test Creep test is a term used to describe a variety of methods employed to determine the deformation developed in a material under a prolonged application of static tensile or compressive loading at some predetermined temperature. Creep tests are usually long-time tests, where it is not practical or economical to conduct the test to rupture.

Diamond pyramid hardness (Vickers) test The Vickers hardness test is a common measure of the indentation hardness for metals. It is an index of the resistance to surface penetration of a square-based, 136° pyramidal-shaped diamond indenter subjected to a specified load (usually 1 kg). The Vickers hardness number is calculated by dividing the applied load by the calculated area of the indentation.

Drop ball impact test The drop ball impact test is used to determine the amount of energy absorption required to fracture a material. It is conducted by dropping a metal ball of known weight on a part or specimen from

559

regularly increasing heights. The intervals between height increases should be small, within the practical limitations of the test procedure. The height at which the material fails is used as a figure of comparison.

Durometer hardness The durometer is used to measure the indentation hardness of soft materials such as rubber. A flat annular presser foot is brought into contact with the specimen to be tested. The durometer reading indicates the degree to which a spring-loaded hardened steel indenter penetrated the specimen. The Durometer scale is based on 100, which is the reading for a piece of plate glass. A reading of 0 indicates that the indenter extends 0.100 in. (2.5 mm) beyond the foot. Therefore, a higher durometer hardness number indicates greater resistance to penetration and a greater indentation hardness.

Fatigue test A fatigue test is used to determine the maximum number of cycles of repeated loading that a material can withstand before failure. Typical data taken are number of cycles and stress applied. Fatigue limit, fatigue strength, and fatigue life can be found from the stress-cycles diagram.

Fatigue loads can be applied axially, torsionally, or flexurally and in one direction, completely reversed, or partially reversed. Most fatigue tests are flexural tests. These fatigue tests are generally conducted using either rotating-beam or vibrating-type machines. In the rotating-beam machine, a load is suspended from the specimen, and it is rotated rapidly, with each rotation completing one stress cycle for the specimen. Vibratory machines flex a specimen either by magnetism or centrifugal force. Specimens vary according to the type of material and the type of test procedure used.

In determining fatigue failure, one criterion is the complete separation of the specimen into two pieces. Another criterion is a specified reduction in the stiffness of the material. The fatigue strength of a material is reported as the maximum stress that can be developed by a material for a specified number of cycles under specified testing conditions.

File hardness test The file hardness test is a measure of the ability of a material to resist the abrasion of a file that is run over the material's surface. The visual quality of scratches made is used as comparative data. This test is generally conducted only on metals.

Flexure test The flexure test is used to determine the behavior of a material subjected to bending stresses. The specimen is supported as a beam and subjected to a gradually increasing bending load at the center of the beam. Data are taken concerning the applied load and deflection of the beam. The flexure test can be applied to a variety of materials and is highly reliable and sensitive to slight variations in specimens.

Impact test The impact test is a destructive test in which the material is subjected to one or more blows of specified magnitude. The energy absorbed in rupturing the specimen is measured or determined using test data.

Izod impact test In the Izod impact test, a rectangular or cylindrical specimen is supported vertically as a cantilever beam and struck transversely. The energy required to break the specimen is either read directly or determined by calculation. The Izod impact test can be used to determine the toughness of a variety of materials. Different materials have different specifications regarding notch, cross section, and hammer weight used.

Repeated bend test The repeated bend test is used as a measure of the ductility of relatively ductile metals. It uses a specimen slightly more than 1 in. (25 mm) wide and at least 6 in. (150 mm) long. The specimen is clamped at one end into a holding fixture. The specimen is then bent 90° degrees; then 180° bends are made until a crack appears or the material breaks. Data taken are visual (quality of fracture or cracks and location and the number of cycles required to rupture the specimen).

Rockwell hardness test The Rockwell test measures the indentation hardness of a material. It indicates the resistance of a material to surface penetration by an indenter under a specified load. Higher Rockwell hardness numbers indicate greater indentation hardness. Rockwell hardness numbers are read directly from the dial. Different scales are used, depending on the indenter and the load applied.

The two general types of indenters used in Rockwell hardness testing are the brale diamond, a 120° cone-shaped indenter with a 0.2-mm radius spherical tip, and the hardened steel ball indenter, 1/16 to 1/2 in. (1.6 to 13 mm) in diameter. Depending on the material tested, different major loads and indenter sizes and type are used. Typically, different combinations are shown on the test machine.

Scleroscope hardness Scleroscope hardness is a measure of the hardness or impact resilience of materials, primarily metals. A graduated scale shows the height of rebound for a standard diamond-tipped hammer falling freely on the part or specimen from a fixed height. The scale is calibrated so that a reading of 100 represents the average for quenched high-carbon steel. Higher scleroscope readings are interpreted as higher impact resilience, or lower energy absorption. Scleroscope data are subject to variations in surface uniformity.

Shear test The shear test is used to show the resistance of a material to the application of a shearing load. A piece of the material to be tested, usually round, is placed in a suitable die, and a gradually increasing load is applied. Maximum shear load is recorded, and maximum shear stress and strength are calculated.

Tension test The tension test is used to determine the behavior of a material under tensile loading. A specimen of known cross section having a known gage length is pulled apart at a slow, steadily increasing rate. Measurements are taken on load applied and extension developed. Tensile stress and strain are calculated, and stress-strain curves are developed. Using the test data and the curve plotted, the elastic limit, percent elongation, modulus of elasticity, proportional limit, percent reduction in area, tensile strength, yield point, and yield strength can be determined. Specimens vary depending on the material under test. Specimens are typically round or flat. Depending on the grippers used, the ends may be plain, threaded, or shouldered. Specimens are typically "dog-bone-shaped," with a reduced center portion where the gage length is marked.

Torsion test A cylindrical specimen with a reduced center section and gage length marked is twisted to rupture. Typically, data are taken regarding the angle of twist and torque applied. A torque-twist diagram or a stress-strain diagram can be plotted. Properties that can be determined are elastic limit, modulus of elasticity, proportional limit, torsional strength, yield point, and yield strength.

APPENDIX D
Glossary of Selected Terms

ABS A polymer consisting of acrylonitrile, butadiene, and styrene.

Absolute humidity The mass of water vapor per unit volume of air.

Accuracy The degree to which a measured value represents the true value for that test.

Acid A chemical compound containing a nonmetal, hydrogen, and sometimes oxygen that neutralizes bases to form salts. Inorganic acids contain H^+ ions. Organic acids contain the carboxyl group.

Strong Acids in Water

Hydrochloric	HCl
Hydrobromic	HBr
Hydroiodic	HI
Nitric	HNO_3
Perchloric	$HClO_4$
Chromic	H_2CrO_4
Permanganic	$HMnO_4$
Sulfuric	H_2SO_4

Weak Acids in Water

Hydrofluoric	HF
Hydrocyanic	HCN
Acetic	$HC_2H_3O_2$
Formic	$HCHO_2$
Benzoic	$HC_7H_5O_2$
Nitrous	HNO_2
Phosphoric	H_3PO_4
Ascorbic	$HC_6H_7O_6$ (vitamin C)

Acrylic resin Any of a group of thermoplastic resins that are formed through polymerization of acrylic acid, methacrylic acid, esters of these acids, or acrylonitrile. Acrylic products are lightweight, weather-resistant, and clear.

Advanced composite One that contains continuous fiber reinforcement.

Age-hardening Also known as precipitation-hardening, because one phase of a two-phase system precipitates and evenly distributes into the other phase. This precipitation and dissolution may be accelerated with heat treatment.

AI Asphalt Institute.

AISI American Iron and Steel Institute.

Alcohols A group of organic chemicals that contain the -OH group, such as methanol (CH_3OH) and ethanol (C_2H_5OH).

Aliphatic compounds Organic chemicals that contain long-chain carbons and not ring-type structures.

Alloy A substance with metallic properties that comprises two or more elements that do not separate upon mixing. Most alloys consist of two or more metals, such as iron and carbon, that form the alloy steel.

Alumina Term used for pure aluminum oxide.

Amorphous Noncrystalline-structured materials, such as some glass.

Anisotropic materials Materials that exhibit different properties along different axes.

Annealing Heat-treatment procedure that converts metal to its softest condition.

Anode The pole or electrode in a plating cell or electrolytic cell that provides electrons. The anode is the negative post on a battery or the positive electrode in a plating cell.

Anodizing Process through which an anode material is coated with ions of the electrolyte to form a hard, chemical-resistant outer surface.

ANSI American National Standards Institute.

Antiknock compounds Substances used to raise the antiknock properties of fuels, particularly gasoline, expressed by octane number. Tetraethyl lead was primarily used as an antiknock agent, but it has been replaced due to Environmental Protection Agency (EPA) regulations. Replacements include tertiary butyl alcohol (TBA) and methyl tertiary-butyl ester (MTBE).

API American Petroleum Institute.

ASM American Society for Metals.

Asphalt Bituminous material of high molecular weight that occurs naturally or as a distillation product of crude oil. It is primarily used as a binding agent in road construction and in roofing products.

Attenuation Loss or weakening effect, particularly used in reference to sound.

Austempering Heat-treatment process by which steel is quenched at a rate that misses the knee of the TTT curve until just above the M_s line and held there until transformation is complete.

Austenite Face-centered cubic structure of steel.

Austenitic stainless steels Nonmagnetic stainless steels whose composition holds them in a face-centered cubic structure at room temperature.

Average The sum of the data in a set of numerical data divided by the number of data taken.

Base A substance that contains an excess of hydroxide (OH^-) ions. Bases neutralize acids to form salts and water.

Benzene Aromatic hydrocarbon, which consists of six carbon atoms and six hydrogen atoms arranged in a hexagonal ring structure.

Bessemer converter Machine that blows air up through the bottom of the melt to convert iron into steel.

Brass Alloy of copper and zinc.

Brittleness A lack of ductility. Brittle fractures occur through rupture of the interatomic bonds. Brittle fractures occur more readily in tension than shear; therefore, brittle materials have high compressive strengths and lower tensile strengths. Some typically brittle materials are stone, concrete, brick, ceramics, and some metals, such as high-carbon steels and cast iron.

Bronze Alloy of copper and anything else, traditionally tin.

Calorie Also known as the gram calorie, the heat required to raise the temperature of 1 g of water by 1° C.

Carbonitriding Heat-treatment process through which steels are case-hardened by heating them in a gaseous environment rich in both carbon and nitrogen.

Carborundum Silicon carbide (SiC) abrasive.

Carburizing Case-hardening treatment of steels in which the metal is submerged in a high-carbon atmosphere and the carbon is allowed to diffuse into the outer layers of the metal.

Case-hardening Process through which the outer layers of a steel are significantly harder than the inner core.

Cast iron Iron product that contains a carbon content greater than 2%.

Cathode Electrode in an electrolytic cell that accepts electrons; the negative electrode in a plating cell or the positive post on a battery.

Cation An element or radical that is positively charged.

Cement factor The number of bags of cement required per cubic yard of concrete.

Cementite Iron carbide.

Ceramic Compounds of metallic and nonmetallic elements that exhibit properties different from either, including stone, clay, and cement.

Cermets Ceramics that have metals as an ingredient.

Cetane number Measure of the ignition quality of diesel fuel. It is expressed as the percentage of cetane that must be mixed with liquid methylnaphthalene to produce the same ignition performance as the fuel being tested. Higher cetane numbers designate cleaner-burning fuel and better performance.

Charpy test Impact test that uses a swinging pendulum to strike a notched specimen. The specimen is held on both ends and struck in the middle.

Clay A fine-grained soil that consists primarily of hydrous aluminum silicate.

Coke Anthracite or hard coal that has all hydrogen and other atoms removed by heating in an oxygen-free atmosphere.

Composites Two or more components, which, when joined, exhibit properties unlike those available separately.

Concrete Any aggregate bonded together with a cementing agent. Typically, an artificial stone product made of gravel, sand, and a cement.

Condensation reaction The reaction of two compounds that results in a by-product that is a smaller molecule, such as water or alcohol.

Copolymer Polymer in which the mer is made up of two or more different molecules.

Cracking Process of refining petroleum in which large molecules are broken down into smaller units under heat and pressure or in the presence of a catalyst.

Creep Deformation that results from a load (typically tensile) applied over an extended amount of time.

Curing A treatment that is applied to increase the strength properties of a material such as polymers or concrete. The increased strength is the result of additional linkages being formed between molecules.

Cyaniding Process of case-hardening a steel by submerging it in sodium cyanide. The carbon and nitrogen in the cyanide are absorbed into the outer layers of the steel.

Deflection Deformation that results from a bending stress.

Density Mass per unit volume of a material.

Ductility Property of materials that allows them to undergo large permanent deformations prior to fracture.

Durometer Hardness tester used mainly for plastics and elastic materials that measures the elastic deformation of the surface of the material.

Elastic limit the greatest amount of stress that a material can develop without taking a permanent set.

Elasticity Property of materials that allows them to deform under load and return to their original condition when the load is removed.

Elastomer A synthetic material with elastic properties, such as rubber.

Electric arc converter Machine that uses large carbon electrodes to carry current to pig iron to burn out excess carbon and convert iron to steel.

Entrainment Dispersion of an undissolved liquid or gas in a liquid or gaseous medium.

Eutectic Composition of a system that yields the lowest melting temperature. Also, the composition of a system containing two or more components that allows the system to go directly from a liquid to a two-phase solid without passing through the two-phase region.

Extrusion Process of pushing a billet of material through a die to produce pieces that have uniform cross-sectional forms, such as beams and tubing. Used for aluminum and various polymers.

Fatigue Failure of materials due to cyclic strain.

Fatigue limit Maximum stress below which a material will not fail regardless of the number of cycles.

Ferric materials Materials composed of iron with a valency of $+3$.

Ferrite Alpha iron or a solid solution containing less than 0.025% carbon in iron. Generally considered pure iron.

Ferritic stainless steels Low-carbon steels that contain 20% to 37% chromium; designated as the 400 series of stainless steels.

Ferrous materials Materials composed of iron with a valency of $+2$.

Flame-hardening Hardening of the surface of a metal by heating it with a flame and then quenching it. Only the surface is heated into the austenitic range; therefore, only the surface goes into martensite.

Flash Extra material that flows into the joint between mold cavity halves, especially in compression molding. Flash must be removed after molding.

Flash point The temperature at which an oil will ignite but not maintain a flame.

Fluid Any material that flows in the liquid, vapor, or gaseous form.

Gage length The length over which the strainometer measures the strain or deformation.

Galvanizing The process of coating iron or steel with zinc to protect the base material from oxidation.

Gamma iron Austenite.

Glass The amorphous structure of silicates and certain oxide formations. It is generally considered a super viscous liquid.

Hardness The ability of a material to resist surface penetration. Different tests involve indenters, plastic deformation, elastic deformation, and resistance to surface abrasion.

Hardwoods Woods that come from deciduous trees.

High polymers Also called high-molecular-weight polymers; materials that have combined so many of the same molecules that the structure has an extremely high molecular weight.

Holography The use of lasers to produce interference patterns of true 3-D.

Hybrid Composite material that contains two or more different reinforcing fibers.

Hydraulic cement Term used in the past to distinguish between true cements and those of hydrated lime. Cements will harden under water, whereas lime is water-soluble and will dissolve if submerged under water for an extended period of time.

Index of refraction The ratio of the speed of light in a vacuum to the speed of light within a substance. It can be calculated by dividing the sine of the angle of incidence by the sine of the angle of refraction.

Inorganic compounds Compounds that do not include hydrocarbons and their derivatives. However, some simple compounds such as carbon dioxide and metallic carbonates are classified as inorganic compounds.

Intermetallic Material that is a combination of two or more different metal types.

Interply knitting The process of sewing together different layers or plies of reinforcing fibers. If the fibers are of different types, the product is termed an interply hybrid.

Interstitial An atom or particle not on a lattice point in a crystal; an imperfection.

Intraply hybrid A composite with different fiber types within each layer or ply.

Ion Electrically charged atom or group of atoms that has lost or gained electron(s).

Isomer Molecules that have the same number and types of atoms which are arranged differently. Isomers have the same molecular weight but exhibit different properties.

Isotropic material A material that exhibits different properties in different axes.

Izod test An impact test that uses a swinging pendulum to strike a notched specimen that is held firmly at one end.

Jominy test An end quench test for steel to determine the hardening depth.

Joule Unit of energy (in SI units) equal to the work done when a force of 1 N is exerted over a distance of 1 m.

Lay-up In composite manufacture, the position and direction of the reinforcing material.

Lime Term used for quicklime (CaO) and hydrated lime [Ca(OH)$_2$].

Linseed oil Vegetable oil derivative of flax. Due to oxidation, it darkens and thickens when exposed to air.

Liquid Material state in which the shape of the material depends on the vessel in which it is contained but its volume does not.

Maraged steel A high-nickel-content, low-carbon steel that is heated to 1500°F (822°C), air-cooled, and then aged at 900°F (486°C). Maraged steel is generally considered a superalloy.

Martempering A heat-treatment process for steels in which steel is heated into the austenitic temperature range, held at about 500°F (262°C) until it is uniformly heated throughout, and then quenched to room temperature.

Martensite The body-centered tetragonal structure of steel. It is the hardest form of steel, produced by rapidly quenching a steel that has been heated into the austenitic temperature range.

Martensitic stainless steels Stainless steels that can easily be hardened by quenching in oil. Produced from the 400 series of stainless steels, they are very resistant to attack by weak acids and bases.

Mechanical properties The properties of materials that refer to the material's behavior under applied loads.

Mer The basic molecule of polymers.

Miller index The method used to designate a specific plane or group of planes in a crystalline structure.

Modulus of elasticity Also known as Young's modulus; calculated by finding the slope of the stress-strain curve for a given material within the range of its linear proportionality between stress and strain.

Modulus of rupture The maximum nominal tensile stress developed in a specimen subjected to a bending load when the specimen fails in tension, calculated in the elastic region.

Monel An alloy of copper and nickel.

Monomer A chain molecule without crosslinkages, as opposed to a polymer.

Multigrade oil Engine oil that meets the requirements for more than one SAE viscosity grade classification. For example, 10W-40 grade motor oil exhibits 40-weight properties at higher temperatures and 10-weight properties at lower temperatures. The W means winter grade.

Nitriding Case-hardening process in which a steel is placed in a high-nitrogen atmosphere, allowing the nitrogen to diffuse into the outer layers of the steel.

Nominal stress and strain Nominal stresses and nominal strains are based on the original dimensions of the specimen. "True" stresses and "true" strains are computed based on instantaneous values while the specimen is subjected to given load conditions.

Nondestructive testing Processes used to determine properties of materials that do not render the material unsuitable for its intended purpose.

Normalizing A heat-treatment process in which steel is heated into the austenitic temperature range [about 100°F (38°C)] and then cooled to room temperature in still air.

Octane number Value that expresses the antiknock properties of gasoline, referenced to a standard fuel reference. The value cited at the gasoline dispensing pumps is an average of the research octane number (RON) and the motor octane number (MON).

Open hearth converter A method for changing iron into steel by blowing air across the top of the melt.

Organic chemistry The chemistry of carbon and carbon-containing materials.

Organic compound Substances that contain carbon and hydrogen; they may also contain other elements.

Oxygen lance converter A method of changing iron into steel by using a pipe to blow pure oxygen through the melt.

Pascal A unit of force per unit area or pressure in the metric system. One pascal equals 1 N of force per square meter.

PCB Polychlorinated biphenyls, a class of synthetic chemicals that do not occur naturally in petroleum but have been found as contaminants in used oils. They are classified as health hazards.

Pearlite The eutectoid composition of steel formed of alternating layers of ferrite and cementite.

Percent elongation The total percent strain that a specimen develops during testing.

Percent reduction in area The difference between the original and final cross-sectional areas of a test piece, expressed as a percentage.

Phase diagram Graph that illustrates the temperature and alloy composition of a material. Used to show the relationship between the temperature and alloy content of a material and the points at which changes in state occur.

Physical properties Properties of a material that are affected by heat, light, magnetism, electricity, and other such forces. Physical properties include other properties, such as mechanical, optical, electrical, and thermal.

Plastic deformation The permanent deformation that occurs after a material has been strained beyond the elastic limit.

Plasticity Property of a material to deform at a constant rate without fracture.

Plastics Term used for a variety of materials that are easily formed. Often used synonymously with polymers, although not all polymers exhibit plastic behavior.

Poisson's ratio While the test specimen is within the elastic range with loading taking place along a single axis, the ratio of lateral to axial strain.

Polymer A large molecule formed by linkage between two or more simple, unsaturated molecules (monomers).

Portland cement A material used to glue aggregates together to form concrete. When reacted with water, it produces a hard, rocklike material.

Precision The degree to which repeated readings agree with each other.

Proof stress The nominal stress that produces a corresponding specified permanent strain, typically 0.1% or 0.2%. The proof stress is used in calculations for materials that deform plastically without a definite yield point.

Proportional limit The greatest stress that a material can develop without deviating from linearity between stress and strain.

Reaction injection molding (RIM) The mixing of components in the proper ratio in a high-pressure chamber and forcing of the material into a mold. This process is used for polyurethanes, epoxies, and other liquid chemical mixtures.

Refractory materials Materials that can withstand very high temperatures.

Reliability The repeatability of a test, or the closeness of sets of data.

Resin transfer molding (RTM) The transfer of a catalyzed resin into a mold in which reinforcement has been placed.

Rheology The study of the deformation and flow of materials in terms of stress, strain, temperature, and time.

Rupture or breaking strength The stress applied at rupture or the load applied at rupture divided by the original cross-sectional area.

SAE Society of Automotive Engineers.

Sand A naturally occurring substance that is composed primarily of silica having a grain size between 0.004 in. (0.1 mm) and 0.375 in. (10 mm).

Saybolt furol viscosity The time in seconds required for 60 mL of a petroleum product to flow through a calibrated orifice under controlled conditions. The term *furol* is a contraction of fuel and road oils. The Saybolt furol viscometer has a larger orifice than the Saybolt universal viscometer.

Saybolt universal viscosity The time it takes for 60 mL of a petroleum product to flow through a calibrated orifice under controlled conditions.

Silica Silicon oxide (SiO_2).

Sintering The fusing together of particles under pressure at temperatures below their melting temperatures.

Slip, or slippage The movement of parallel planes within a material in parallel directions. These material planes slip or slide past each other, causing a fracture along that plane.

Specific gravity The density of a material divided by the density of water.

Spheroidizing A heat-treatment process for steel that causes the cementite in the steel to form small spheroids, reducing the brittleness of the steel.

Stainless steels Steels that have high chromium or nickel content that retards oxidation.

Steel An alloy of carbon and iron containing less than 2% carbon.

Strain The change per unit length in the linear dimensions of a body accompanying a coincident change in stress.

Stress The intensity of the internally distributed forces or components of forces that resist a change in the form of a body. Commonly measured in units dealing with a force per unit area, such as pounds per square inch (lb/in.²) or megapascals (MPa). The three basic types of stress are tension, compression, and shear. The first two—tension and compression—are called direct stresses.

Superalloys Alloys of chromium, nickel, titanium, and other metals with iron. When properly heat-treated, these alloys can withstand tensile stresses exceeding 300,000 lb/in.² (2.069 GPa).

Superconductivity The property of a material or alloy that allows the resistance to approach zero at reduced temperatures.

Tempering A secondary heat-treatment process in which steel is reheated to about 700°F (374°C) and then cooled. Tempering relieves the internal stresses in the steel, making it tougher.

Tensile strength The maximum tensile stress that a material is capable of developing during a test.

Terra cotta Burned clay used in applications such as tile and ornamental work.

Thermoplastic materials Polymers that soften on heating or reheating.

Thermosetting materials Polymers that harden and char or burn when heated or reheated.

Toughness The ability of a material to withstand applied stress or the amount of energy required to cause rupture.

TTT curve A graphical representation of the time and temperatures at which changes occur in a particular steel as it changes from austenite to pearlite or martensite. The cooling rates for these structures can be taken directly from the curve.

Ultimate strength The greatest stress that a material can withstand prior to failure.

Valency The combining power of an atom or the capacity of an atom to form chemical bonds with other atoms.

Validity The ability of a test to measure what it is supposed to measure.

Viscosity Measurement of the resistance of a fluid to flow due to the internal friction caused by the cohesive forces in the material. Typically used to describe the flow characteristics of materials such as motor oils.

Vulcanizing The addition of sulfur to the polymerization of rubber.

Warp The longitudinal direction of the reinforcing fabric in composites.

Weft The transverse direction of the reinforcing fabric in composites.

White metals Low-melting-point metals, including antimony, bismuth, cadmium, lead, and tin.

Wrought iron Iron with a very low carbon content (less than 0.1%) that has a fibrous structure. Often used in decorative work.

Yield (concrete) The quantity of concrete obtained in cubic feet or yards per bag of cement.

Yield point Two critical points within the yield range for ductile materials are the upper yield point (UYP) and the lower yield point (LYP), which are used to characterize the plastic range of a material.

Yield strength The stress at which a material exhibits a specified limiting permanent set.

Yield stress The minimum stress at which strain occurs without a further increase in stress. The additional strain is the result of plastic deformation. Most alloys do not exhibit distinct yield stresses; therefore, a proof stress must be used.

Young's modulus See *Modulus of elasticity.*

APPENDIX E

Answers to Selected Questions and Problems

CHAPTER 1

5. Organic compounds are substances that contain carbon and hydrogen and may include other elements. Inorganic compounds do not contain hydrocarbons or their derivatives.

6. a. hydrogen: $1s$
   b. helium: $1s^2$
   c. sodium: $2s^2$, $2p^6$, $3s$
   d. copper: $2s^2$, $2p^6$, $3s^2$, $3p^6$, $3d^{10}$, $4s$
   e. iron: $2s^2$, $2p^6$, $3s^2$, $3p^6$, $3d^6$, $4s^2$
   f. gold: $2s^2$, $2p^6$, $3s^2$, $3p^6$, $3d^{10}$, $4s^2$, $4p^6$, $4d^{10}$, $4f^{14}$, $5s^2$, $5p^6$, $5d^{10}$, $6s$
   g. silver: $2s^2$, $2p^6$, $3s^2$, $3p^6$, $3d^{10}$, $4s^2$, $4p^6$, $4d^{10}$, $5s$

7. Ionic, covalent, metallic, and van der Waals forces

8. n: principal quantum number
   l: angular momentum
   M: energy level
   M_s: spin direction

9. a. Isotope: atoms that vary from the normal atomic mass found in naturally occurring forms of the element.
   b. Ion: versions of the original element that deviate from the normal neutral electrical state.
   c. Allotrope: alternative crystalline structures.

10. (1) cubic, which includes simple cubic, body-centered cubic, and face-centered cubic; (2) tetragonal, which includes simple tetragonal and body-centered tetragonal; (3) orthorhombic, which includes simple orthorhombic, body-centered orthorhombic, base-centered orthorhombic, and face-centered orthorhombic; (4) rhombohedral; (5) hexagonal, which includes hexagonal close-packed and face-centered cubic close-packed or complex cubic structure, also referred to as the diamond structure; (6) monoclinic, in the simple and base-centered varieties; and (7) triclinic.

CHAPTER 2

1. Steel contains a certain percentage of carbon.
2. Cast iron has a higher carbon content.
3. a. 1020: 0.20% C, carbon only
   b. 1060: 0.60% C, carbon only
   c. 1340: 0.40% C, manganese
   d. 3350: 0.50% C, nickel and chromium
   e. 4120: 0.20% C, chromium and molybdenum
   f. 4340: 0.40% C, nickel, chromium, and molybdenum
   g. 6140: 0.40% C, chromium and vanadium
   h. 8640: 0.40% C, nickel, chromium, and molybdenum
4. Plain-carbon steel: carbon is the primary alloying element.
   Low-alloy steel: low-alloy structural steels have less alloy content than most other alloy steels.
   High-alloy steel: high percentage of alloying elements.
5. Major ores are (1) magnetite, (2) hematite, and (3) taconite.
6. Coke: anthracite or hard coal that has been converted by heating in the absence of oxygen.
   Slag: impurities that are drawn off the top of the melted iron.
12. Common steel shapes include the following:

Angles:	Angles can be of equal or unequal leg length. Equal-leg lengths range from 1×1 up to 8×8 in. Unequal-leg lengths go up to 9×4 in.
Bars:	Solid shapes, hot- or cold-drawn, in sizes ranging from ¾ to 12 in. thick.
Beams:	Standard I and H beams. I beams range from 3×2.25 in. up to 24×8 in. H, or wide-flange, beams range from 8×5.25 in. up to 36×16.5 in.
Billets:	Section of ingot suitable for rolling.
Blooms:	Slab of steel, where width and depth are generally equal.
Channels:	Sizes range from 1.25×3 in. up to 4×8 in. They are U-shaped in cross section.
Plates:	Large, flat slabs thicker than ¼ in.
Sheets:	Generally less than ¼ in. in thickness, hot- or cold-rolled, and available in coils as well as flat sheets.
Tubing:	Square, rectangular, and round tubing and pipe can be made from sheet stock. Rectangular tubing ranges in size from 2×3 in. to 12× 20 in. Square tubing ranges from 2×2 in. to 16 ×16 in. Round tubing and pipe is available from ¼ in. up to several feet in diameter in several thicknesses.
Wires:	Drawn from bars that have been rolled down to small diameters. Steel wire is further fabricated into nails, cable, fence, screens, and similar products.

13. As the carbon content of the steel increases:
   The steel becomes more expensive to produce.
   The steel becomes less ductile, i.e., more brittle.
   The steel becomes harder.
   The steel loses machinability.
   The steel has a higher tensile strength.
   The melting point of the steel is lowered.

The steel becomes easier to harden.

The steel becomes harder to weld.

14. Aluminum: used to produce killed steel, promotes toughness
Manganese: promotes strength, hardenability, and hardness
Boron: promotes hardenability
Copper: promotes corrosion resistance
Chromium: promotes corrosion resistance and hardenability
Niobium: increases tensile strength
Titanium: high strength at high temperatures
Tungsten carbide: high hardness
Vanadium: toughness and impact-resistance

15. Steel that has a percentage of aluminum as an alloying element; this helps deoxidize the melt and results in high uniformity.

16. Steel loses its strength at high temperatures due to impurities in the melt, such as sulfur, which accumulate at the grain boundaries.

17. Ferrite is the BCC version of iron and its solid solutions.

18. Austenite is the FCC version of iron and its solid solutions.

19. Pearlite is the euctectoid composition of ferrite and cementite.

20. Cementite is iron carbide.

21. Martensite is the hard and brittle product formed when steel is heated into the austenitic range and rapidly cooled. It is composed of ferrite saturated with dissolved carbon.

22. A grey cast iron, where all of the carbon content is in graphite form, appears grey and dull. A white cast iron, with the carbon as iron carbide or cementite, appears bright and silvery at the fracture.

26. Ferritic stainless steels, austenitic stainless steels, and martensitic stainless steels

27. High-strength steels, also referred to as superalloys or maraging steels, contain 18% to 25% nickel, 7% cobalt, and small amounts of other elements, such as molybdenum, titanium, and zirconium.

28. Two different metals in contact with an electrolyte, typically water, comprise an electrolytic cell or galvanic cell. One metal acts as the positive electrode, or anode, whereas the other acts as the negative electrode, or cathode. This process erodes the anode. The oxide or iron forms a larger crystal than the steel itself, causing the rust to buckle away from the surface of the metal and to flake off, eroding the surface.

35. a. 1020
b. 1340
c. 4120
d. 4340
e. 2540

36. 60:40:10 ductile cast iron

39. A finite period of time is required for full conversion.

CHAPTER 3

1. Nonferrous metals are those that do not contain iron as a principal alloying element.

2. Copper-zinc alloys are brasses; copper and any other alloys are generally bronzes.

4. The triple-plating process is used to plate steels. The steel is first degreased and cleaned well. It is then etched with nitric acid to roughen the surface of the steel, giving the plating

something to hold onto. The steel is first given a thin layer of copper. After washing, the copper is then covered with a thin layer of nickel. The part is then washed again, and a final layer of chromium is deposited on the surface of the part. Coats of about 0.0002 in. provide a shiny, decorative surface, whereas platings of 0.05 in. are used to provide wear resistance. Chromic acid is used as the electrolyte plating solution.

6. a. Tin
    b. gold
    c. silver
    d. copper
    e. chrome
    f. nickel
7. Refractory metals, by definition, have melting points above 3600°F (2000°C), some of them reaching as high as 6150°F (3426°C). These metals include iridium, osmium, and ruthenium, which are in the platinum group. Common refractory metals include chromium, columbium, hafnium, molybdenum, niobium, rhenium, tantalum, tungsten, and vanadium.
8. As an alloying element in steels
9. Phosphorous
10. a. 65% Cu, 35% Zn
    b. 60% Cu, 40% Zn
    c. 70% Cu, 30% Zn
    d. 95% Cu, 5% Al
    e. 88.5% Cu, 10% Ni, 1.5% Fe
11. The flotation process involves grinding the ores into powder and placing the powder in water. A foaming material such as soap is added to create a froth that brings the copper ore to the surface. The ore is then skimmed off, leaving the undesirable materials in the water.
12. After refining by the flotation process, the ore mixture is roasted in an oven. After roasting, the ore mixture contains copper oxides, iron sulfides, copper sulfide, iron oxides, iron sulfates, silicates, and other impurities. The ores are then placed in a smelting furnace, where they are melted at temperatures of about 2600°F (1438°C). Once melted, the mixture is called *matte copper,* which contains about 30% copper.
13. After matte copper is produced, the mixture is then placed in a converter with a flux (usually silica), where air is blown through the melt. The oxygen in the air, much as in the steel-conversion process, reacts with the sulfur in the melt and produces reasonably pure copper. The copper is then drained off and poured into ingots. The sulfur dioxide produced in the conversion process bubbles out as the copper cools. This gives the copper the name *blister copper.* This copper is 98% to 99% pure.
14. Superconductors have very little resistance to current flow. A current started in a superconducting circuit will continue to flow almost indefinitely.
15. The tin in solders allows the solder to "wet," or bond with, the metals to be joined.
16. Refractory metals are used in high-temperature applications, where they exhibit high strength; they oxidize rapidly at low temperatures.
17. The Hall method involves the electrolysis of a molten solution of alumina in cryolite or sodium aluminum fluoride (Na_3AlF_6) at temperatures around 1745°F (950°C). Once in solution, the aluminum is separated by electrolysis. The Hall method brought the price of aluminum down to as low as 15¢ per lb, or 0.033¢ per kg.
18. Pure aluminum oxide is extracted using the Bayer process. Bauxite is open-pit or strip mined near the surface. The ore is crushed and washed to remove undesirable materials and dried. The dried powder is then mixed with soda ash (Na_2Co_3), lime (CaO), and water, after which it forms sodium aluminate ($Na_2Al_2O_4$). After filtering, aluminum hydrate

[AlO(OH)] is precipitated from the solution and heated to temperatures of about 2000°F (1100°C) to form aluminum oxide (Al_2O_3). This product is approximately 99.6% pure.

19. The anodizing process involves placing an anode of aluminum in an electroplating cell with oxalic, sulfuric, or chromic acid as the plating solution or electrolyte. Current is applied to the solution, and the anode is plated with an oxide coating, which is hard and wear-resistant. Anodized coatings give the metal a better appearance and may be colorized by using different metals in the electrolyte or by dying the metal.

20. Carbon monoxide gas is washed over the heated ore, which converts the nickel to nickel carbonyl [$Ni(CO)_4$]. Nickel carbonyl is a volatile compound that turns from a solid directly into a gas at a temperature of about 1783°F (973°C), which causes it to decompose into metallic nickel dust and carbon monoxide.

21. Lead is versatile and is used for its high density, ductility, low melting point, corrosion resistance, and ability to lubricate. Common uses include (1) corrosion-resistant tank linings; (2) lead-acid storage batteries; (3) type metal for the printing trade; (4) bearing metals; (5) biological shields against radiation; (6) pipes and pipe fittings; (7) cable shielding; (8) low-melting solders; (9) alloying element for various metals, and (10) thermal fuses.

22. For example, the galvanizing process is used to provide cathodic protection. As long as there is zinc available, the zinc will give up electrons to the iron, which protects the iron from corrosion. This process is termed cathodic protection.

CHAPTER 4

2. Normalizing involves heating the metal well into the austenite range, letting it remain there for 1 hour per inch of thickness, and then letting it cool in still air at room temperature. Normalized metals have an even grain size that makes them easier to machine. Normalizing is often used prior to other heat treatments to provide good machinability.

3. The slow cooling of a steel from the austenitic temperature range is called annealing. This leaves the steel in its softest possible condition.

4. Case-hardening, also referred to as surface-hardening, involves one of four different methods: carburizing, nitriding, cyaniding, or carbonitriding. Case-hardening is used on parts such as gear teeth, cutting wheels, and tools.

5. Parts are heated below the 1333°F (723°C) lower transformation temperature and cooled slowly. The cementite in the steel is spheroidized, producing a tougher steel.

6. Tempering (sometimes called drawing, supposedly because it "draws" the hardness from the metal) is the process of reheating the metal immediately after hardening to a temperature below the transformation temperature and cooling to increase the ductility and toughness of the steel. After hardening, the steel is heated to between 700° and 800°F (375° and 430°C) and is allowed to remain there for 1 hour per inch of thickness. The steel is then allowed to cool. Tempering softens the steel slightly.

7. Martempering is similar to general tempering, except the part is quenched to a temperature just above the M_s line [between 500° and 600°F (260° and 316°C)] for a few seconds to allow the temperature throughout the part to stabilize.

8. Austempering resembles martempering, except after leveling the temperature at about 700°F (375°C), it is held there for a longer period of time while it passes through the P_s and P_f lines. At this temperature, bainite is formed.

10. When quenched, there are four separate actions that occur: (1) vapor formation, (2) vapor covering, (3) vapor discharge, and (4) slow cooling.

23. An interstitial is a particle not located at a lattice point in a crystal; therefore, it is an imperfection in the lattice.

CHAPTER 5

1. Single covalent bonds between atoms do not provide for additional atoms to be added; they are said to be saturated. Saturated molecules have strong intramolecular bonds but weak intermolecular bonding. When carbon and hydrogen form unsaturated molecules, such as ethylene and acetylene, the molecules form double or triple covalent bonds. An unsaturated molecule doesn't have the necessary hydrogen atoms to satisfy the outer shell of the carbon atoms. Many molecules form many double bonds and are referred to as polyunsaturated compounds.

2. Two molecules with the same composition that form two different configurations having different properties, such as propyl (1-propanol) and isopropyl (2-propanol) alcohol, are called isomers.

3. Addition polymerization involves bonding many of the same mer by simply splicing them together. In copolymerization, more than one molecule makes up the mer. Condensation polymerization involves chemical reactions that form new molecules and generally produce a by-product such as water. A catalyst may be required to induce condensation polymerization.

8. Some of the more common production methods are casting, blow molding, compression molding, transfer molding, injection molding, extrusion, lamination, vacuum forming, cold forming, filament winding, calendaring, and foaming.

10. Plastic or plasticity is a term used to describe the mechanical property of a material whereby it will undergo a significant deformation or elongation without breaking. Polymer is the proper term applied to those materials commonly thought of as plastic.

11. Intramolecular and intermolecular bonding

13. Various materials can be added to polymers to improve certain properties, reduce the cost of the polymer, improve the formability of the material, and/or colorize the material. These additives are termed fillers, plasticizers, coloring agents, and lubricants. They are often divided into two groups: performance- or processing-related additives, depending on their intended purpose.

14. Noncrystalline state

15. Isotropic materials exhibit the same properties regardless of viewing plane; anisotropic materials' properties depend on the viewing angle.

16. Crystallinity: formation of crystals; crystallization: the ability to form crystals

19. The study of the deformation and flow of materials in terms of stress, strain, temperature, and time

20. A combined elastic and plastic response to applied loads

21. There are primarily two principles used in the foaming process: physical foaming and chemical foaming.

22. Physical foaming requires the injection of a gas during the molding process. When this combination is heated, the gas expands and causes the combination to expand. Expanded polystyrene is an example of a polymer that has been physically foamed.

Chemical foaming is the result of the chemical reaction that occurs when two chemicals are mixed. One part of the reaction is the base polymer. The other part of the reaction is a catalyst or initiator and a chemical that reacts with the polymer to form a gas. When these two parts are combined, they produce a gas that causes the material to expand. Polyurethane foam is a chemically foamed polymer.

23. Polystyrene and polyurethane

25. A synthetic material with elastic properties, such as rubber

26. Vulcanization is produced by adding sulfur to the mix. The sulfur produces cross-linkages between the chains of the rubber molecules, which make the rubber stiffer and harder. These cross-linkages reduce the slippage between chains, increasing the elastic properties of the material. In the vulcanizing process, some of the double covalent bonds between molecules are broken, allowing the sulfur atoms to attach themselves and form cross-linkages between molecules.

28. Elastomer molecules form chains, similar to polymers. These chains may take one of two forms, either the cis- form or the trans- form. The chains of the elastomer rotate and spin endlessly around single carbon-to-carbon bonds, such as the case with isoprene. When these single bonds become double bonds, they are no longer free to rotate. Looking at the molecules that make up the elastomer, if the two end groups are on the same side of the enclosed double bond, the molecule is in the cis- form. If the end groups are on opposite sides of the double bond, it is in the trans- form.

30. For elasticity and not plasticity

CHAPTER 6

3. Due to climate trends, a tree grows more rapidly in the spring than in the summer. Cells that grow in the spring tend to be longer, with thinner walls, whereas the cells that grow in the summer are short, with thicker walls. Because of this, summerwood tends to be denser, stronger, and usually darker in color than springwood.

5. Hygroscopy is the property of a material in which it absorbs water from the air when it is present in the surrounding environment and gives up water when the surrounding air is less moist than the material.

8. 0.085%

9. Collapse, pitch, knot, chipped grain, loosened grain, raised grain, torn grain, skips in surfacing, sizing variations, saw burns, and gouging

10. Hardwood grading is based on the percentage of the piece of lumber that can be cut into smaller pieces of standard size and maintain one face clear of defects and the other face sound in composition. Based on this system, the standard grades for hardwood lumber are firsts, seconds, selects, no. 1 common, no. 2 common, sound wormy, no. 3A common, and no. 3B common.

11. Softwood lumber is graded according to use, size, and method of manufacture. Lumber classified according to use includes yard lumber, structural lumber or timber, and factory, or shop, lumber. Softwood lumber comes in three grades: select, standard, and common. Select lumber is free of all major defects and has at least 75% of the strength of a clear test specimen. Standard lumber may contain minor defects but must maintain at least 60% of the strength of a clear test specimen. Common lumber may contain defects and is used where strength is not a primary concern. Common lumber must have at least 30% of the strength of a clear test specimen.

12. Several adhesives are used to produce plywood, particleboard, and hardboard and to assemble veneers. These adhesives are primarily animal glues, blood-albumin glues, fish glues, glues made from starch, casein, and synthetic resin glues.

13. The term *veneer* refers to a thin sheet of material or a form of protective facing. In the wood industry, veneers are used as decorative or protective facings and are rotary-cut, sliced, or sawed from a log, or flitch.

15. Standard preservatives are grouped into three categories: creosote, oil-based preservatives, and water-based preservatives.

16. Pressure-impregnation consists of placing the wood in a pressure vessel and forcing the preservative into the pores of the wood.

17. Mechanical protection involves erecting barriers to attack, such as painting, ventilation, and water drainage. Soil poisoning presents a chemical barrier to insect attack. Liquid chemical preservatives are applied by brushing or spraying to protect the wood's surface. Dip diffusion involves dipping the wood into a tank of preservative and allowing the wood to soak up the preservative, thus permeating the wood.

CHAPTER 7

1. Soda-lime glass is still the most predominant and cheapest glass. It is used in the manufacture of windowpanes, bottles, light bulbs, and other similar items.

2. Tempered glass is often used in such applications as swinging doors, eyeglasses, sliding glass doors, such as patio doors, and safety glass.

3. Borosilicate glass (trade names include Pyrex, Kimax, and others) exhibits excellent thermal qualities and is, therefore, used in the production of ovenware and the heat-resistant glass commonly used in scientific and technical products. High-content silica glass contains approximately 96% silica. Typically, it is made from 96.5% silica, 3% boron oxide, and 0.5% aluminum oxide. Applications of high-content silica glass include space-shuttle windows, missile nose cones, laboratory glassware, and heat-resistant coatings.

4. Fused silicates, or quartz glasses, are produced from 100% fused silica having approximately the same properties as quartz. It has a very high optical transparency and very high softening point, and it will transmit ultraviolet waves (unlike most glasses). Fused silica glass consists entirely of silica in a noncrystalline state. It is the most expensive glass to produce, is the most difficult to fabricate, and can withstand greater temperatures. It is, therefore, used in severe conditions where there is no plausible substitute, such as telescopes and melting crucibles.

5. Rare-earth glass contains no silicates at all. It contains approximately 28% lanthanum oxide (La_2O_3), 26% thorium oxide (ThO_2), 21% boron oxide, 20% tantalum pentoxide (Ta_2O_5), 3% barium oxide, and 2% barium tungstate ($BaWO_4$). It has the highest refractive index of any class. Because of its clarity and high refractive index, it is used almost exclusively in the manufacture of lenses and optical applications.

7. 6.975° or about 7°

8. 1.33

9. 11°

10. Ceramic materials are complex compounds and solutions that contain both metallic and nonmetallic elements and are typically heated at least to incandescence during processing or application. They are typically hard and brittle and exhibit high strength, high melting points, and low thermal and electrical conductivity.

11. Drying, firing, and glazing

12. Some materials must be fired at elevated temperatures to provide fusion and cause chemical reactions in the material that produce desired properties. The fusion process is called

sintering. As a result of sintering, the boundary edges of the individual particles fuse together to form bonds. This bonding gives ceramics their strong, rigid, brittle nature.

13. Spalling is defined as the thermal cracking of ceramic materials. Most ceramics will crack if rapidly heated or cooled.

15. The general categories of silica glass include soda-lime glass, borosilicate glass, lead oxide glass, aluminosilicate glass, high-content silica glass, fused silica glass, and pyrocerams.

16. Laminated, tempered, and wired

17. Glass should, theoretically, have an incredibly high tensile strength and be highly resilient to breakage. In actuality, the surface of the glass has minute cracks, which develop during the cooling period. These cracks concentrate the stress. When an object hits the glass, these cracks open up and break the bonds that hold the glass together. Bond after bond breaks, allowing the crack to propagate through the glass.

18. Limestone, sandstone, shale, slate, granite, and marble (synthetic and natural)

19. Aluminous clay, ball clay, corundum, cristobalite, feldspar, flint clay, kaolin, mullite, and vermiculite

20. There are four basic methods associated with clay manufacturing: (1) slip casting, (2) soft-mud process, (3) stiff-mud process, and (4) dry-press process. These four processes vary in the amount of water used to temper the clay mixture.

21. Acidic, basic, and neutral

22. These materials include the aluminas (Al_2O_3), chromic oxides (Cr_2O_3), iron oxides (Fe_2O_3), magnesias (MgO), silicas (SiO_2), zirconia (ZrO_2), zircon ($ZrO_2 \cdot SiO_2$), and various other materials, including nitrides, carbides, borides, and graphite.

23. Cermets are composed of ceramic particles and powdered metal, which have been fused.

26. Aluminum oxide and silicon carbide are two examples of abrasives. Aluminum oxide is preferred for harder materials, because it wears away faster, thus exposing new cutting surfaces more frequently. Silicon carbide is used for cast irons and softer nonferrous metals.

CHAPTER 8

1. Cement is an adhesive used to bond the ingredients in concrete. Cements, as in organic rubber cements or inorganic portland cements, are adhesive materials that coat and bind the aggregate (bulk or filler material). Concrete has been defined as the product of bonding any aggregate together with a cementing agent that hardens into a solid mass.

2. Ferric oxide serves as a flux.

3. Type I: Normal
   Type II: Modified
   Type III: High early strength
   Type IV: Low heat
   Type V: Sulfate resistant

4. Portland cement is made from limestone and clay, with ferric oxide added as a flux. These materials are blended together in a powdered form and heated in a kiln to approximately 2900°F (1606°C). This drives all the water from the mixture. This mixture is drawn from the kiln in the form of a black slag called clinker.

5. Hydraulic cements, such as portland cements, set and harden by taking up water in complex chemical reactions called hydration.

6. The primary factor in strength for concrete is the water-to-cement ratio, measured in gallons per sack of cement.

7. Total weight $= 180\ \text{lb} + 30\ \text{lb} + 16\ \text{lb} = 226\ \text{lb}$
   Volume produced $= (226\ \text{lb})/144\ \text{lb/ft}^3 = 1.57\ \text{ft}^3$
   Yield $= (1.57\ \text{ft}^3 \cdot 94\ \text{lb/bag})/(30\ \text{lb}) = 4.92\ \text{ft}^3/\text{bag}$
   Cement factor $= (27\ \text{ft}^3/\text{yd}^3)/(4.92\ \text{ft}^3/\text{bag}) = 5.49\ \text{bags/yd}^3$

8. Assuming a 1:2:4 ratio:
   Water $= 0.24 \cdot 5\ \text{yd}^3 = 1.2\ \text{yd}^3$
   Cement $= 0.18 \cdot 5\ \text{yd}^3 = 0.9\ \text{yd}^3$
   Aggregate $= 0.58 \cdot 5\ \text{yd}^3 = 2.9\ \text{yd}^3$
   Specific gravities are:
   Water $= 1.00$
   Cement $= 3.10$
   Sand $= 2.65$
   Gravel $= 2.65$
   Densities are:
   Water $= 62.4\ \text{lb/ft}^3$
   Cement $= 94\ \text{lb/ft}^3$
   Sand $= 105\ \text{lb/ft}^3$
   Gravel $= 96\ \text{lb/ft}^3$
   Bulk weight of each constituent is:
   Water $= 1.2\ \text{yd}^3 \cdot 27\ \text{ft}^3\text{yd}^3 \cdot 62.4\ \text{lb/ft}^3 = 2022\ \text{lb}$
   Cement $= 0.9\ \text{yd}^3 \cdot 27\ \text{ft}^3/\text{yd}^3 \cdot 94\ \text{lb/ft}^3 = 2284\ \text{lb}$
   Aggregate $= 2.9\ \text{yd}^3 \cdot 27\ \text{ft}^3/\text{yd}^3 \cdot 2.65 \cdot 62.4\ \text{lb/ft}^3 = 12{,}948\ \text{lb}$
   Total weight $= 2022 + 2284 + 12{,}948 = 17{,}254\ \text{lb}$
   Water $= (2022\ \text{lb})/(8.345\ \text{lb/gal}) = 242.3\ \text{gal}$
   Cement $= (2284\ \text{lb})/(94\ \text{lb/bag}) = 24.3\ \text{bags}$
   Aggregate $= (12{,}948\ \text{lb})/(2000\ \text{lb/ton}) = 6.48\ \text{ton}$ (2.16 tons sand and 4.32 tons gravel)

9. W/C ratio $= (242.3\ \text{gal})/(24.3\ \text{bags}) = 9.97\ \text{gal/bag}$

10. $144\ \text{lb/ft}^3 \cdot 27\ \text{ft}^3/\text{yd}^3 = 3888\ \text{lb/yd}^3$

11. Area $= 24 \cdot 28 \cdot 0.5 = 336\ \text{ft}^3$
    $(336\ \text{ft}^3)/(27\ \text{ft}^3/\text{yd}^3) = 12.45\ \text{yd}^3$
    $12.45\ \text{yd}^3$ of concrete

12. a. Tricalcium silicate $(3\ CaO \cdot SiO_2)$
    b. Dicalcium silicate $(2\ CaO \cdot SiO_2)$
    c. Tricalcium aluminate $(3\ CaO \cdot Al_2O_3)$
    d. Tetracalcium aluminoferrite $(4\ CaO \cdot Al_2O_3 \cdot Fe_2O_3)$

13. A standard cone is filled with concrete and then removed, leaving a standing cone of concrete, which will tend to fall and spread out. The difference between the top of the concrete while in the cone and the top of the concrete once it has fallen, or slumped, is the slump value for that batch of concrete.

14. A silt test is made by filling a clear container with about 2 in. of aggregate. The container is filled about three-quarters full with water and shaken vigorously for about a minute. The mixture is then allowed to sit on a level surface for about an hour. If a layer of silt more than 1/8 in. settles on the aggregate, the aggregate is rejected or washed prior to using it in the concrete.

15. A 500-mL clear container is filled about two-thirds full with a 3% lye solution for the calorimetric test. This test is conducted to check for organic matter in the aggregate. Lye is sodium hydroxide (NaOH). The mixture is shaken for approximately 1 min and allowed to sit for about 24 hours. The lye above the aggregate at the end of 24 hours should be clear or a light yellow. The darker the color, the more organic matter that the aggregate contains.

16. Accelerators and retarders are also concrete additives, which advance or delay the initial set of a concrete. An accelerator can be used in cold weather for early removal of forms or to protect the concrete. Calcium chloride in amounts up to 2% is often used. Retarders are often used in hot weather to allow sufficient time to work the concrete. The retarder will delay hydration of the cement, which retains the water for workability. Starch or sugar in amounts up to 0.05% are used to retard hydration for up to 4 h.
17. Air-entraining additives are used to reduce the effects of the freeze-thaw cycle.
18. One method involves placing reinforcing bars or a reinforcing grid in the center of the concrete as it is poured. The concrete is poured around the reinforcement. The purpose of the reinforcement is to absorb the stresses in the concrete.
19. In prestressing, the concrete is kept in compression. One method of prestressing concrete requires placing steel rods or cable under tension in position as the concrete is poured over them. After the concrete has set up, the tension on the rods is released. This places the concrete in compression. The tensile stress in the prestressed rods in the concrete exerts a compressive force on the concrete. For the concrete to fail in tension, the tensile load on the concrete must exceed the compressive force exerted by the tension in the rods.
20. Exerting a tensile force on the reinforcement after the concrete has set up or has cured

CHAPTER 9

1. Composites are composed of two or more combined materials; they exhibit improved properties over their individual components.
2. All composites generally have one thing in common: a matrix or binder, along with a reinforcing material. If the fibers are directionally oriented and continuous, the material is termed an advanced composite.
3. Composite materials, matrix, reinforcement material, reinforcement placement, reinforcement type, aspect ratio, reinforcement volume, and manufacturing method used
4. Intraply, interply, selective placement, and interply knitting
5. Alternating layers are bonded together in plies.
6. Sandwich structures have thin layers of facing materials over a low-density honeycomb core, such as a polymer foam or expanded metal structure. A familiar sandwich-structured composite is corrugated cardboard.
7. Hand lay-up, injection molding, lamination, filament winding, and pultrusion
8. A hybrid is a combination of fibers, giving the best of each to the composite. Hybrids are available in various forms: intraply, interply, selective placement, and interply knitting.
9. The direction in which the stress or load is applied is called the warp; the direction perpendicular to the stress is called the weft. Between the warp and the weft are the bias directions.
10. The length-to-diameter, or aspect, ratio of the fibers used as reinforcement influences the properties of the composite. The higher the aspect ratio, the stronger the composite. Therefore, long, continuous fibers are better for composite construction.

CHAPTER 10

1. a. The synthetic binder generally placed on one or both of the surfaces to be bonded
   b. The material or materials to be bonded

     c. A conversion that involves cross-linking by polymerization

     d. The term used to describe the process of a liquid adhesive converting to a hardened condition

     e. The property of an adhesive that causes the surface coated with an adhesive to form a bond on contact with the other surface to be joined; the stickiness of the adhesive

     f. The measured resistance of an adhesive joint to stripping

2. Wettability refers to the ability of a material to wet the surface of the adherend materials. This happens when forces between adhesive molecules and base molecules are greater than between the adhesive molecules. Ideally, all of these forces should be the same (within the adhesive, between the adhesive and the adherend materials, and within the base material).

4. Pressure-sensitive adhesives do not wet the surface to which they adhere. Tape is one example of a pressure-sensitive adhesive. The adhesive used remains permanently tacky and does not set. On the back side of the tape is a release coat. This coating allows the tape to be unrolled without stripping the adhesive from the tape itself.

5. Bond strengths are directly proportional to the width of the lap joint but are independent of the length or overlap. Joint stresses are generally concentrated in the first inch of overlap; as the length of the lap increases, the unit strength decreases.

6. a. Glue is a term generally applied to materials that occur naturally, such as those from animal, vegetable (starch), casein, and soybean origins.

     b. Cements are used for a specific purpose, such as rubber cement, plastic cement, and glass, stone, leather, or other cements.

     c. Pastes are generally water-soluble products used for paper and paper products.

11. Contemporary sealants include polysulfides, silicones, polyurethanes, butyl, polybutene, polyisobutene, acrylic rubbers, and polychloroprene.

12. Organic coatings may contain the following: (a) a binder or vehicle; (b) coloring agents or pigments; (c) solvents and additives to control viscosity; and (d) additives to act as inhibitors, stabilizers, or thickeners and to alter physical or chemical properties.

13. a. Protect materials from corrosion, exposure and weathering

     b. Improve visibility through luminescence or reflectivity

     c. Provide electrical, thermal, or acoustic insulation

     d. Improve appearance through decorative effects

     e. Enhance marketability and advertise the product

     f. Improve safety (warning)

     g. Inform users (traffic lanes, road signs, and so on)

     h. Camouflage or hide certain features (hunter's clothing)

     i. Identify persons or products (color coding)

14. There are several types of enamels currently available, including acrylic, alkyd, epoxy, phenolic, silicone, urethane, and vinyl.

15. a. Paints are a mixture of a binder and pigment that typically hardens by cross-linking.

     b. Enamels are a blend of a paint and a varnish. They harden through cross-linking, where the agent is oxygen available in the environment.

     c. Lacquers contain a binder dissolved in a solvent that dries by evaporation of the solvent to produce a thermoplastic finish.

     d. Varnishes form a thermosetting film but contain no pigment.

16. There are three types of solvents used in binders: solvents, diluents, and thinners. Active solvents are used to dissolve the binder and lower the viscosity of the coating.

17. a. Active solvents are used to dissolve the binder and lower the viscosity of the coating.

   b. Diluents are used to weaken the active solvent and increase the overall amount of coating.

   c. Thinners are used to increase the amount of coating material without reducing the effectiveness of the solvent.

18. Latex paints are water-based paints. Similar to aerosol paints, where a liquid is dispersed in a gas, a latex emulsion is a suspension of fine particles of spherical organic resin particles in water. When latex paints are applied, the water evaporates, leaving the resin particles to form a film coating. Latex paints form a film through coagulation of the particles.

19. Five basic types of inorganic films are anodizing, chromizing, siliconizing, and oxide and phosphate coatings.

20. Inorganic coatings or films react with the surface to which they are applied and become a part of that surface.

21. a. The anodizing process is used to convert the surface of a product to an oxide. The metal product is immersed in a bath of electrolyte, such as boric acid, chromic acid, oxalic acid, or sulfuric acid. When anodizing a metal product, the part is lowered into a lead-lined tank containing the anodizing solution. The lead lining in the tank acts as the cathode (–) for the reaction. The solution becomes the anode (+). An electric current is applied at voltages of 12 to 24 V. The surface of the part is typically oxidized to thicknesses up to approximately 0.005 in. (0.13 mm).

   b. Chromizing involves heating a ferrous metal part in a closed container in which chromium and hydrogen are present. The metal part absorbs approximately 30% of the chromium to depths of 0.0005 to 0.005 in. (0.013 to 0.13 mm). Chromizing produces a corrosion-, abrasion-, and mildly acid-resistant surface.

   c. Oxide coatings are used as protection against corrosion. They are produced by exposing surfaces to oxidizing gases or solutions under elevated temperatures. The gases or solutions oxidize the surface of the metal part. Common oxide coatings are black oxide, dichromate coatings, and gun metal finishes.

   d. Siliconizing is basically the same process as chromizing, but it is done in the presence of silicon carbide and chlorine. It typically penetrates to a depth of 0.100 in. (2.5 mm).

23. a. Hot dipping is also a process where one metal is dipped into a bath of plating material. The part that is dipped must have a higher melting point than the plating material.

   b. Cold dipping is a process by which the plating material displaces the base metal on the surface of the part when it is dipped into a bath of plating material. This is possible when the electrochemical potential of the base metal is higher than the material used to plate the part.

   c. Surfacing is used to build up worn parts or to increase the abrasion and corrosion properties of the part. Once the buildup of the surface reaches more than 3% of the base metal, the process is called cladding.

   d. Metal spraying is performed by blowing an atomized metal against a surface to be coated. As the sprayed particles hit the surface, they flatten out into flakes. These flakes lock into the irregularities in the surface of the part. The surfaces to be coated are often sandblasted or gritblasted prior to spraying to roughen the surface and provide better bonding. The depth of the sprayed surface does not usually exceed 0.005 in. (0.13 mm). Metal-sprayed parts can be fired after spraying to diffuse the flakes and produce a better finish. Any material that can be melted or atomized can be sprayed.

   e. Fluidized bed coating processes involve immersing heated parts into a tank filled with the coating material, typically a thermoplastic polymer such as PVC. The polymer coating is in a granular form and can be colorized. In the container of coating material, pressurized air is blown to keep the coating stirred during application. The part is heated prior to application, then lowered into the container full of coating, and the air is applied. Any material that can be heated can be coated with the fluidized bed process.

f. Galvanized steel is a zinc-coated steel produced by the hot dip process or by electroplating. The zinc coating protects the base steel from corrosion.

g. Plastisol and organisol dipping involve immersing the part to be coated in a bath of coating compound. This process is used to coat such items as tool handles and kitchen utensils. The part may be heated and then dipped into the thermoplastic (PVC) coating. The longer the part remains in the bath, the greater the buildup of the coating.

CHAPTER 11

1. When a substance burns, the oxygen in the air combines with the fuel. This combination process is known as oxidation.

2. For fuels to burn, they must first be at or raised to a temperature that breaks the molecular bonds holding the complex molecules together, forming simpler molecules such as CO_2 and H_2O. As the bonds are broken, energy is released in the forms of heat and light.

4. The combustion equation of the materials must balance out.

5. The unit most commonly used to describe heat energy is the British thermal unit (Btu). One Btu is the heat energy required to raise the temperature of 1 lb of water 1°F. The heat-energy value or heating value for solid fuels such as coal is measured in Btu per pound (gram). The heat-energy value released by liquid fuels such as gasoline is measured in Btu per gallon (liter). Finally, the heat energy value obtained from gaseous fuels such as natural gas is measured in Btu per cubic foot (cubic meter).

6. One Btu is the heat energy required to raise the temperature of 1 lb of water 1°F.

12. The first petroleum company was formed in 1854 in Titusville, Pennsylvania, by George Bissel.

13. It is generally believed that petroleum comes from the decomposed remains of animals and plants that were covered millions of years ago. At that time, the earth was covered by a large percentage of water, containing a variety of marine and plant life. The remains of those plants and animals settled on the bottom and were covered with sediments dumped there by different tributaries. Cut off from air, the remains decomposed and were transformed into petroleum and gases through biochemical processes involving the aerobic and anaerobic bacteria and the enzymes present.

16. In some situations, the pores of the rock formation are too small to allow the oil to flow to the surface. In these situations, the rock formations can be broken up by explosives (when in sandstone) or dissolved by acids (when in limestone); the rock layers are then opened up, allowing the oil to flow to the surface.

17. Petroleum is produced by high temperatures and pressures, which act on decomposed organic material (plants and animals) over very long periods of time.

19. Ethyl alcohol (grain alcohol), or ethanol, may also be used as a motor fuel, especially when blended with gasoline (sometimes called gasohol). Five to twenty-five percent of anhydrous alcohol (alcohol with all water removed) may be used in different blends. Advantages include an increased octane rating, cleaner burning, and cooler burning temperature. Disadvantages stem from the lower heat-energy value of the alcohol, which means that more fuel must be burned or the burn rate must be increased to produce the same power.

20. Viscosity is a measure of a liquid's resistance to flow.

CHAPTER 12

1. Strength, in the broadest sense of the term, refers to the ability of a material or group of materials to resist the application of a load or loads without failure.

2. Mechanical properties are those that deal directly with the behavior of materials under applied forces.

3. Materials testing is an area of research that deals primarily with the methods and procedures used to measure and reliably, precisely, and accurately determine the mechanical properties of materials. Primarily, these are (1) a determination of the applied force or load, and (2) the test specimen's change in length based on the applied force.

4. Stress is defined as the intensity of the internally distributed forces or components of forces that resist a change in the form of a body. Strain is defined as the change per unit length in the linear dimensions of a body accompanying a coincident change in stress.

5. a. Stiffness deals with the resistance of a material to deformation under load while in an elastic state.
   b. Hardness is a measure of a material's resistance to indentation or abrasion of its surface.
   c. Elasticity is the ability of a material to deform without taking a permanent set when the load is released.
   d. Plasticity refers to a material's ability to deform outside of the elastic range and yet not rupture.
   e. Energy capacity is a function of a material's strength and stiffness. It deals with the ability of a material to absorb energy or have work done on it.

9. The stresses and strains previously discussed are called nominal stresses and nominal strains because they are based on the original dimensions of the specimen. "True" stresses and "true" strains are computed based on instantaneous dimensions when the specimen is subjected to given load conditions.

10. While the test specimen is within the elastic range, the ratio of lateral to axial strain, with loading taking place along a single axis, is called Poisson's ratio. Poisson's ratio is found by dividing the lateral strain by the axial strain.

11. The elastic limit is the greatest amount of stress that a material can develop without taking a permanent set. The proportional limit is the greatest stress a material can develop without deviating from linearity between stress and strain. The yield strength, or yield point, is most often determined by the offset method, in accordance with ASTM standards, which define yield strength as the stress at which a material exhibits a specified limiting permanent set.

18. Three general failure indications are slippage, separation, and buckling. However, a combination of these fractures may occur during actual testing.

20. Once a dislocation reaches the surface of the crystal, it produces a permanent change in the size of the spaces between atoms. This process is called slip, because it appears that one part of the crystal has slipped over another. Once at the surface, the original dislocation is finished. Under increasing loads, new dislocations are generated. This process can be observed in the plastic action of real materials under applied loads. This increasing stress is required to drive the dislocations through the material to the surface. Once the load is removed or held constant, the dislocations cease to move through the material.

CHAPTER 13

8. The term precision is used to indicate the repeatability of a measure or measurement, whereas the term accuracy is used to signify how close a measurement or measure is to the "true," or correct, value.

9. Test significance is a measure of the extent to which the information obtained through the test procedure is a predictor of the performance of the same material in service. If the information obtained cannot be used to predict actual performance of a material in service, the test has little or no significance.

CHAPTER 14

7. a. The point where the stress-strain curve first deviates from proportionality is called the proportional limit.
   b. The elastic limit is the point beyond which an increase in stress induces a nonproportional increase in strain.
   c. The yield strength, or yield point, is most often determined by the offset method in accordance with ASTM standards, which define yield strength as the stress at which a material exhibits a specified limiting permanent set.
   d. This peak is the maximum stress that can be developed in the material. This point is known as the ultimate tensile stress of the material.
   e. Breaking strength is the applied stress required to break the specimen.
   f. The modulus of elasticity is the ratio of the incremental increase in stress to the coincident incremental increase in strain.
   g. Toughness refers to the difficulty in breaking a material or its ability to withstand shock loading.
   h. Stiff materials do not exhibit much deformation under a load.
   i. Ductility is the ability of a material to undergo permanent deformation prior to failure.

9. Data and calculations for tensile testing usually include elastic limit, percent elongation, modulus of elasticity, proportional limit, percent reduction in area, tensile strength, yield point, yield strength, and other related properties.

10. Stress = force/area
    $= (2500 \text{ lb})/[\pi \cdot (0.2525 \text{ in.})^2] = 12{,}482 \text{ lb/in.}^2 \ (86.1 \text{ MPa})$

11. Stress = force/area
    $= (5000 \text{ lb})/[\pi \cdot (0.5 \text{ in.})^2] = 6366 \text{ lb/in.}^2 \ (43.9 \text{ MPa})$

12. Strain $= (l_f - l_o) + l_o$
    $= (4.55 - 4.5) + 4.5 = 0.011, \text{ or } 1.11\%$

13. Original area $= 0.200 \text{ in.}^2$
    Final area $= 0.198$
    % reduction in area $= [(A_o - A_f) + A_o] \cdot 100 = 1\%$

14. Stress = force/area
    $= (8000 \text{ lb})/(3 \text{ in.} \cdot 5 \text{ in.}) = 533 \text{ lb/in.}^2 \ (3.7 \text{ MPa})$

15. Stress = force/area
    $12{,}000 \text{ kg} \cdot 9.8 \text{ m/s}^2 = 117{,}600 \text{ N}$
    $0.03 \text{ m} \cdot 0.05 \text{ m} = 0.0015 \text{ m}^2$
    Stress $= (117{,}600 \text{ N})/(0.0015 \text{ m}^2) = 78.4 \text{ MPa}$

16. Stress = force/area
    $(100{,}000 \text{ lb})/(2 \text{ in.} \cdot 3 \text{ in.}) = 16{,}667 \text{ lb/in.}^2$

100,000 lb/in.$^2 \cdot$6 in.^{2} = 600,000 lb
17. Stress = force/area
    Yield strength = 700 MPa (101,523 lb/in.^{2})
    700 MPa$\cdot$(0.1 m$\cdot$0.1 m) = 7000 N
    (7000 N)/(9.8 m/s^2) = 714 kg
    10,000 lb/in.^{2} = 69 MPa (Yes)
18. 18$\cdot$0.01 = 0.18 in.
    l_f = 18.18 in.
19. Stress = force/area
    100,000 lb/in.^{2} = force/[$\pi \cdot$(0.375 in.)^2]
    Force = 226,354 lb

CHAPTER 15

1. Creep is the plastic deformation resulting from the application of a long-term load.
2. Will be accelerated.
5. Applied load, test period, resultant strain, and environmental conditions.

CHAPTER 16

2. Concrete is strongest in compression and is used in situations requiring greater compression strength.
4. The specimens required for compression testing tend to be short and larger in diameter. The longer the compression specimen is, the greater the risk of buckling and of column action. Column action is an instability that develops in the specimen. This column action leads to a buckling of the material, where the material bulges out to the side due to an elastic instability.
    The ends of the specimen should be flat, parallel, and perpendicular. Otherwise, stresses will be concentrated at the ends. The ends should also be perpendicular to the loading axis to help prevent bending due to eccentric loading. Eccentric loading occurs when the load is applied in a direction that is not perpendicular to the axis of the specimen. In general, the more rapid the rate, the higher the indicated strength. For example, the indicated strength of a specimen loaded at 1000 lb/in.^{2}/min may be as high as 5% greater than one loaded at 100 lb/in.^{2}/min, for instance.
5. Compressive strength = (5000 lb)/(2 in.$\cdot$2 in.)
                        = 1250 lb/in.^{2} (8.6 MPa)
6. Area = πr^2
    (105,000 lb)/[$\pi \cdot$(3 in.)^2] = 3714 lb/in.^{2} (25.6 MPa)
7. Area = πr^2
        = 28.3 in.^{2}
    Compressive strength$_1$ = (85,000 lb)/(28.3 in.^{2}) = 3004 lb/in.^{2}
    Compressive strength$_2$ = (87,500 lb)/(28.3 in.^{2}) = 3092 lb/in.^{2}
    Compressive strength$_3$ = (93,000 lb)/(28.3 in.^{2}) = 3286 lb/in.^{2}
    No, reject the batch.
8. 500 lb/4 = 125 lb/leg
    Stress = force/area

$= (125 \text{ lb})/[\pi \cdot (0.75 \text{ in.})^2] = 70.7 \text{ lb/in.}^2$

9. Stress = force/area
   Area = force/stress
   $= (200,000 \text{ lb})/3200 \text{ lb/in.}^2$
   $= 62.5 \text{ in.}^2$, or a square of 7.9 in. on a side
   Area = force/stress
   $= (400,000 \text{ lb})/3200 \text{ lb/in.}^2$
   $= 125 \text{ in.}^2$, or a square of 11.2 in. on a side

10. Area $= \pi r^2$
    $(62.5 \text{ in.}^2)/\pi = 19.89 \text{ in.}^2 \rightarrow$ a radius of 4.5 in.
    $(125 \text{ in.}^2)/\pi = 39.79 \text{ in.}^2 \rightarrow$ a radius of 6.3 in.

11. $(60,000 \text{ lb})/4 = 15,000 \text{ lb/column}$
    Stress = force/area
    Area = force/stress
    $= (15,000 \text{ lb})/2750 \text{ lb/in.}^2 = 5.45 \text{ in.}^2$, or a square of 2.34 in. on each side

12. Stress = force/area
    $(1500 \text{ lb/in.}^2) \cdot [\pi \cdot (0.75 \text{ in.}^2)] = 2651 \text{ lb max. each leg}$
    $(60,000 \text{ lb})/(2651 \text{ lb/leg}) = 22.6$, or 23, legs

13. Area $= 28.3 \text{ in.}^2$
    $4500 \text{ lb/in.}^2 \cdot 28.3 \text{ in.}^2 = 127,350 \text{ lb (No)}$

14. Area = force/stress
    $= (60,000 \text{ lb})/(3500 \text{ lb/in.}^2) = 17.14 \text{ in.}^2$
    Area $= \pi r^2$
    $(17.14 \text{ in.}^2)/\pi = r^2 = 5.46 \text{ in.}^2$
    $r = 2.34 \text{ in.}$
    The maximum is a specimen of 4.5↔9 in.

15. Area $= 28.3 \text{ in}^2$
    Force = stress·area
    $= (3200 \text{ lb/in.}^2) \cdot (28.3 \text{ in.}^2) = 90,560 \text{ lb}$

CHAPTER 17

3. Stress = force/area
   Shear strength $= (1500 \text{ lb})/[\pi \cdot (0.25 \text{ in.})^2] = 7639 \text{ lb/in.}^2$

4. Shear strength $= (3500 \text{ lb})/[\pi \cdot (0.375 \text{ in.})^2] \cdot 2 = 3961 \text{ lb/in.}^2$

5. $(5000 \text{ kg}) \cdot (9.8 \text{ m/s}^2) = 49,000 \text{ N}$
   Area $= \pi \cdot (0.05 \text{ m})^2 = 0.0079 \text{ m}^2$
   Shear strength $= (49,000 \text{ N})/(0.0079 \text{ m}^2) = 6.24 \text{ MPa}$

6. Shear strength $= (49,000 \text{ N})/(0.016 \text{ m}^2) = 3.12 \text{ MPa}$

7. $E = 2G(1+n)$
   E is the modulus of elasticity for the material.
   G is the shear modulus for the material.
   n is Poisson's ratio for the material.
   $2G = E/(1+n)$
   $= 2 \times 10^7/(1.35) = 1.48 \times 10^7$
   $G = 7.4 \times 10^6$

8. $E = 2G(1+n)$

$= 2 \cdot (10 \times 10^6) \cdot 1.30 = 2.6 \times 10^7$

9. $G = \tau / \gamma$
$= (5000 \text{ lb/in.}^2)/(\tan 5°) = 5.72 \times 10^4 \text{ lb/in.}^2$

10. $3500 \text{ kg} \cdot 9.8 \text{ m/s}^2 = 34{,}300 \text{ N}$
$(34{,}300 \text{ N})/(\tan 3°) = 654{,}483 \text{ MPa}$
$E = 2G(1+n)$
$= 2 \cdot 654{,}483 \text{ MPa} \cdot 1.28 = 1.68 \text{ GPa}$

11. Torque $=$ force $\cdot$ distance
$= 35 \text{ lb} \cdot 1 \text{ ft} = 35 \text{ lb-ft}$

12. $25 \text{ kg} \cdot 9.8 \text{ m/s}^2 = 245 \text{ N}$
$245 \text{ N} \cdot 0.25 \text{ m} = 61.25 \text{ N} \cdot \text{m}$

13. $S = 16T/\pi d^3$
$= (16 \cdot 3000 \text{ lb-in.})/[\pi \cdot (0.75 \text{ in.})^3]$
$= 36{,}217 \text{ lb/in.}^2 \ (250 \text{ MPa})$

14. $S = 16Td_o/\pi(d_{o4}{}^{-d}{}_{i4})$
$= (16 \cdot 25 \text{ N} \cdot \text{m} \cdot 0.05 \text{ m})/\pi [(0.05 \text{ m})^4 - (0.035 \text{ m})^4]$
$= 424 \text{ MPa}$

15. $S = 16T/\pi d^3$
$= (16 \cdot 600 \text{ lb-in.})/[\pi \cdot (0.25 \text{ in.})^3]$
$= 195{,}570 \text{ lb/in.}^2$

16. $S = 16T/\pi d^3$
$d^3 = 16T/S\pi$
$= (16 \cdot 125/(100{,}000 \text{ lb/in.}^2 \cdot \pi)$
$= 0.006 \text{ ft, or } 0.076 \text{ in.}$
$d = 0.424 \text{ in.}$

17. $\varepsilon = (\theta \cdot d \cdot \pi)/360 \cdot L$
ε is the torsion strain (in in./in. or m/m)
θ is the angular displacement (in degrees)
d is the diameter (in in. or m)
L is the gage length over which the strain is measured (in in. or m)
$\varepsilon = (5° \cdot 0.5 \text{ in.} \cdot \pi)/(360 \cdot 3 \text{ in.})$
$= 0.0073$

18. Assume a torsional stress of 500 in.-lb.
$S = 16T/\pi d^3$
$= (16 \cdot 500 \text{ in.-lb})/[\pi \cdot (6.5 \text{ in.})^3]$
$= 20{,}732 \text{ lb/in.}^2$
$S = 16Td_o/\pi(d_{o4}{}^{-d}{}_{i4})$
$= (16 \cdot 500 \text{ in.-lb} \cdot 0.5 \text{ in.})/\pi [(0.5 \text{ in.})^4 - (0.49 \text{ in.})^4]$
$= 262{,}416 \text{ lb/in.}^2$
Given the same torsional stress, the pipe would develop the greater stress. However, the strength of the bar would be greater, because it develops less stress. To equal the strength of the bar, the pipe would also have to be a bar.

19. $\varepsilon = (q \cdot d \cdot \pi)/360 \cdot L$
$= (7° \cdot 0.75 \text{ in.} \cdot \pi)/(360 \cdot 18 \text{ in.})$
$= 0.0025$
$\varepsilon = (q \cdot d \cdot \pi)/360 \cdot L$
$= (7° \cdot 1 \text{ in.} \cdot \pi)/(360 \cdot 24 \text{ in.})$
$= 0.0025$
They exhibit the same strain.

CHAPTER 18

1. Flexural strength $= 3FL/2wh^2$
   F = applied load (in lb or kg)
   L = distance between two outer points (in in. or cm)
   w = width of specimen (in in. or cm)
   h = height of specimen (in in. or cm)
   F.S. $= (3 \cdot 1500\ \text{lb} \cdot 4\ \text{in.})/[2 \cdot 3\ \text{in.} \cdot (0.25\ \text{in.})^2]$
   $= 48,000\ \text{lb/in.}^2$

2. 331 MPa

3. Flexural strength $= 3FL/2wh^2$
   $F = (\text{F.S.} \cdot 2 \cdot wh^2)/3L$
   $= [50,000\ \text{lb/in.}^2 \cdot 2 \cdot 6\ \text{in.} \cdot (0.5\ \text{in.})^2]/(3 \cdot 12\ \text{in.})$
   $= 4167\ \text{lb}$

4. Flexural strength $= 3FL/2wh^2$
   $h^2 = 3FL/(\text{F.S.} \cdot 2 \cdot w)$
   $= (3 \cdot 25,000\ \text{lb} \cdot 24\ \text{in.})/(55,000\ \text{lb/in.}^2 \cdot 2 \cdot 12\ \text{in.})$
   $= 1.36\ \text{in.}^2$
   $h = 1.17\ \text{in. minimum}$

5. $F = (\text{F.S.} \cdot 2 \cdot w \cdot h^2)/(3L)$
   $= [3,000\ \text{lb/in.}^2 \cdot 2 \cdot 36\ \text{in.} \cdot (12\ \text{in.})^2]/(3 \cdot 72\ \text{in.})$
   $= 12,000\ \text{lb}$

CHAPTER 19

1. BHN $= (2 \cdot 3000\ \text{kg})/\pi \cdot (10\ \text{mm})(10\ \text{mm} - (10\ \text{mm})^2 - (4\ \text{mm})^2$
   $= 229\ \text{BHN}$

2. BHN $= (2 \cdot 500\ \text{kg})/p \cdot (10\ \text{mm})(10\ \text{mm} - (10\ \text{mm})^2 - (2.5\ \text{mm})^2$
   $= 3.29\ \text{BHN (very low)}$

3. DPH $= 1.8544P/d^2$
   $= (1.8544 \cdot 10\ \text{kg})/(0.2\ \text{mm})^2$
   $= 464\ \text{DPH}$

4. DPH $= 1.8544P/d^2$
   $= (1.8544 \cdot 10\ \text{kg})/(0.33\ \text{mm})^2$
   $= 170\ \text{DPH}$

5. KHN $= 1.43P/L^2$
   $= (1.43 \cdot 0.5\ \text{kg})/(0.05\ \text{mm})^2$
   $= 286\ \text{KHN}$

6. KHN $= 1.43P/L^2$
   $= (1.43 \cdot 10\ \text{kg})/(0.075\ \text{mm})^2$
   $= 2542\ \text{KHN (very high)}$

7. 283 BHN
8. Can't
9. HR_B
10. HR_C
11. HR_A

12. HR_F
14. Durometer, scleroscope
15. 6.5

CHAPTER 20

1. $E = W \cdot D$
   $= 1.25 \text{ lb} \cdot 3 \text{ ft}$
   $= 3.75 \text{ ft-lb}$
2. $E = W \cdot D$
   $= 10 \text{ kg} \cdot 9.8 \text{ m/s}^2 \cdot 1 \text{ m}$
   $= 98 \text{ N} \cdot \text{m}$
3. $E = m \cdot r \cdot (\cos \text{ß} - \cos \alpha)$
   $= 60 \text{ lb} \cdot 2.625 \text{ ft} \cdot (\cos 27° - \cos 76°)$
   $= 102.23 \text{ ft-lb}$
   $60 \text{ lb} = 27.24 \text{ kg}$
   $27.24 \text{ kg} \cdot 9.8 \text{ m/s}^2 = 267 \text{ N}$
   $31.5 \text{ in.} = 80 \text{ cm} = 0.8 \text{ m}$
   $E = 267 \text{ N} \cdot 0.8 \text{ m} \cdot (\cos 27° - \cos 76°)$
   $= 138.64 \text{ N} \cdot \text{m}$
4. $E = m \cdot r \cdot (\cos \text{ß} - \cos \alpha)$
   $= 60 \text{ lb} \cdot 2.625 \text{ ft} \cdot (\cos 0° - \cos 76°)$
   $= 119.40 \text{ ft-lb}$
   $60 \text{ lb} = 27.24 \text{ kg}$
   $27.24 \text{ kg} \cdot 9.8 \text{ m/s}^2 = 267 \text{ N}$
   $31.5 \text{ in.} = 80 \text{ cm} = 0.8 \text{ m}$
   $E = 267 \text{ N} \cdot 0.8 \text{ m} (\cos 0° - \cos 76°)$
   $= 161.93 \text{ N} \cdot \text{m}$
5. $E/mr + \cos \alpha = \cos \text{ß}$
   $\cos \text{ß} = (75 \text{ ft-lb})/(60 \text{ lb} \cdot 2.625 \text{ ft}) + \cos 76°$
   $\cos \text{ß} = 0.7181$
   $\text{ß} = 44.10°$
6. $E/mr + \cos \alpha = \cos \text{ß}$
   $30 \text{ kg} \cdot 9.8 \text{ m/s}^2 = 294 \text{ N}$
   $\cos \text{ß} = (120 \text{ N} \cdot \text{m})/(294 \text{ N} \cdot 0.8 \text{ m}) + \cos 76°$
   $\cos \text{ß} = 0.7521$
   $\text{ß} = 41.23°$
7. $E = 60 \text{ lb} \cdot 2.625 \text{ ft} \cdot (\cos 12° - \cos 30°)$
   $= 17.66 \text{ ft-lb}$
8. $E = 30 \text{ lb} \cdot 2.625 \text{ ft} \cdot (\cos 11° - \cos 25°)$
   $= 5.93 \text{ ft-lb}$
   $30 \text{ lb} = 13.62 \text{ kg}$
   $13.62 \text{ kg} \cdot 9.8 \text{ m/s}^2 = 133.5 \text{ N}$
   $31.5 \text{ in.} = 0.8 \text{ m}$
   $E = 133.5 \text{ N} \cdot 0.8 \text{ m} \cdot (\cos 11° - \cos 25°)$
   $= 8.04 \text{ N} \cdot \text{m}$

9. $E/mr + \cos \alpha = \cos \beta$
   $\cos \beta = (105 \text{ ft-lb})/(60 \text{ lb} \cdot 2.625 \text{ ft}) + \cos \alpha$
   $= 0.9086$
   $\beta = 24.69°$
10. $E = 16 \text{ lb} \cdot 2.5 \text{ ft} \cdot (\cos 15° - \cos 90°)$
    $= 38.64 \text{ ft-lb}$
    $16 \text{ lb} = 7.26 \text{ kg}$
    $7.26 \text{ kg} \cdot 9.8 \text{ m/s}^2 = 71.15 \text{ N}$
    $30 \text{ in.} = 0.762 \text{ m}$
    $E = 71.15 \text{ N} \cdot 0.762 \text{ m} \cdot (\cos 15° - \cos 90°)$
    $= 52.37 \text{ N} \cdot \text{m}$

CHAPTER 21

4. $\sigma = 10.18 \cdot LF/d^3$
   $= 10.18 \cdot (3 \text{ in.} \cdot 1500 \text{ lb})/(1.375 \text{ in.})^3$
   $= 17{,}622 \text{ lb/in.}^2$
5. $L = sd^3/10.18F$
   $= (55{,}000 \text{ lb/in.}^2) \cdot (1.5 \text{ in.})^3 + (10.18 \cdot 5000 \text{ lb})$
   $= 3.65 \text{ in.}$

CHAPTER 22

2. Radiographic tests include those tests that use short electromagnetic waves, such as X or gamma rays. Using these waves, it is possible to record features in the interior of solid objects and obtain information about their size and other aspects. Radiography can be used for a wide range of applications by varying the voltage, current, film type, and other parameters of the test.
3. Computerized axial tomography (CAT) uses the concept of measuring the radiation intensity across an X-ray image, digitizing that image, and analyzing the data to reconstruct an image based on the data. CAT scans are widely used in medical applications but are finding wider uses in industrial applications. The object is held firmly in place while the source and detectors are rotated around the specimen, or the object is rotated between the source and detector.
5. Eddy current inspection uses alternating magnetic fields and can be applied to any electrical conductor. Leakage flux inspection uses a permanent magnetic or dc-generated electromagnetic field and is used only for ferromagnetic materials. In eddy current testing, the alternating fields set up eddy currents in the specimen around flaws, which can be detected. Any flaw that produces lines of leakage flux can be detected using the leakage flux technique. These techniques are often combined with other nondestructive techniques, such as ultrasonics.

Index